Geomorphology and Hydrology of Karst terrains

Geomorphology and Hydrology of Karst Terrains

William B. White

New York   Oxford
OXFORD UNIVERSITY PRESS
1988

Oxford University Press

Oxford   New York   Toronto
Delhi   Bombay   Calcutta   Madras   Karachi
Petaling Jaya   Singapore   Hong Kong   Tokyo
Nairobi   Dar es Salaam   Cape Town
Melbourne   Auckland

and associated companies in
Berlin   Ibadan

Copyright © 1988 by Oxford University Press, Inc.

Published by Oxford University Press, Inc.,
200 Madison Avenue, New York, New York 10016

Oxford is a registered trademark of Oxford University Press

All rights reserved. No part of this publication may be reproduced,
stored in a retrieval system, or transmitted, in any form or by any means,
electronic, mechanical, photocopying, recording, or otherwise,
without the prior permission of Oxford University Press.

Library of Congress Cataloging-in-Publication Data

White, William B. (William Blaine), 1934–
  Geomorphology and hydrology of karst terrains.

  Includes index.
  1. Karst. 2. Hydrology, Karst. I. Title.
GB600.W47   1988   551.4′47   87-23996
ISBN 0-19-504444-4

9 8 7 6 5 4 3 2 1

Printed in the United States of America
on acid-free paper

To Bette
Steadfast companion in many dark places
of the world

Preface

Karst terrains have been of interest to a few fanatics for a long time, mostly because of the dramatic landscapes and the challenge of cave exploration. Investigations of karst were often in remote and rural areas far from the hassle of the human scene. Indeed, many karst researchers found the subject interesting because it allowed them to avoid the human scene. All that has changed. Urbanization has crawled over the mountains and into once rural limestone valleys. The geomorphology of karst landscapes and the hydrology of karst drainage systems are now subjects of interest to water-supply specialists, urban planners, and environmental engineers. One of the primary objectives of this book is to address karst hydrology and geomorphology from this new point of view.

Hydrology has long been a quantitative science. The physics of fluid flow in porous media is described by Laplace's equation, a much-studied differential equation whose solutions give rise to the well-known flow nets and isopotential surfaces of groundwater hydrology. Likewise, the very subject matter of rainfall, runoff, and well yields demands expression in exact numbers. Unfortunately, the neat mathematical formalisms do not work well in karst, and it has taken some time to sort out the best way to blend standard hydrology with what is known of karst from direct physical exploration.

Traditionally, geomorphology has been a qualitative science. Landforms are named and the processes that sculpture them are identified and argued about. Much of the effort during the past several decades has been to make geomorphology more quantitative, usually following one of two approaches to the baffling complexity of the earth's landscape. The megathinkers place their faith in statistics. By measuring size and shape of the landscape and temperature, rainfall, runoff, flood frequency, and other "process" parameters, they hope that multivariate statistics aided by big, number-crunching computers will filter out the general laws of landscape evolution from the background noise of local detail. In contrast, the microthinkers attempt to reduce the processes of landscape evolution to physics and chemistry. Their approach is to strip away the local detail until the chemical and mass transport processes are amenable to analysis by the ordinary laws of physical chemistry, fluid mechanics, and thermodynamics. Their final model, however, still must contain some remnant of the character of the real world. The author, as will be quickly discovered, belongs to the microthinker school. After all, karst does develop predominantly by chemical processes. If geomorphology is to be reduced to physics and chemistry anywhere, karst ought to be one of the better places.

The first four chapters are mainly descriptive. They paint a picture of surface and underground landforms in karst regions, and name a lot of names. Chapter 5 provides a summary of carbonate geochemistry as the subject is understood at present. Chapter 6 does the same for karst hydrology. It begins with groundwater in ordinary porous media aquifers, introduces some principles of fluid mechanics, and then outlines some of the current ideas on the hydrology of carbonate rocks.

Chapters 7, 8, 9, and 10 are the heart of the subject. These chapters discuss the chemistry of karst waters, the processes of sedimentary in-filling, the origin of caves, and the evolution of karst systems down through geologic time. The material in these chapters is drawn mainly from contemporary research.

Karst is not limited to carbonate rocks. Chapters 11 and 12 briefly introduce karst in evaporite rocks, which are more soluble than the carbonates, and karst in such rocks as granites and quartzites, which are generally regarded as insoluble.

The last two chapters are devoted to environmental problems in karst, loosely separated into land-use problems (Chapter 13) and water resources problems (Chapter 14). In reality, the two problems cannot be separated. If urbanization spreads into karst regions, the subject of these two chapters is what it is all about.

People who are interested in caves and karst are a diverse group from widely varying backgrounds. I do not assume my readers have a deep knowledge of geology, because many people concerned with karst are not geologists; I do, however, assume some knowledge of chemistry and mathematics, at about the level of a good second-year college curriculum. I find quantitative statements, as either equations or data plots, to be more satisfying than verbal armwaving.

This is my only excuse for the rather large number of equations scattered through the text.

Terminology of karst terrains is a continuing and vexing problem. Because karst landforms intrude themselves into human affairs, a tremendous variety of local names, in many langauges, have accumulated. Some of these have been transformed into formal scientific usage by careful analysis and precise definition; others have been adopted, helter-skelter, and entered the literature through a sort of osmosis. Repeated attempts by international bodies such as the Karst Commission of the International Geographical Union have not done much to clear up the tangle. It has long been a popular practice to borrow Slavic terms—doline, uvala, polje, among others—and, unfortunately, to apply them to areas where the landforms and the geologic setting are quite different from those of the Adriatic karst. A recent tendency to use English terms defined in context, rather than force the landforms into a complex polyglot of international terms, has been a welcome relief. I have followed this practice as much as possible.

Two instances warrant mention. Small closed depressions are a dominant landscape element in nearly all karst areas. I have used the terms *sinkhole* and *doline* interchangeably (the latter term, in its original Slavic usage, simply means "little valley," not necessarily a closed depression). For features of intermediate size, I use *compound sink* or *valley sink* rather than *uvala;* the uvalas of the Adriatic karst are similar to compound sinks. The largest closed depressions are simply called *poljes* because the most detailed descriptions are from the Adriatic karst, and these are the poljes against which other large depressions are compared.

More controversial, perhaps, is my use of *cutter* rather than the British term *grike* for the long trenchlike solution features that crisscross many limestone surfaces beneath the soil mantle, and which are of critical importance in many land-use investigations. Thus, we use the term cutter-and-pinnacle topography rather than grike-and-clint topography, in spite of the promotion of "grike" as an international term by British colleagues. "Cutter" is widely used in the United States in the engineering literature; "grike" is more widely used in the geomorphological literature. Maybe it is just a perverse belief that Tennessee phosphate miners have equal standing with Yorkshire sheepherders.

July 1987   W. B. W.

Acknowledgments

My interests in karst research originated in the exploration of caves. Over the decades I have accumulated an enormous debt of gratitude to the many cavers who have been with me in the field, have provided me with maps, samples, and observations, and who have pulled me out of a variety of tight spots. Much credit is also due to the National Speleological Society, whose leaders, particularly in the early days, had the intelligence to encourage cave science at a time when no one else cared.

Much of my field work has had the support of the Cave Research Foundation (CRF). It is no accident that many of the illustrations come from the great cave systems of Mammoth Cave National Park, where many of the CRF activities have been undertaken.

One's ideas are enlarged, modified, and refined by constant interaction with colleagues. I have had many such opportunities. At the risk of offending the others, some who have particularly helped me get my act together are Rane L. Curl, George H. Deike III, Derek C. Ford, Arthur N. Palmer, and James F. Quinlan.

I owe a great deal of thanks to former and present graduate students who had the nerve to take on a karst-related thesis: Evan T. Shuster, Henry W. Rauch, John W. Hess, Bruce E. Gaither, Janet S. Herman, Joseph W. Troester, John R. Kastrinos, Daniel L. Chess, Ira D. Sasowsky, and George Veni.

I owe a special debt to Richard A. "Red" Watson, who talked me into starting this book many years ago and has played an important role in encouraging book writing on karst subjects.

Much of my own karst research has been done in collaboration with my wife, Dr. Elizabeth L. White. She also played a role of operations manager, and without her efforts the present book would likely never have been finished.

Contents

1 Introduction, 3
2 Surface landforms in karst regions, 8
3 Underground landforms in karst regions, 60
4 Karst landscapes, 103
5 The chemistry of carbonate dissolution, 119
6 Karst hydrology, 149
7 Geochemistry of karst waters, 193
8 Soils, sediments, and depositional features, 220
9 Theories, models, and mechanisms for the origin of caves, 264
10 Karst evolution and Pleistocene history, 302
11 Evaporite karst, 328
12 Karst and karst-like features in slightly soluble rocks, 340
13 Land use and land management problems in karst, 355
14 Water resources problems in karst, 380

References, 406
Illustration Credits, 446
Index, 447

Geomorphology and Hydrology
of Karst terrains

1

Introduction

1.1 THE PROBLEM OF KARST LANDSCAPES

In many regions of the earth, usually where carbonate rocks underlie the surface, there occurs a landscape known widely and loosely as *karst*. Karst landscapes and their underlying caves are created by the chemical solution of the bedrock. The characteristic landforms of karst regions are:

    Closed depressions of various size and arrangement
    Disrupted surface drainage
    Caves and underground drainage systems

The degree of development of these landforms varies greatly from one region to another. Some karst terrains are a rough and jumbled land of deep depressions, isolated towers, and pointed hills. Others may be gently rolling plains, soil-covered, with perhaps only the gentlest of depressions to label them as karst.

    The aspect of karst that distinguishes it from any other landscape is the dominance of solution as a geomorphic agent. Karst occurs on carbonate rocks, gypsum, and to a minor extent on certain other rocks. Although mass transport by mechanical erosion must play some role in the removal of the insoluble components of the bedrock, solution and solutional transport (as distinguished from solutional weathering followed by mechanical transport) is a dominant process in landform sculpturing.

    In much English-language literature, karst terrain is pictured as an area dotted with sinkholes and sinking streams. Landform development, in short, is required to reach extreme and dramatic stages before the special denotation karst is applied. East European and particularly Soviet writers take a much broader view and consider all landforms produced by the solution process to be karst—regardless of scale, surface expression, or rock type. We shall adopt this broader definition here.

Gvozdetskii (1967) estimated that some 50,000,000 km^2 of land surface could qualify as karst. This staggering number represents about 20 percent of the earth's land surface. Karst landscapes, in the broad sense that the Soviet writers use the term, are of major importance to geomorphology and need to be more widely understood.

1.2 HISTORICAL DEVELOPMENTS

The study of karst is a blend of many elements and interests, of which the most important are cave science, geomorphology of surface landforms, and karst hydrology. The oldest of these is probably the study of caves. Because of their role as shelters and places of worship, and their association with the underworld, caves have been of interest since the dawn of history. Geomorphology is a relatively recent offshoot of the geological sciences, but geomorphologists have been interested in the dramatic landforms of karst regions since the beginning of the science. Hydrology, the third scientific element in the study of karst, is the most recent. The mysterious underground rivers of the karst have excited curiosity for a long time, but only in the past few decades has this interest been integrated into a broader view of the behavior of surface and groundwater in karstic drainage basins.

Of the three components to the study of karst, only speleology, the study of caves, would claim, perhaps illegitimately, to be a science in its own right. The speleologist, an isolated and somewhat provincial character, has been tied most closely to biologists (specializing in cave habitat and ecology) and archeologists. Speleology also has strong connections with cave exploration and sport caving.

The first step in any aspect of geomorphology is to visualize the shape and form of the landscape. Traditionally maps, and more recently aerial photography, infrared imagery, LANDSAT, and other high technology techniques have accomplished this objective on the earth's surface. Caves cannot be so easily visualized. One cannot see a cave. One can see, at any moment, only the short segment of passage in which one is standing. The visualization of a cave can only be accomplished by painstaking station-by-station surveying, with little help from modern technological devices. Those who explore caves can document their discoveries only by preparing maps. The strong tradition of cave surveying, often with high standards of cartographic excellence, has resulted in a large data base on which the cave scientist can draw.

Sport cavers in the past decade have developed new exploration techniques, such as single-rope methods for descending long vertical drops and the increased use of wet suits and diving gear for exploring very wet or underwater caves. These contributions, combined with a certain competitive urge to find the "longest" or "deepest," have also greatly extended the data base available for scientific study.

A new factor has entered the study of karst terrains in the past few decades—our use of the karst. Karst landforms are intrusive in human affairs—lost sheep in the clints of Yorkshire, flooding of entire farms in the poljes of Yugoslavia, and leaky dams in the Tennessee Valley. The karstlands of eastern United States are predominantly rural. Some limestone regions are farmed, others used only for pasture, where the pervasive cutter and pinnacle topography frustrates the plow. The rural character of the eastern American karstlands is rapidly changing. Urban development is sprawling from the East Coast megalopolises to the limestone valleys of the West. Solution cavities in the bedrock lie in wait for the builder. Subsoil pinnacles that will support a barn or farmhouse perform less effectively under the weight of a ten-story apartment building. Foundation engineering takes on new importance. Water supplies become more critical because pollution is easily injected into the limestone aquifers and spreads effectively by little-known routes. Water supply development and protection has become a fourth element in the study of karst, which gives the subject an entirely new importance.

1.3 THE KARST LITERATURE

The diversity of viewpoints that have converged to make up the present-day study of the geomorphology and hydrology of karst is reflected in an equally diverse literature. Those who wish to study the subject in depth can expect to devote considerable time to tracking down some obscure and exotic literature sources.

The geomorphology of karst landscapes has been investigated mainly by professional geologists in the United States and by physical geographers in Great Britain and much of Europe. Results are reported in journals such as *Zeitscrift für Geomorphologie* and *Earth Surface Processes and Landforms* and in the mainstream journals such as the *Geographical Journal, Bulletin of the Geological Society of America,* and the *Journal of Geology.*

For a time the International Geographical Congress had a Commission on Karst Phenomena. Several meetings of the commission identified such research problems as the role of climate in determining karst landforms and debated an international nomenclature for karst features. Two meetings produced collections of papers summarizing the state of knowledge at the time (Lehmann, 1960; Sweeting and Pfeffer, 1976).

Books in English describing karst from a primarily geomorphological point of view include Jennings (1985), Sweeting (1972), and Jakucs (1977). Herek and Stringfield's (1972) *Karst: Important Karst Regions of the Northern Hemisphere* provides excellent reviews with references to other literature for the principal countries of Europe and the USSR.

The hydrology of carbonate terrains has been studied in the United States by hydrologists and hydrogeologists, who often had little interest in karst phenomena as such. The word *karst* seldom appears in the titles of articles, frustrating keyword and computer literature searches. Many papers of interest appear in *Water Resources Research, Journal of Hydrology*, and *Ground Water*. The International Association of Hydrogeologists (IAH) has had a long-standing interest in karst aquifers, and many of the Memoirs of the IAH contain papers on karst. Many papers on the chemistry of carbonate dissolution equilibria and kinetics appear in the geochemical literature, particularly *Geochimica et Cosmochimica Acta*. Here again, the most important papers were not written by scientists with a primary interest in karst.

Various symposia have reviewed the hydrology of karst terrains, beginning with the 1965 Dubrovnik Symposium on fractured rocks (IASH, 1967). A second international symposium was held in Yugoslavia in 1975 (Yevjevich, 1976), and a third at Western Kentucky University in 1976 (Dilamarter and Csallany, 1977). The 12th Congress of the International Association of Hydrogeologists was devoted entirely to karst hydrogeology (Tolson and Doyle, 1977).

Few books in English are entirely devoted to karst hydrology. Zötl (1974) describes his elaborate water-tracing experiments in the Austrian Alps. Milanović's (1981) *Karst Hydrogeology* provides insight into the problems of engineering works in karst and gives considerable detail about karst hydrology in Yugoslavia.

The speleological component of karst research is a different beast. Few papers that deal specifically with caves are published in the mainstream literature. In each country there is one or more caving societies, and most of them publish journals. These specialty journals contain most of the papers about caves. The National Speleological Society is the national caving organization in the United States and publishes the *NSS Bulletin*. In addition to national organizations, there are many local chapters (called "grottos" in the United States) and independent caving clubs that have their own publications. Some, such as the Association for Mexican Cave Studies, produce high-quality publications including many cave descriptions and maps. Much of the descriptive information from these publications is collected in the *Speleo-Digest,* published by the National Speleological Society.

Most valuable for international communication has been the International Speleological Congresses, held under the auspices of the International Speleological Union at roughly four-year intervals. The congress Proceedings contain summary papers from many countries, and these in turn give access to the large but obscure national literatures.

Many books have been published on cave exploration, usually from the sport caver's point of view. Likewise, many collections of cave descriptions are available (see Chapter 3). Ford and Cullingford (1976) and Bögli (1980) describe karst science from a speleological viewpoint.

2
Surface Landforms in Karst Regions

2.1 KARST AND DRAINAGE

2.1.1 *Drainage Basins*

Rain runoff on the land surface is aggregated first into tiny rivulets, then into small streams and gullies, which join to form larger and larger streams. The stream drainage network is a highly ordered structure from the tiny feeder streams in the headlands to major rivers that drain eventually to the sea.

The land area that drains runoff into any selected tributary can be outlined with a drainage divide; thus, rain falling within the divide flows into the selected tributary, whereas rain falling outside the divide flows into some other stream. The area within the divide becomes the drainage basin or catchment area of the selected tributary. In practice, the spot marking the downstream end of the tributary is often the location of a stream gage and may be called a gage point. Many drainage divides can be drawn, and smaller basins can be nested within larger basins as the gage point is moved downstream. Every point on the land surface belongs to some drainage basin.

It was Horton (1945) who found that the arrangement of tributaries was highly ordered, and his concepts have been expanded by others, particularly by Strahler (1952, 1964). Suppose the smallest streams that appear on a given map are labeled "first order." When two first-order streams join, the resulting stream is labeled "second order," and so on. In the Strahler ordering system, all unbranched tributaries are first order, and junction of two streams of order μ becomes a stream of order $\mu + 1$, counted only from the junction point downstream. The junction of a lower-order stream with one of higher order does not change the order of the higher-order stream. The number of stream segments of any given order becomes smaller as the order increases. This is expressed quantitatively as the bifurcation ratio

$$R_b = \frac{N_\mu}{N_{\mu+1}} \quad (2.1)$$

where N_μ is the number of segments of order μ, and $N_{\mu+1}$ is the number of segments of next higher order. The bifurcation ratio can be considered a property of the drainage basin. The basin itself can be assigned an order, which is the order of the trunk stream that drains the basin at the gage point. The order of nested basins, therefore, becomes higher and higher as the gage point moves downstream. The characteristic ratios, however, remain constant and are expressed as a series of laws. Horton's law of stream numbers is

$$N_\mu = R_b^{(k-\mu)} \quad (2.2)$$

where k is the order of the trunk stream. Horton's law of stream lengths is

$$\overline{L}_\mu = R_L^{(\mu-1)}\overline{L}_1 \quad (2.3)$$

where $\overline{L}_\mu$ is the mean length of streams of order μ, R_L is the length ratio, and $\overline{L}_1$ is the mean length of first-order streams. The law of stream areas is (Schumm, 1956)

$$\overline{A}_\mu = R_A^{(\mu-1)}\overline{A}_1 \quad (2.4)$$

where the area terms have the same definition as the length terms.

The characteristic features of the drainage basin are its pattern, its area, its relief (often defined as the elevation at the highest point on the drainage divide minus the elevation at the gage point), and some measure of shape. The discharge of surface water leaving the basin at the gage point is related to area by an empirical equation of the form

$$Q = jA^m \quad (2.5)$$

where the values of j and m are characteristic of both the basin itself and the frequency and intensity of storms.

2.1.2 Transitions Between Fluvial and Karst Drainage

A characteristic feature of karst regions is the diversion of surface drainage into underground routes. The highly ordered pattern of the surface drainage basin is disrupted, tributary streams end abruptly in swallow holes, and high-order streams emerge abruptly from karst springs.

Large drainage basins in karst areas are likely to have both fluvial and karst (underground) drainage components. The relative importance of the two is determined by the thickness of soluble rocks compared with the total thickness of rocks exposed, and with the areal fraction of exposed carbonate rocks compared with the total drainage basin area. In a few regions, such as some of the Caribbean Islands and parts of the Adriatic karst, thick limestones are exposed over entire large drainage areas, so that all drainage is karstic (Fig. 2.1). Such regions of completely karstic drainage and landforms are rare, and Cvijič (1960) assigned special significance to them, calling them *holokarsts*. Karst regions with mixed karstic and fluvial characteristics, called *fluviokarsts* by other writers, he called merokarsts.

Sometimes the fluvial pattern developed on overlying clastic rocks is superimposed on underlying carbonates as the landscape is degraded. The superimposed drainage system gradually loses first its water, then the stream channels, and finally the form of the drainage pattern itself, as interior drainage becomes dominant over surface drainage. Likewise, there is usually a borderland on nonkarst rocks from which surface streams flow on to the karst and are lost at swallow holes. Downstream, the subterranean drainage reappears as spring-fed tributaries of surface streams.

As degradation of the land continues, sooner or later the limestone is destroyed and the landscape returns to a completely fluvial regime. Remnants of karst remain for quite a while as isolated fragments of underground streams and sculptured residual hills. This may be old age in some sense, but the process is not part of any sort of "karst cycle." Karstification is a process of destruction, and it runs on a one-way track from the time the limestone is first exposed to weathering until the last of it disappears down the river.

There is another process which may return a fluvial landscape even before the limestone is lost. Cvijič's holokarsts develop best in very pure limestones. Rocks are then transported entirely by the solution process. Most limestones are not pure, or at least not over large stratigraphic intervals. There is always a fraction of clay, sand, or chert that must be removed by mechanical transport. Likewise, the purer limestones may be sandwiched between beds of shale or sandstone, the weathering products of which must be transported mechanically. Further, if the drainage from the borderlands must flow through the interior drainage of the karst, the eroded products of the borderland rocks must also be transported through underground routes. When the load of clastic material becomes too massive for the subsurface streams to handle, the underground channels are blocked and drainage eventually returns to the surface. Again, fluvial erosion occurs in the karst, with the streams flowing over a pad of insoluble alluvium.

Figure 2.1 Holokarst. The mountains of Montenegro, west of Titograd, Yugoslavia, are a gently chaotic terrain with no integrated surface drainage.

2.1.3 Karst Valleys

The transition from a fluvial valley to a karst valley takes place as the valley becomes underdrained because both trunk and tributaries lose their flow to the developing underground drainage system. Consider the annual runoff past the gage point (termed an annual hydrograph; Fig. 2.2). Surface runoff is flashy, and individual storm events appear as sharp peaks on the hydrograph. During long dry periods, the flow in the stream is reduced to the base flow obtained from groundwater discharge along the stream banks. Imagine that the surface valley has just deepened far enough to allow the stream to contact the carbonate rocks. The hydrostatic head provided by the normal valley gradient is the driving force for percolating water in the carbonate rock under the valley. When the interior drainage system has developed a bit, some of the stream flow diverts into it along with the groundwater that would otherwise have nourished the stream during low flow periods (Q_1 in Fig. 2.2). Base flows are reduced, but the stream still flows all year. As the underground system becomes still larger, its capacity increases until it can carry the entire base flow (Q_2). At this time, the surface stream goes dry during the low flow periods of the summer. As the underground conduits continue to enlarge, they begin to engulf first the smaller and later

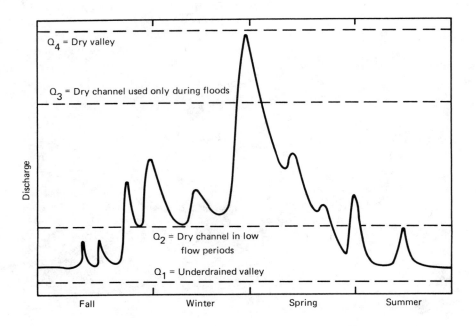
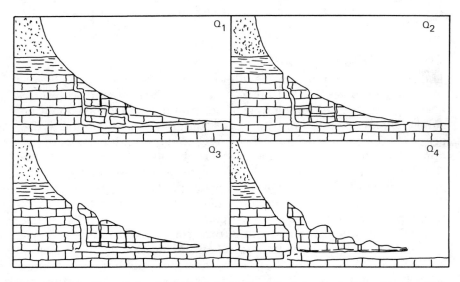

Figure 2.2 *Top:* Schematic drawing of annual hydrograph of a surface stream basin. The dashed lines indicate the carrying capacity of the evolving underground drainage system. *Bottom:* The evolving underground drainage system from an underdrained valley (Q_1) through a dry channel during low flows with well-defined swallow hole (Q_2), through the development of an incised upstream channel and swallow hole (Q_3), to the complete loss of the surface channel with concurrent development of a blind valley upstream and breakup of the valley profile through doline development downstream.

Surface landforms in karst regions / 13

the larger storm runoff discharges, so that flow in the surface stream (Q_3) occurs only during spring peak runoff. At the same time the smaller tributaries are completely lost to underground routes. Throughout this sequence of development, the valley retains its fluvial shape, its gradient, and the surface channel is clearly visible, although it may frequently be dry. Such valleys may be termed underdrained (Fig. 2.3).

When all of the discharge from the basin is diverted underground the valley becomes a *dry valley* (Q_4). The surface stream may flow only during exceptional floods. Under these conditions, the channels become degraded by bank caving and growth of plants, and all traces of them may be lost. Localized solution within the valley floor may cause sinkholes to develop, eventually disrupting the valley profile. If the soluble rocks are thick and their areal extent large, all traces of valley morphology could disappear, to be replaced with an undulating surface of sinkholes. Around the margins are the borderland surface streams that sink where they contact soluble rock.

Blind valleys are an integral part of the evolution of karst areas with nonkarstic borderlands. As the karst area develops into mature forms, surface streams gradually disappear underground. However, the tributary streams that flow onto the karst from the borderlands

Figure 2.3 The dry bed of the Elk River, Randolph County, West Virginia. The valley is completely underdrained where the river crosses the Greenbrier limestone.

are still downcutting. This leads to the formation of tributary valleys with blind footwalls, under which the tributary streams disappear. Likewise, the stream emerging from the karst area may retreat headward, incising a valley into the karst uplands. The emerging stream flows from a blind headwall, sometimes called a steephead or pocket valley.

Blind valleys are also associated with the later degrading stages of karst development. During this period karst streams are often seen emerging from springs at the blind headwall of a valley, only to sink again under a blind footwall. The streams develop a segmented pattern with some reaches of underground flow and some reaches of surface flow. The segments of subsurface drainage become shorter and shorter as the land is cut down and, just before the end, a feature known as a natural bridge may appear. This is a short segment of cave passage, usually highly modified by frost pry and other forms of surface weathering.

2.1.4 Form and Process in Karst Valleys

Baker (1973) traced many small first-order surface streams underground in the Helderberg plateau karst of New York. He was able to connect the surface with the underground drainage and then retrace it to the surface through the karst springs. Horton's law of stream number (Eq. 2.2) can be rewritten as

$$\log N_\mu = (k - \mu) \log R_b \qquad (2.6)$$

Baker's least-squares fitted values for the branching ratios were 2.4 for Skull Cave, 2.7 for Howe's Cave, and 3.3 for McFail's Cave. These values are above the theoretical lower limit of 2.0 but are less than typical range of 3.0 to 5.0 for surface streams (Strahler, 1964). The lower than average values for the underground drainage system may mean that not all tributaries were counted in the analysis. More examples are needed to evaluate the significance of the parameters.

Sinking stream subbasins can be ordered using the swallow hole as the gage point ["swallet ordering" in Williams's (1966a) terminology]. The number of such basins of given order falls off exponentially with increasing order (Fig. 2.4). In the Kentucky example, streams of first to third order flow northwestward from the clastic rocks of the Glasgow Upland and sink along the contact with the Mississippian St. Louis limestone. These streams aggregate in an unknown way and flow northward to discharge through large springs on Green River in Mammoth Cave National Park. Those streams flowing into the Parker Cave system (Quinlan and Rowe, 1978) are known to join into a master trunk. If the downstream drainage is assumed to be fourth order (by the joining of at least two

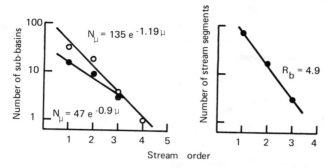

Figure 2.4 *Left:* Distribution of the number of subbasins of given order as a function of the order for sinking streams in Ingleborough (England) *(open circles)* and the Pennyroyal Plain, Kentucky *(solid circles)*. Ingleborough data are from Williams (1966a). Equations are least-squares fitted exponential functions. *Right:* Distribution of number of stream segments of various orders as a function of order for sinking streams along the southern margin of the Sinkhole Plain, Kentucky.

third-order streams), fitting the data in Fig. 2.4B gives a branching ratio of 4.9—close to ratios frequently observed in surface streams.

The length of the main channel of a drainage basin, when extended to the basin divide, is known to relate to the basin area by a power function. Hack (1957) found

$$L = 1.4A^{0.6} \qquad (2.7)$$

for a group of basins in the Shenandoah Valley of Virginia. White and White (1979) compared the drainage characteristics of 62 karstic basins in the Appalachian Highlands with 71 basins on noncarbonate rocks. They found that

$$L = 1.56A^{0.59} \qquad (2.8)$$

for the noncarbonate basins and

$$L = 1.51A^{0.61} \qquad (2.9)$$

for the carbonate basins. L = length of the longest collector, in miles; A = basin area in square miles. These results reflect the generally fluviokarst character of the Appalachian karst. Examination of the length of the longest collector for sinking streams, L_s, in the 62 karstic basins produced

$$L_s = 2.29A_s^{0.85} \quad r^2 = 0.88 \qquad (2.10)$$

showing that the truncation of the sinking streams by swallow holes somewhat changes the length–area relationships.

Total length of all the streams is also related to area. Comparisons of the control set of noncarbonate basins with a subset of 38 carbonate basins containing 30 to 100 percent of the basin area on carbonate rock produced

$$L_T = 1.35A^{1.03} \quad r^2 = 0.95 \quad (2.11)$$

for noncarbonate basins and

$$L_T = 2.39A^{0.98} \quad r^2 = 0.85 \quad (2.12)$$

for the karstic basins, again an almost identical result. The karstic basins differ from the noncarbonate basins mainly in the goodness-of-fit of the data. The lower values for the correlation coefficients of the karstic basin data reflect the more disordered character of these basins, but what appears superficially to be extensive karst development has relatively little effect on the properties of the surface channel system. Clearly, such relations cannot continue to hold in karst basins where all drainage is subsurface and remnants of surface channels are lost.

Stream channels are steep in the headwaters and flatten in the downstream reaches. The slope varies continuously. There has been a long and continuing debate about the appropriate mathematical description of the longitudinal profile. The two leading candidates are an exponential form and a logarithmic form. The exponential form can be written as

$$(E - E_{\text{ref}}) = E_O e^{-Kl} \quad (2.13)$$

if the profile is drawn in the downstream direction (White and Hess, 1982). E_{ref} is the elevation of the reference datum and E_O is the elevation of the origin with respect to the reference datum. K is the slope constant and has units of reciprocal length. Using elevation of the stream mouth as the zero of elevation creates artifacts when the profile is drawn close to the downstream end. The logarithmic form can be written as

$$E_D - E = A - K \ln l \quad (2.14)$$

following Hack (1957, 1973). E_D is the elevation at the divide, and l is the distance downstream measured from the divide. This equation creates artifacts when the profile is close to the divide. Shepherd (1985) claimed that the exponential form is preferable for profiles of streams draining into aggradational areas, whereas the logarithmic function serves better for stream profiles that terminate upstream in scarps or steep divides.

Fitting functions are useful for extrapolating stream profiles and for correlating fragments of terrace level and cave passage elevations with river profiles.

Stream channels and channels in dry valleys should be related to present and past base level streams. A study of various tributaries of the Obey River in northcentral Tennessee (White and White, 1983) is suggestive. Sunk Cane Branch (Fig. 2.5) drains from the margin of the Cumberland Plateau into the large closed depression known as Big Sunk Cane. The valley of the west branch of the Obey River cuts nearly 100 m below the broad-rolling, doline-pocked floor of the depression. From the margin of the depression, a dry stream channel continues into a deep sinkhole near the center. When fitted with a logarithmic function, the active upstream portion of the stream forms a straight-line segment from the head of the stream to the edge of the closed depression; no breaks occur in the slope at any of the bedrock contacts. The slope does break at the edge of Big Sunk Cane, and the dry channel continues with a gentler slope for 3 km across the floor of the depression. When this segment is fitted and extrapolated across the boundaries of the depression, it reaches the Obey Valley close to the Highland Rim elevation, a major erosion surface

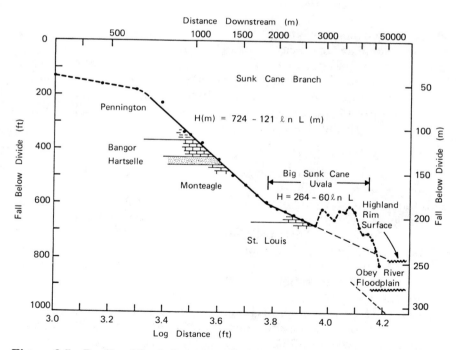

Figure 2.5 Profile of Sunk Cane Branch, northcentral Tennessee. Equations show least-squares fit to Hack's (1973) logarithmic function. The rock units shown are the Mississippian age limestones, Hartselle sandstone, and Pennington shale that crop out along the western margin of the Cumberland Escarpment. [Adapted from White and White (1983).]

along the western margin of the Cumberland Mountains. The active stream segment extrapolates to the river valley below the present-day Obey River floodplain.

The underground courses of karst streams appear to maintain, on the average, the profile and gradients they would have had if they had been flowing in normal surface channels. The details of the subsurface drainage are controlled by the structural and stratigraphic setting, but the overall pattern overrides these controls, and streams emerge from the karst at elevations appropriate to the hydraulic characteristics of the drainage basins that feed them. The dry underdrained channels in fluviokarst may be related to river terraces and erosion surfaces in the same way that dry upper-level cave passages can be.

2.1.5 *Swallow Holes*

The term swallow hole (or swallet) is used in a loose sense to indicate the place where a sinking stream goes underground. The equivalent Slovene term is *ponor*.

Swallow holes come in many sizes and shapes. Some are places where major streams abruptly go underground, either vertically through their beds or laterally into their banks. Some swallow holes are pits, some are open cave entrances, and some are choked; others are simply reaches of stream bed where water is lost. Upstream from the swallow hole the stream flows at its full volume; downstream the bed is dry. In between is an intermediate reach, where the water is lost gradually in the stream bed alluvium. Often there is no "hole" associated with the swallow.

2.1.6 *Karst Springs*

The rise points where karst waters return to surface routes are frequently springs of majestic proportions. Waters from very large catchment areas are localized in a single point of discharge. Some of the largest springs in the U.S. are found in Missouri (Vineyard and Feder, 1974) and in Florida (Ferguson et al., 1947).

Some karst springs flow from open cave mouths without any significant hydrostatic head. These are referred to as gravity springs (Fig. 2.6). Others are quiet upwellings of water from a shallow depth below the level of the surface stream that carries water from the spring. Mostly these are alluviated springs. The water must rise and force its way out against the damming effect of rockfalls, glacial alluviations, backwash sediments from surface rivers, and related barriers. Large springs in the limestone areas of New York have been plugged with glacial debris, and the present-day springs rise through channels between the bedrock walls and the glacial fill. A similar situation prevails in the south central Kentucky karst, where surface streams were alluviated by periglacial outwash material.

Surface landforms in karst regions / 19

Figure 2.6 A representative large gravity spring and blind valley headwall: Mammoth Spring at the head of Honey Creek, Mifflin County, Pennsylvania.

Waters that rise under artesian heads are known as vauclusian springs, after the Fountain of Vaucluse in France. Such springs are often structurally controlled, for example by a fault. The waters of vauclusian springs may be under considerable hydrostatic pressure, and a pronounced boil or fountaining may occur at times of high discharge. Many of these springs have open orifices and a few have been explored by divers to depths of 100 m or more.

An intermediate type are the near-shore and offshore springs that occur in many coastal karst areas. The offshore springs are rise areas of fresh water amid the salt water. Some are fed by groundwater flow in permeable or fractured rock. Others—many of which are large springs—are related to karst feeder systems that developed during the Pleistocene sea level minima and have since been flooded with salt water as a result of the post-Pleistocene rise in sea level.

2.2 LANDFORMS WITH NEGATIVE RELIEF

On most terrains the land slopes away continuously from the mountains to the sea. One can always walk downhill, arriving eventually at a stream that can be followed to a river, which can be followed to the ocean. A characteristic feature of karst landscapes is that the

20 / Geomorphology and hydrology of karst terrains

land usually slopes down into closed depressions from which the only exit is underground.

Features of negative relief can be described by their width, w, length, l, and depth, d. Families of landforms can be differentiated by the ratios of these parameters (Table 2.1). Although it is convenient to apply names to landforms based on their size and geometry, the underlying mechanisms are strongly interrelated.

2.2.1 Dolines, Shafts, and Related Features

The terms doline and sinkhole (used interchangeably) refer to relatively shallow, bowl-shaped depressions ranging in diameter from a few to more than 1000 m. Depths typically range from less than a meter, in shallow swales barely discernible to the eye and not represented on topographic maps, to depths of hundreds of meters. Also found in karst areas are shafts or pits of various kinds and depths, many of which open from the bottom of closed depressions. They may connect with caves below, or they may narrow to a tangle of narrow fissures or be blocked with rubble. Dolines and pits have traditionally been discussed separately, in part because dolines can be inspected casually in the field or on air photographs or topographic maps. Pits require specific exploration and mapping, often by people with special equipment and special training. However, they are closely related to the same processes of vertical solution and transport.

Most reviews and textbooks speak of two mechanisms for the formation of dolines: dissolution of the bedrock to form a funnel or bowl-shaped depression, and collapse of shallow cavern roofs to form a funnel or bowl-shaped depression. The real situation is considerably more complicated. Every closed depression has three components:

1. A drain—some high-permeability pathway that carries the internal runoff collected by the sinkhole into the subsurface.

Table 2.1
Geometrical Classification of Karst Landforms[a]

	$l/w \simeq 1$	$l/w \gg 1$
$w/d \geq 1$	Dolines (sinkholes) Compound and valley sinks Poljes	Cutters Solution corridors Solution canyons
$w/d \ll 1$	Solution chimneys Vertical shafts Subsidence shafts	Solution fissures

[a]Key to symbols: w = width, l = length, and d = depth; all are in the same units.

2. The solutionally modified zone at and just below the bedrock surface.
3. A cover of soil, colluvium, glacial drift or moraine, volcanic ash, or other unconsolidated material that makes up the land surface. In some landscapes, the cover may be absent.

Three processes act within these three components. Dissolution of carbonate bedrock is either restricted to the near-surface region or continues into the vadose zone, with enlargement and modification of the drains. Transport of soil and other insoluble material into the subsurface occurs by piping, in-washing, and gravitational collapse. Bulk bedrock is transported into the subsurface by collapse and upward stoping of preexisting solution cavities. These processes operate on widely different time scales. Dissolution of the bedrock is very slow; soil arch failure can be catastrophically fast.

Consider first the drains. High permeability pathways through the vadose zone are provided by fracture concentrations, by fracture intersections, and along bedding planes in steeply dipping limestones. As these networks of fractures are enlarged by solution, the flow may be concentrated along a single pathway, which can develop into either a solution chimney or a vertical shaft (Fig. 2.7).

Solution chimneys are structurally controlled and irregular in shape and ground plan. They follow structurally dominated pathways, which may include sloping or horizontal components in addition to vertical segments. The cross section is often in the shape of a fissure. Some solution chimneys are of considerable size, an outstanding example being the 320-m-deep Sotanito de Ahuacatlan in Mexico, which is mostly developed along a single fracture (Raines, 1972). A solution chimney in Big Ridge Cave, in central Pennsylvania, is about 100 m deep in vertically bedded Helderberg limestone.

Vertical shafts, in their ideal form, are right circular cylinders whose vertical walls cut the bedding regardless of the inclination of the beds. Pohl (1955) provided the definitive description. Their shape, controlled by the hydraulics of fast-moving water films, is largely independent of structure and bedding. Vertical shafts form in specific hydrogeologic environments, most importantly where there is a solution-resistant capping bed that retards vertical solution and focuses the water on specific points of attack. The base of the vertical shaft is a bedrock basin from which water drains through small channels. Abandoned drains are found higher on the walls of shafts. Basin and drain are often obscured by breakdown and rubble in shafts that open to the surface. The classic shafts were described from the central Kentucky karst, but identical features are found in the Appalachian plateaus of West Virginia, Tennessee, northern Alabama, and northwest Georgia. The distinction between

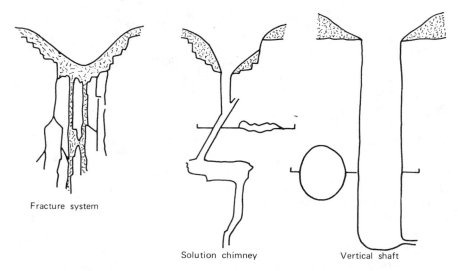

Figure 2.7 Three developments of drain systems for closed depressions: a solutionally widened fracture zone with enough permeability to permit soil transport to the subsurface; a solution chimney, which is essentially a vertical cave developed by selection of one pathway through the fracture system; and a vertical shaft.

vertical shafts and solution chimneys is based on the balance between hydraulic control and structural control of shaft morphology; many shafts take on aspects of both forms. Vertical shafts may be completely roofed over and are discovered through their drains as part of cave systems. Collapse of roof material leaves some shafts open to the surface, where they may or may not be located in a large closed depression. Some open shafts act as inlet points for sinking streams. Some shafts are completely filled with soil and rubble, with a small closed depression as their only surface expression.

Solution sinks are depressions in the bedrock surface. The bowl-shaped depression is deepened and widened by continuous dissolution of the bedrock as water moves down the sides of the sink toward the drain. Solution sinks continue to enlarge until the boundary of the sink meets the boundary of an adjacent sink coming the other way. Generally, solution sinks are roughly circular, although some elongate along lines of major fractures. Solution sinks may drain into solutionally enlarged zones of fractures, into solution chimneys, or occasionally into vertical shafts.

In arctic or alpine regions, the walls of solution sinks are often bare bedrock, sometimes with a bit of soil at the bottom (Fig. 2.8A). Many temperate climate solution sinks are completely mantled with soil, so that the bedrock form of the sinkhole is not visible (Fig. 2.8B). A relatively small drain may become plugged with soil. If the residual soil is an impermeable clay, the bottom of the sinkhole may be sealed so that water accumulates, forming a *sinkhole pond* (Fig. 2.8C). Sinkhole ponds are very common in sinkhole areas, but they are ephemeral features of the landscape because the plug may give way emptying the pond in a few minutes or hours.

Closed depressions also form by collapse within the bedrock. Collapse sinks (Fig. 2.9A) are shallow features. As the land surface is lowered by solution and erosion, the underlying bedrock, perhaps weakened by vertical solution along joints and fractures, comes within the influence of the stress pattern in the roof overlying a preexisting cave passage or solution cavity. Ceiling beams are weakened and cut by advancing surface erosion, and the roof rock collapses into the cavity. If the cavern into which collapse occurs contains a stream, the rock and soil can be removed and a vertical walled sink is preserved (Fig. 2.8D). Steep-walled dolines, providing access to underground streams, are known as karst fensters. In the absence of a stream, continuing erosion causes walls to slump, and soil drapes over them. Cave passages can sometimes be entered through the bottom of a collapse sink, in which case the origin of the feature is obvious. In many cases, however, the collapse is so completely mantled by soil and debris that its form is not distinguishable from sinkholes of other origins.

Collapse can also originate at some depth in the bedrock, if it is assisted by fracture zones or by a poorly consolidated bedrock material. Subsidence shafts are deep-seated collapse features. They originate from the collapse of a cave or solution cavity that stopes its way to the surface. If circulating groundwater is available to dissolve and remove fallen blocks, and a fracture zone provides a line of structural weakness, the upward stoping process can proceed for a long distance before the shaft breaks open to the surface. A subsidence shaft accessible from the surface may bottom out in collapse rubble high above the cave passage that gave birth to it (Fig. 2.9C).

Subsidence shafts may either be roughly cylindrical or open directly into large cave rooms. The entrance room of Hellhole Cave, West Virginia (Davies, 1958), and Devil's Sinkhole, Texas (White, 1948), are of the latter type. Vertical subsidence shafts are represented by the Empalme entrance to the Rio Camuy System, Puerto Rico (Gurnee, 1967), and the deep Mexican shafts of El Sotano and Sotano de las Golandrinas, the former, at more than 400 m, being the deepest shaft now known in North America (Raines, 1968).

Deep pits sometimes form by upward subsidence through resistant or nonsoluble beds. One example is Dante's Descent on the Coconino Plateau in northern Arizona. There has been extensive col-

Surface landforms in karst regions / 25

Figure 2.8 Some representative dolines: (A) large alpine doline with exposed bedrock walls and accumulation of soil at the base; (B) soil-mantled sinkholes in southcentral Kentucky; (C) series of sinkhole ponds along major fracture of southcentral Kentucky. Note that the dry sink in the background is deeper than the water-filled ones; (D) Smullton Sink, a collapse sink and karst fenster in Brush Valley, Pennsylvania; (E) a soil piping or cover collapse sink near State College, Pennsylvania. In the background, a previous piping failure is now rubble-filled; (F) closeup showing a sharp break of the soil rim resulting from shear fracture during collapse of soil arch.

lapse of the resistant Supai sandstone into a cavernous zone in the underlying Redwall limestone. The resulting pit is 120 m deep with vertical to overhanging walls. A second example is the Big Hole near Braidwood, New South Wales (J.N. Jennings, 1966), where the entire exposed 100 m is Devonian sandstone.

Upward stoping initiated by deep-seated solution cavities, where

26 / *Geomorphology and hydrology of karst terrains*

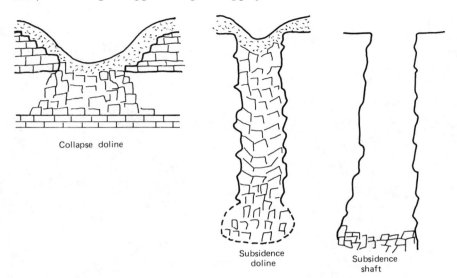

Figure 2.9 Three types of collapse and bedrock subsidence feature: collapse doline formed by ceiling fracture of a shallow cave passage; bedrock subsidence doline formed by upward stoping of a solution cavity at depth; subsidence shaft formed by upward stoping of material into a cavern and its concurrent removal.

transport processes are inadequate to remove fallen material, produces a subsidence doline at the surface (Fig. 2.9B). These look much like other dolines, but the depressions may be underlain by tens or hundreds of meters of broken rock.

A related vertical karst feature is the cenote described in Yucatan (Cole, 1911; Pearse et al., 1936). Cenotes are karst features that lead to water, vital to the Mayan culture because the Yucatan has no surface streams. The depressions in the flat karst plain of northeastern Yucatan appear to be of at least four types (Reddell, 1977): caves, solution dolines, vertical shafts, and stoping chambers with bell-shaped and small surface openings.

If the soil cover is thick, vertical transport of soil into the subsurface through the solutionally widened drain may produce sinkholes at the surface that do not require depressions in the underlying bedrock. Characteristic vertical walled sinkholes in the soil caused by soil piping and collapse are sometimes called cover collapse sinkholes. Alternatively, the soil may slump continuously into the solutional openings in the bedrock, always stoping upward to the surface to form cover subsidence sinkholes (Beck, 1984).

Figure 2.10 illustrates the stages in the development of a cover

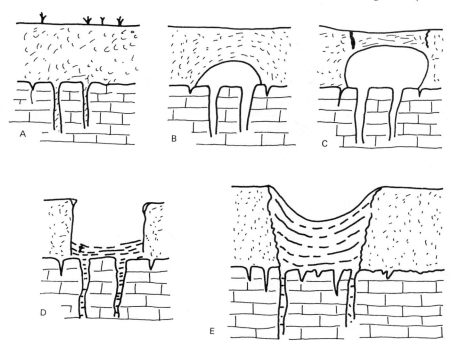

Figure 2.10 The development of two types of soil piping doline: (A) initial conditions: thick soils (or other unconsolidated cover) overlie vertical solution pathways in the bedrock; (B) flushing of the drain and initial formation of void beneath a soil arch; (C) critical void when the width of a void and thickness of the remaining soil can no longer sustain the arch; (D) arch failure with formation of a cover collapse doline; (E) final form of cover subsidence doline in which continuous ravelling and piping of soil have created a surface depression without formation of a large void and soil arch.

collapse sinkhole. Infiltrating water washes soil into the subsurface. Flow is concentrated into a single solution tube or widened fracture that acts as a drain. There must be some mechanism in the subsurface for disposing of the in-washed soil; a cave with a flowing stream is the most effective. The first stage is to clear the drain of sediment, producing a highly efficient pathway for sediment transport. If soils are thick and cohesive, a cavity forms, roofed with a soil arch between the bedrock and remaining soil. These cavities can be stable for long periods of time, and they grow gradually as storms provide water for further soil flushing and cavity enlargement. As the cavity becomes larger, and the roof becomes thinner, a critical thickness is reached when the arch can no longer support its own weight. Collapse is abrupt, forming a steep-walled sink (Fig. 2.8E,F). The in-falling soil usually blocks the drain, temporarily deactivating the sink. Then bank caving and slope wash eventually close it up again until fresh soil flushing repeats the cycle.

The formation, collapse, and refilling of cover collapse sinkholes is a natural process. New sinks dot the landscape in most karst regions, and human intervention is unnecessary, although as will be discussed in Chapter 13, these sinks are particularly sensitive to human activities on the karst land surface. Because frozen ground makes a good protective roof for the enlarging cavity, cover collapse sinkholes tend to form more commonly in the spring when high rates of internal runoff from snow melt or early rains transport the soils and the thawing ground triggers collapse.

Whether soil piping activity generates a cover collapse sink or a cover subsidence sink depends on the nature of the drain and the thickness and cohesiveness of the soils. Cover subsidence sinks (also known as raveling sinks in the engineering literature) form where the cover material is not sufficiently cohesive to maintain a mechanically stable arch above the developing void. The effect of continued soil piping is transmitted rapidly to the surface, where it is manifested by gradual settling of the ground. Only a gentle sag may appear at first, detectable only by the collection of circular puddles after rainstorms. At later stages circular shear cracks may appear around the perimeter of the sink, as the ground settles further within the circle. Subsidence, along with possible widening of the sink, continues at variable speeds over a period of hours, days, or weeks, depending on how efficiently the underlying soil is being piped away.

The processes of bedrock dissolution, piping of unconsolidated material, and bedrock collapse and stoping lead to five categories of doline and three categories of open shaft. Most real examples have been influenced by more than one process, and there is a continuum of landforms between these end members. Because solution dolines develop slowly, and collapse dolines degrade slowly after the initial collapse, these may be regarded as permanent features of the landscape on a human time scale. Cover collapse dolines form abruptly, and cover subsidence dolines may make themselves felt at the surface in a matter of hours or days.

2.2.2 Closed Depression Features of Intermediate Size

Closed depressions range in size from meters to kilometers (Fig. 2.11). As the closed depressions enlarge, they generally develop more than one inlet point and may develop a subsidiary set of surface drainage channels; mass transport processes may operate differently at different locations in the same depression.

Caprock-protected sinks are large closed depressions developed

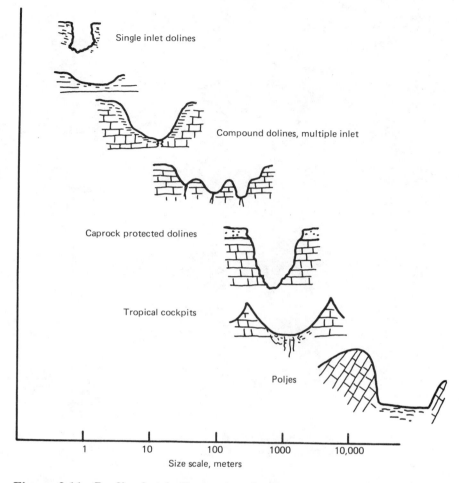

Figure 2.11 Profile sketch illustrating the size scale and types of closed depression features.

in the margins of plateaus. In the Cumberland Mountains of the southern Appalachians, the mountain flanks are underlain by cavernous limestone, but the tops of the ridges are protected by a thick cap of sandstone. If for some reason the sandstone caprock is breached some distance back from the plateau margin, internal drainage is initiated. Solution beneath the sandstone, collapse of the caprock, and vertical shaft development combine to form large, closed depressions (Fig. 2.12). The rim of the depression is protected by the sandstone so that these sinkholes tend to be very deep and steep-sided.

30 / *Geomorphology and hydrology of karst terrains*

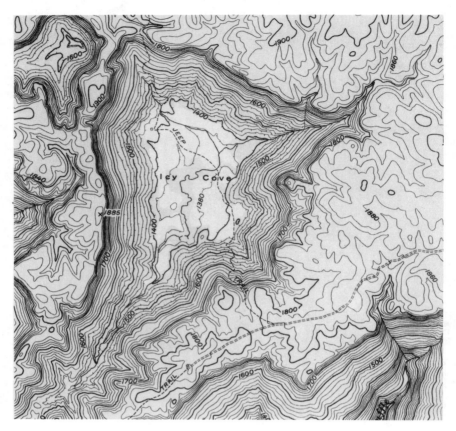

Figure 2.12 Icy Cove in the Calfkiller River Basin, Tennessee, is an example of a caprock-protected doline. The high area is part of the Cumberland Plateau, supported by the Pennsylvanian Pottsville sandstone. Lower portions of the cove are in Mississippian limestone. Four-square-mile segment of U.S. Geological Survey, Monterey Lake Quadrangle.

Compound sinks are formed when individual sinkholes grow and coalesce to form large closed depressions with multiple points of infiltration (Fig. 2.13) Compound sinks are common in the doline karst plains of the Appalachians, in southern Indiana, in central Kentucky and Tennessee, and in southern Missouri. Compound sinks are generally shallow but may occupy areas of several square kilometers. These features are called uvalas in the Slovene karst (Terzaghi, 1958).

Valley sinks are closely related to compound sinks (Fig. 2.13). They form in the last stages of degradation of underdrained valleys. After the main drainage has gone underground, continued runoff into the dry valley and sinking tributary streams along the sides of the valley create sinkholes. These deepen and widen until they also coalesce into large closed depressions occupying the space where the valley used to be.

Cockpits are large sinkholes found typically in thick limestones in tropical climates. The depressions are bowl-shaped, often up to a kilometer or more in diameter. The type locality is the Cockpit region of Jamaica, where a 1000-m-thick band of limestone has been eroded into a karst belt 30 km wide and 90 km long (Sweeting, 1958). Cockpit is a local name assigned because of the resemblance of the bowl-shaped sinks to cock-fighting arenas. Cockpits are often sufficiently large that subsidiary channel systems develop on the walls. The map

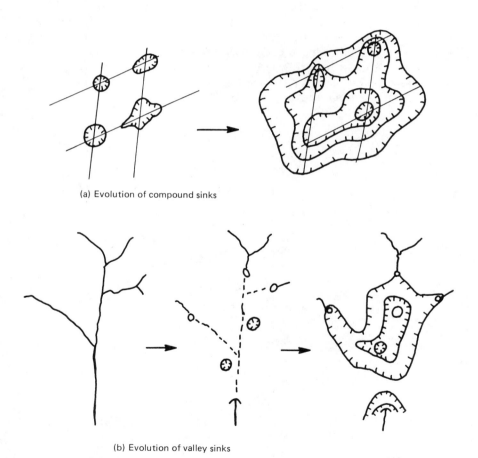

(a) Evolution of compound sinks

(b) Evolution of valley sinks

Figure 2.13 The evolution of (a) compound sinks (uvalas) and (b) valley sinks.

of individual cockpits is therefore serrate or star-shaped rather than circular or elliptical, as is the case of smaller dolines.

Cockpits often join along a rather sharp ridgeline, and the residual mass of rock at the intersection of three or four cockpits may be a pyramidal hill. Williams (1971) drew the drainage divides, which essentially follow the ridgelines between closed depressions for several karst areas in New Guinea. He showed that this allowed the entire area to be patterned (or tiled) with rough polygons of various sizes and numbers of sides; this he termed *polygonal karst* (Fig. 2.14). Laying out the polygonal pattern allows the closed depression surface to be reduced to a geometric pattern, which is then amenable to further analysis of polygon size and topology.

The surface channel pattern found within larger cockpits and other large closed depressions can be given a Horton- or Strahler-type stream ordering (Fig. 2.15) (Williams, 1971). A zero-order depression is defined to include the case of those cockpits or dolines with no channel pattern. These closed depression orders do not seem to follow a Horton law relation for the New Guinea examples, but closed depression orders measured in the Jamaican cockpits have a linear frequency distribution (Fig. 2.16) (Day, 1976).

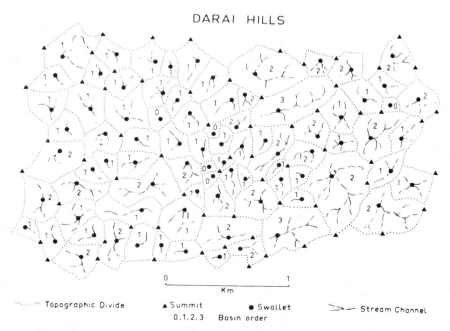

Figure 2.14 Polygonal karst from the Darai Hills, New Guinea. [From Williams (1971).]

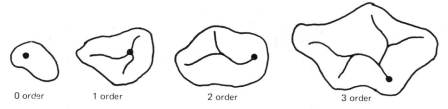

Figure 2.15 Sketch defining the order of internal drainage in a doline, following the suggestion of Williams (1971).

2.2.3 Measures for Dolines, Cockpits, and Uvalas

The area of a doline can be defined by a contour drawn at the break in slope at the edge of the doline. If the doline is not circular, a long axis, l, can be drawn and a longest distance, w, perpendicular to it can be drawn. To the extent that the doline approximates an ellipse, the area $A = (\pi/4)lw$. The mean doline area is obtained by averaging areas of individual dolines over the total number of dolines observed:

$$\overline{A}_D = \frac{1}{N_D} \sum_i A_{D,i} \qquad (2.15)$$

The ratio, l/w, is an aspect ratio that describes the eccentricity of the doline. The orientation of the long axis can be related to structural controls. An aspect ratio for more asymmetric dolines is given by Williams (1972a).

Three measures have been devised to describe the arrangement of dolines with respect to each other: the depression density, the doline area ratio, and the area of internal drainage (Table 2.2). The depression density is obtained by counting all dolines within the karst area and dividing by the area. All depressions are counted,

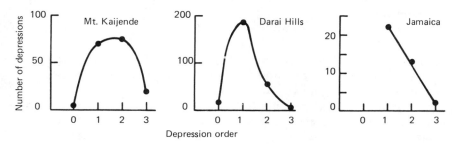

Figure 2.16 Distribution of closed depressions of various orders as a function of order for three tropical karst areas: Mt. Kaijende and the Darai Hills in New Guinea (Williams, 1971) and Jamaica (Day, 1976).

Table 2.2
Measures of Doline Development

Measure	Definition	Units
Depression density	$D_D = \dfrac{N_D}{A_K}$	L^{-2} (km^{-2})
Doline area ratio	$R_D = \dfrac{1}{A_K}\sum_i A_{D,i}$	Dimensionless
Area of internal runoff	$A_I = \sum_i A_{I,i}$	L^2 (km^2)

[a]Key to symbols:
$A_{D,i}$ = area of individual dolines
$A_{I,i}$ = catchment areas of individual dolines
A_K = area of karst
N_D = total number of dolines in karst area

including those that are nested within larger complex closed depressions. Depression density measures the number of entry points of internal runoff water into the subsurface. The doline area ratio is determined by summing the areas of individual dolines over all dolines in the karst region and dividing by the area of karst. This parameter is unity in polygonal karst. It is the reciprocal of the index of pitting, as defined by Williams (1966a). The area of internal runoff is determined by delineating all catchments that drain internally into dolines and summing these over the entire basin. Both depression density and doline area ratio can be subdivided among different parts of a karst region or among the outcrop areas of different rock types. Area of internal runoff is determined by local divides for overland flow and may actually include the runoff area from nonkarstic rocks, if overland flow from these areas drains into sinkholes.

To a first approximation, doline depths follow an exponential distribution

$$n = N_0 e^{-Kd} \qquad (2.16)$$

Troester et al. (1984) examined doline depth distributions for four temperate karst regions and two tropical karst regions. In all cases the depth distribution was approximately exponential, but with different values of the exponential coefficient, K. If all data are normalized by plotting n/N_0 (Fig. 2.17), the patterns displayed seem to relate to the internal relief within the karst. The Florida karst is very flat, with broad, shallow dolines. The Missouri, Kentucky, and Appalachian karsts are developed on Paleozoic limestones and dolomites and have an intermediate range of relief. The two tropical

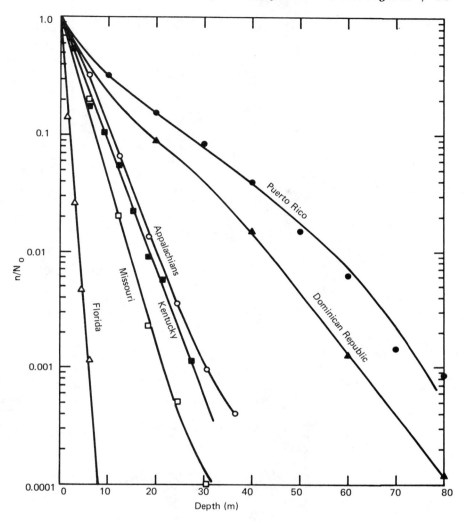

Figure 2.17 Sinkhole frequency–depth distributions for six karst regions. These data were fitted to the equation $n = N_0 e^{-Kd}$ to obtain the fitting coefficients listed in Table 2.3. [Adapted from Troester et al. (1984).]

karsts have the greatest internal relief; they also deviate the most from simple exponential form. These deviations imply that more complicated distribution functions are needed.

The K coefficients are listed in Table 2.3. Exponential curves have a scale factor that requires that $x = 1$ when $y = 1/e$. Thus, the exponent, $Kd = 1$ when the population of sinkhole depths falls to $1/e$ of the initial value. This, in turn, allows $1/K$ to be defined as a charac-

Table 2.3
Least-Squares Fitting Coefficients for Exponential Depth Distribution and Other Properties of Doline Populations

Karst region	Total number of sinkholes	Depression density (km^{-2})	Mean depth $\bar{d}$ (m)	d_e (m)	N_0	K (m^{-1})	r^2
Temperate karst regions							
Appalachians							
Mississippian limestone	2,182	1.45	7.3	4.62	4,276	0.22	0.98
Ordovician limestone	1,506	1.31	8.3	4.62	4,000	0.22	0.98
Ordovician dolomite	1,472	1.00	8.0	4.18	4,190	0.24	0.98
All dolines	5,160	1.25	7.8	4.48	12,608	0.22	0.99
Kentucky	830	5.41	5.4	4.02	892	0.25	0.99
Missouri	2,217	—	6.8	3.23	9,789	0.31	0.99
Florida	3,395	7.94	—	0.85	12,299	1.18	0.99
Tropical karst regions							
Puerto Rico	4,308	5.39	19	11.35	6,876	0.088	0.99
Dominican Republic	7,205	5.71	23	8.93	69,153	0.11	0.99

Source: Troester et al. (1984).

teristic depth for each population. These depths, labeled d_e in Table 2.3, are different from the mean depths, $\bar{d}$ which are also listed. The coefficient N_0 is the number of sinkholes of zero depth if the exponential distribution function is valid over the entire range. The implication is that the catchment area for internal runoff is finite at the time of sinkhole initiation.

Regression analysis of a population of 127 sinkholes on the Highland Rim sinkhole plain (Mills and Starnes, 1983) showed that sinkhole area was strongly related to drainage basin area (area of internal runoff, as defined in Table 2.2), but the correlation with sinkhole depth was weak at best. The relation between doline area and area of internal runoff is what would be expected if sinkhole size is indeed related to the available volume of water funneled into it. The lack of relationship between sinkhole area and sinkhole depth is in agreement with the extrapolation of sinkhole depth distributions to zero depth.

The spatial distribution of dolines can be investigated by examining depression density, doline areas, the relative spacing of nearest-neighbor dolines, and other spatial measures. Figure 2.18 shows the distribution of depression densities measured on 1-km grids drawn on 42 quadrangles in the Highland Rim and Pennyroyal Plain doline karsts of southcentral Kentucky and northcentral Tennessee (Kemmerly, 1982). Because of the large sample—nearly 25,000 dolines—two populations show up quite clearly, confirming an earlier discovery of two doline populations in the Mendip karst of England (Drake and Ford, 1972). The population distribution was independent of the bedrock and was not a result of a mixture of solution dolines and collapse dolines. The populations are explained by a "multigenerational diffusion and competition process (MDCP) model." The essential part of this model, proposed by Ford in 1964, is that the development of first-generation dolines produces a local drawdown in groundwater levels, and the induced hydraulic gradient helps flush soil cover and clear sediment out of the joint and fracture system. These factors induce a second generation of doline development in the region of influence of the primary doline. Support for the model was found in a pronounced clustering of nearest-neighbor dolines.

Kemmerly and Towe (1978) calculated the rate of doline development directly for an area in the Highland Rim sinkhole plain in northcentral Tennessee. They compared the sizes of some 18 dolines as they appeared in air photographs taken in 1937 with their sizes in photographs taken in 1972. These dolines were carefully selected to avoid interference from human activities. The directly determined growth rates were 0.4, 0.7, and 1.0 $m^2 a^{-1}$ for lossial, clayey residual, and silty colluvial soil covers respectively. Extrapolating backward

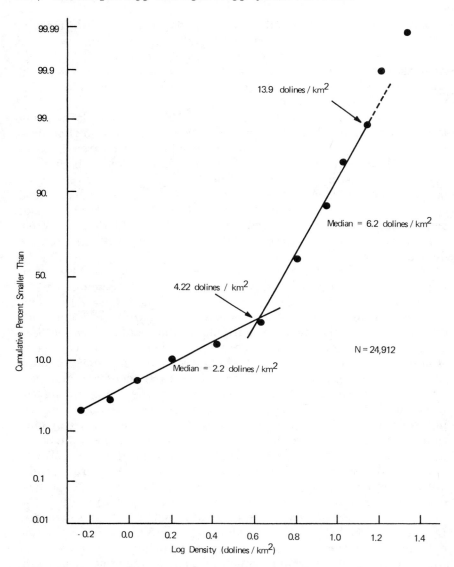

Figure 2.18 Distribution of depression densities in the Highland Rim and Pennyroyal Plain doline karsts. [Adapted from Kemmerly (1982).]

to the initiation of the dolines, ages ranging from 25,000 to 65,000 years were obtained, in agreement with some paleontological and pollen evidence.

An alternative approach was used by Palmquist et al. (1976) in the karst of northeastern Iowa. They examined doline development

Surface landforms in karst regions / 39

in regions that were thinly covered with glacial drift of various but known ages. The depression density at first increases rapidly with age of the drift and then levels off. Regression analysis of the data reveals that

$$D_D = 7.26 \log t - 27.9 \qquad (2.17)$$

where D_D is depression density in number per square kilometer and t is the age of the till in years. A curious result from this study was that an initiation time of about 6000 years seemed to occur between the time when the drift cover was deposited and when dolines first became perceptible. The number is comparable to the initiation time for cave passage development (see Chapter 9) and may represent the time necessary to widen the fracture system enough for clastic material to be transported.

Completely stochastic models for karst landform development have met with only marginal success. McConnell and Horn (1972) examined the distribution of closed depressions in the Mitchell Plain of southern Indiana. By showing that the depressions followed a mixed Poisson distribution, they concluded that at least two independent random processes were operating, which they deduced to be cavern roof collapse and dissolution of the bedrock by infiltrating surface water. Hypotheses of doline distribution as a completely random arrangement and of mutual doline influence were rejected.

Using multiple regression analysis, LaValle (1967, 1968) compared various measures of closed depression morphology in the central Kentucky karst with measures that were supposed to describe processes of karst development. These calculations showed that the elongation of closed depressions and the relative orientation with respect to structural features were correlated with various geologic and hydrologic parameters. The correlations were not sufficiently precise to be of much value in hydrogeologic interpretation of the aquifer.

2.2.4 Poljes

The term polje (literally "field" in Slovene) is used in two senses in karst literature; one refers to the large closed depressions of the Adriatic Coast and the other to any large closed depression with a flat floor.

The poljes of the Adriatic karst are rather special features not duplicated anywhere else in the world. They are elongated depressions 1 to 5 km wide and up to 60 km long. Polje floors are flat and alluviated (Fig. 2.19). Terrace levels occur, and more than one drainage system may be associated with the polje. The poljes are surrounded by rugged limestone mountains, usually with a sharp break between the flat polje floor and the mountain slope. Water enters the

Figure 2.19 The western end of Popovo Polje, Yugoslavia. Note the flat agricultural land on the polje floor and steep limestone mountains. The view shows only a small segment of the 60-km-long polje. The Trebišnjica River flows above the polje floor in a concrete-lined channel built to preserve water for irrigation and hydroelectric power (upper right of photo).

poljes from springs and flows across the polje floors as alluvial surface streams that sink again into caves and swallow holes (ponors) on the opposite side. Water flows from polje to polje at progressively lower elevations until it ultimately appears in the form of large springs on the Adriatic Coast (Fig. 2.20).

Once a season, and sometimes more frequently, the poljes flood because of greatly increased flow from the springs and rising water levels in the mountains, which cause the ponors to reverse flow and become springs. In flood time the entire bottom of the polje may become a lake. Most polje floors are farmland, and roads, houses, and villages can be found in the higher part of the polje floor. Floods threaten homes and villages and devastate crops during the growing season. Intensive efforts to manage the floods allow the waters to be controlled for irrigation and hydroelectric power.

The Dinaric poljes occur in massive limestone in a tectonically complex zone. The axis of elongation of the poljes parallels the structural grain from northwest to southeast. Cvijič (1960) argued for a tectonic origin, pointing out that the poljes occur along faults, which provide zones of weakness for the development of uvalas that grow and coalesce to form the poljes. However, there has been a continuing and at times heated debate over the origin of poljes. Flat polje floors that truncate the bedding make it obvious that controls other than structural ones are in effect. Grund (1903) and Terzaghi (1958) argued that the polje floors are planed to the level of the water table, although the floor elevation of various poljes differ. Roglič (1957) and Louis (1956) explain the flat floors as a balance between the amount of clastic material that must be transported from the basin and the amount of water available to do the transporting. The mountains of the Adriatic karst contain beds of flysch, which add to the clastic load. The bottoms of incipient poljes become choked with clastic material, which generates a local perching of groundwater, and growth of the polje proceeds by lateral solution of the polje walls.

Geophysical work done as part of hydroelectric power projects

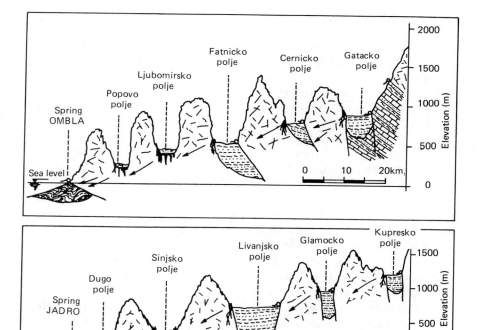

Figure 2.20 Two cross sections of the Adriatic karst showing the stair-stepping of polje levels from the mountainous interior to the coast. Tectonic setting and sedimentary in-filling based on recent test drilling and extensive hydrogeologic mapping. [Adapted from Mijatović (1983).]

strongly supports a mixed tectonic–karstic origin for the poljes (Mijatović, 1983). Some sedimentary in-fillings are shallow, a mere veneer over the solution-planated bedrock floor. Other in-fillings are deep, covering a bedrock floor that may be below sea level. Active faults are abundant. Earthquakes are common. The polje karst of the Adriatic can be seen as the result of intense karstic processes superimposed on a basin and range topography created by block faulting.

The second sense in which the term, polje, is used refers to any large closed depression karst feature with a flat floor. The lower size limit that would separate small poljes from large dolines is thus somewhat arbitrary, and the presence of the flat alluviated floor seems to be the distinguishing characteristic. Gams (1977) writes, "a polje is an extensive (closed) basin with a flat bottom, karstic drainage, and steep slope at least on one side. If the slopes are mostly steep, with a break at the transition to the bottom, and a sinking river, the flat bottom is 400 meters wide at least."

By using only morphometric elements and eliminating the unique structural and stratigraphic setting of the Dinaric karst from the definition, the long-standing controversy over whether poljes exist outside the Dinaric karst can be resolved. Large closed depression features that meet Gams's definition are found in many regions of the world including Jamaica (Sweeting, 1958), Cuba (Gradzinski and Radomski, 1965), the Northwest Territories of Canada (Brook and Ford, 1977), western Ireland (Williams, 1970), and north Borneo (Sunartadirdja and Lehmann, 1960). Many of these poljes have the hydrologic characteristics of the Adriatic poljes, including some degree of seasonal flooding.

2.2.5 Cutters and Associated Features

Solution depressions guided by single joints or fractures seem to have received less attention than the doline-related features. In the middle of the size scale, and perhaps most widely distributed, are the linear slots cut in the bedrock along a guiding structural element. Typically these are fractions of a meter to several meters wide and from a meter to tens of meters deep. Their length is usually indefinite, but many can be traced for tens of meters. Howard (1963) argued that these features, which he termed cutters, were one of the three principal landforms in temperate climate karst, along with sinkholes and caves. Cutter is synonymous with the British term *grike*, which is widely accepted in the English-language literature. Thus cutter and pinnacle topographies and grike and clint topographies are the same. Unlike dolines, cutters do not seem to have well-developed internal drainage. Soil piping and other transport of sediments is less effective through the cutter system than through

Surface landforms in karst regions / 43

dolines. As a result, cutters are usually buried in the regolith and have little surface expression.

Cutters may be seen in profusion in road cuts through the horizontal limestones of Indiana, Kentucky, and Missouri. Cutters in the massive Bedford oolite of Indiana taper to narrow cracks. Cutters in the more heterogeneously bedded limestones of Kentucky, Missouri, and other parts of Indiana tend to bottom out at specific bedding planes (Fig. 2.21). The solution trench has a bedrock floor. Cutters (Fellows, 1965) appear to have a dendritic pattern superimposed on the rectilinear regional joint system and act as collector systems for infiltrating groundwater, which moves laterally under the regolith from minor cutters to master cutters and then underground.

Cutters also occur in regions of folded limestones, where they are usually oriented along the strike and are also soil-filled. Where soils are thin, the intervening pinnacles or ledges of undissolved rock sometimes stand out in relief, although the cutter itself is completely obscured. The resulting long parallel limestone ledges are a common landform in the Appalachian valleys.

Cutters are formed by vertical solution along fractures or are downcut by lateral flow along the bottom of the trench. Solutionally widened joints of fractions of a meter in width are usually classified with other bedrock solutional sculpturing as *kluftkarren* (see Section

Figure 2.21 Cutters overlain by soil mantle in road cut on Interstate Route 71, just east of Louisville, Kentucky.

2.4). In the width range of meters are the cutters. Toward the large end of the size scale are wider and deeper *solution corridors* with widths in the range of tens of meters (Fig. 2.22). These are known as *bogaz* in the Adriatic karst. Similar features in Puerto Rico were described as zanjones by Monroe (1964, 1976).

Still larger are the somewhat enigmatic *solution canyons.* Their widths and depths are on the order of tens to hundreds of meters and their lengths are a kilometer or more. Jennings and Sweeting (1963) described such features from the Fitzroy Basin of western Australia, referring to them as "giant grikes." The rectilinear pattern expected from fracture control is obvious. Other features in the same size range occur in County Sligo, Ireland, where they appear as vertical, walled canyons in limestone with ungraded undulating floors. These "aults" are several hundred meters wide and several kilometers long. There are a large number of them in an *en echelon* pattern. Deep vertical walled canyons are also cut by surface streams, which Roglić (1972) has argued are the true valley form in holokarst. Solution also plays a role in their development.

Completing the pattern set out in Table 2.1 a relatively uncommon landform exists that is the linear equivalent of the shaft. Solution fissures are analogous to the cutters but have depths much greater than their widths. These are found primarily in mountainous regions where high relief and high hydraulic gradients permit their enlargement. Some of these along the Teton shelf in western Wyoming are known to be on the order of 100 m or so deep (Medville and Werner, 1974). Tectonic movements that produce deep, narrow earth cracks in other rock types may also play an important role in the development of solution fissures.

2.3 LANDFORMS WITH POSITIVE RELIEF

Whether a feature of the landscape is positive or negative depends on the point from which one views it. It seems both natural and scientifically useful to view fluvial landscapes as consisting mainly of features of negative relief. The chemical weathering and mass transport, sheet wash, and sediment transport in channels by the action of flowing water all act in a downward direction, and the resulting pattern of slopes and channels is highly structured and of interest to study. The hills, ridges, and mountains become residual objects—fragments of the solid earth not yet attacked by the forces of fluvial erosion. Their pattern is more chaotic, and they are therefore less productive as objects of study.

The creation of karst landforms also takes place by downward-acting forces of solution and mass transport. Karst landforms have interior drainage and the important role of underground drainage

Figure 2.22 Soil-filled solution corridor in the French Pyrenees.

systems. The most interesting positive relief karst features are those developed on thick, massive limestones in tropical areas. In the Caribbean Islands, South China, and New Guinea, for example, occur incredible landscapes of jagged hills rising from flat plains or from a chaotic tangle of narrow gorges not worthy to be called valleys. In these regions features of positive relief completely dominate the landscape.

2.3.1 *Pinnacles*

Occasionally in temperate regions and frequently in arid regions, limestones develop a pinnacle weathering form. Solution along joints and fractures lowers the intervening rock mass and leaves the limestone blocks as isolated spires or pinnacles standing meters or tens of meters above the surrounding land surface. Most pinnacles in both temperate and arid regions occur on steep slopes and cliffs, where they appear to be little different from similar spires that form on other massive rocks. The outlines of pinnacles in temperate and arid regions tend to be smooth; the rock surface follows joint planes with some modification by solution. In tropical climates pinnacles are taller, more spiney, and usually sculptured with complex minor solutional forms (karren), which give the entire rock mass an extremely jagged surface.

There is a continuum in scale from small pinnaclelike solutional sculpturing a few centimeters in height (spitzkarren) through intermediate forms a few meters tall, to significant landforms tens of meters high. One of the most extreme examples of pinnacle development occurs in the Highlands of New Guinea, where interconnected spires on the south flank of Mount Kaijende reach a relief of some 300 to 400 m (Jennings and Bik, 1962).

2.3.2 Residual Hills

A characteristic feature of slope retreat in karst regions is that of internal drainage through vertical solution pathways. When thick layers of limestone are exposed on hill slopes, runoff from higher on the hill or streams collected on clastic caprocks of the hill can make their way to base level more efficiently by dissolving out vertical paths inside the hill. These usually take the form of vertical shafts in regions with a clastic caprock, but solution chimneys and open joints and fractures also exist. As a result, karst regions tend to have steeper, more rugged slopes.

The central Kentucky karst, and many other areas of dissected plateau, has isolated hills locally called "knobs." Knobs are isolated fragments of the plateau, and many retain a fragment of the original caprock that protects them from further erosion. They are frequently ringed with vertical shafts that drain off the precipitation that falls on top of the knob. In the later stages of development, the caprock is eroded away and an isolated limestone hill remains standing in relief above the karst plain.

Isolated hills of limestone standing above nearly flat plains occur in the Adriatic karst, where they are known as hums (Herek and Stringfield, 1972), and in other regions such as the Bom Jesus de Lapa in Brazil, described in detail by Tricart and Cardoso da Silva (1960).

2.3.3 Cone and Tower Karst

A characteristic feature of tropical karst is a residual hill with either vertical or near-vertical sides. The vertical-sided hills are called towers *(Turmkarst)* and the near-vertical hills are called cones *(Kegel-*

karst) (Lehmann, 1936). The detailed form of the cones and towers varies considerably (Fig. 2.23). Figure 2.24 shows a cone karst in Puerto Rico. The cone karst of Java is in the form of sigmoid hills, quite rounded at the top but with steep slopes near the bottom (Tjia, 1969). Some towers such as those of New Guinea (Williams, 1973), which Williams referred to as "broken bottle karst," are nearly vertical and rise to points. Karst towers in South China and north Vietnam are narrow vertical masses of limestone rising several hundred meters above an alluviated plain (Von Wissmann, 1954; Silar, 1965; Chinese Academy, 1976). Karst towers in Cuba, called mogotes, are vertical-walled but have irregular, serrate tops, that are usually nearly flat (Lehmann, 1954a). As Jennings (1972) has pointed out, the detailed forms are complex and variable and there seems to be a gradation between the various styles.

The elevations of the cones and towers vary with the region from a few tens of meters to several hundred meters. Some, such as the mogotes of the Sierra de los Organos, Cuba, are dissected remnants of a structurally controlled line of ridges and show an accordant skyline. Some appear to be the dissected remnants of plateaus with more or less accordant summits. Others have tops set at completely variable elevations.

The internal structure of the cones and towers can be complex. The outer surface is sculptured with various small solution features. The tops, particularly those with relatively flat summits, contain deep pits, some of which may reach the base of the tower. The towers are sometimes riddled with caves. Lehmann (1954a) has called attention to the "foot-cave," a cave opening at the base of the hill at the elevation of the surrounding plain (if there is a surrounding plain). Streams are known to take underground short cuts directly through the cones and towers rather than follow the valleys between them.

The space between the hills is usually irregular and ungraded, unlike the ordered arrangements of tributary valleys in fluvial landscapes. The intervening spaces between towers that rise from alluviated plains often show little evidence of tributary pattern. In other areas, the towers are nestled against each other, separated only by deep, rugged ravines with no trace of tributary pattern.

The cross-sectional profiles provide the best distinction between cone and tower karst and the closely related cockpit karst. In both types of karst there must be a regional system of fractures that permit the development of underground drainage. The underground drainage system must be sufficient to carry all excess precipitation as internal runoff in addition to being competent to carry whatever clastic load may be generated from the solution of limestone. There is a suggestion that the cone and tower karst is the result of cutter

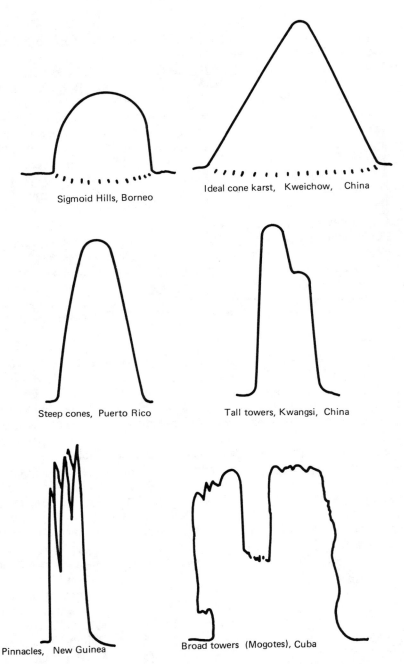

Figure 2.23 Some representative styles of cone and tower karst.

Figure 2.24 Cone karst (kegelkarst) in the northern karst belt of Puerto Rico. The conical hills are underlain by the massive Lares limestone. The low-relief foreground is underlain by the clay-rich Cibao formation.

development carried to an extreme of relief. If solution is uniform along the length of the fractures, and if vertical solution predominates over lateral solution, an interlacing set of deep solution gullies and intermediate residual hills results. If solution is concentrated at fracture intersections and lateral solution tends to predominate, the cockpit is formed. The hillslopes of cone and tower karst are convex hillslopes in cockpit karst are concave.

2.4 SOLUTIONAL SCULPTURING

Direct rainfall, sheet wash, channelized flow, and percolating flow under various kinds of mantle materials produce a myriad of small sculpturings on the bedrock surface of soluble rocks. These forms, as a class, are called karren, a German term that has come into general usage in English. The French equivalent is *lapiez*. The forms are small, often transitional into one another, and much affected by details of the environment in which they are formed (Table 2.4). The names are mainly those of Bögli (1960), whose terminology has come into wide usage. However, following Jennings (1985) and Pluhar and Ford (1970), the preference has been for an anglicized version of Bögli's terminology. An attempt has been made to separate the land-

Table 2.4
Types of Karren

		Hydraulic forms		Etched forms	
		Sheet flow	Channel flow	Structural weakness	Massive bedrock
Bare bedrock surface	↑ Increasing slope	Rill karren (rillenkarren)	Wall runnels (wandkarren)	Solution grooves	Rain pits
		Stepped karren (trittkarren)	Pit and tunnel karren		
			Runnels (rinnenkarren)	Cleft karren (kluftkarren)	Pinnacle karren (spitzkarren)
		Kaminitza	Meandering runnels (meänderkarren)		
Covered surface — Snow pack					Pedistal karren
			Bag-shaped runnels (hohlkarren)	Groove karren	
Soil			Solution wells		
			Rounded runnels (rundkarren)	Ledge pendants (deckenkarren)	Irregular rounded pinnacles

forms into groups: those that are clearly hydraulic forms, those that form by etching a preexisting structural weakness, and those that form by etching the massive bedrock in the absence of obvious structural lines of weakness.

The slope of the exposed rock surface is important in determining the flow regime and the velocities attained. The other variable shown in Table 2.4 is the type of cover. Solution beneath snow pack also produces solutional sculpturing, particularly in the spring when melting snows produce rapid circulation of water between the remaining snow pack and the bedrock surface. Soil mantles reduce velocities to a slow percolating flow. Soil plugs in channels inhibit the free flow of water and can also act as a sponge, holding water long after the storm has ended. Further, biological processes in the soils are sources of carbon dioxide (CO_2), so that the amount of limestone taken into solution beneath the soil may be much larger than the amount dissolved from adjacent rock ledges that are exposed only to rain water.

Sweeting (1972, Chap. 5) lists other environmental and lithologic factors controlling the type and degree of development of karren sculpturing. The amount and distribution of precipitation is important since it controls the runoff regime on the surface to some extent. The lithologic properties of limestone or dolomite are important. The presence of shale partings or chert layers, the thickness of bedding, and the primary porosity of the rock all contribute to the detail of karren development. Karren landforms develop best in massive, thick-bedded, pure limestone. The characteristic forms either do not form or do not persist in thin-bedded, shaley, or chalky limestones. Dolomite exhibits the same range of sculpturing as do limestones, but because the solution process takes place more slowly on dolomite, the forms are usually more subdued, all other things being equal (Pluhar and Ford, 1970).

The European literature on karren landforms (Bögli, 1960, 1980; Sweeting, 1972; Jennings, 1985) is extensive, particularly those papers describing European alpine karst. Corresponding descriptions of landforms from temperate North America are sparse. Pluhar and Ford (1970) investigated dolomite karren on the Niagara escarpment of eastern Canada. Moravec (1974) describes typical karren forms from the southern Appalachians.

2.4.1 *Hydraulic Forms Resulting from Sheet Flow*

On steep slopes or on the tops of sharp-ridged boulders close-spaced solution rills (rillenkarren) are formed. These are parallel troughs, spaced evenly a few centimeters apart, with rounded, parabolic cross sections and sides rising to a sharp cusp between adjacent rills (Fig. 2.25). They form only near the tops of the exposed surface; downslope the rills give way to a flat bedrock surface. The length of the rills increases with increasing slope.

Glew (1977) modeled the formation of rillenkarren using a rainfall simulator on blocks of salt and plaster of Paris. His results indicate that rillenkarren are a stable form that retreats into the rock without change in the overall geometry. The rills terminate in the downslope direction when the flow down the rill system attains a critical thickness. Rillenkarren development is apparently associated with the incipient stages of sheet flow, where the water films are thin and of smaller velocity. Glew varied the slope angle on his models and was able to show that the length of the rills from the crest to the breakup line varied directly with the slope (Fig. 2.26).

Sheets of water flow down sloping surfaces in a pulsating pattern. The pulses tend to generate a stepped surface. Once started, the steps stabilize and enlarge into steplike forms with a nearly flat tread and a nearly vertical riser. Some appear as pockets cut in the rock (e.g., Bauer, 1962), others (Fig. 2.27) as wide steps across the

Figure 2.25 Solution grooving and pitting (modified rillenkarren) under desert weathering conditions. Hakatai Canyon in Grand Canyon National Monument. Note that the rills break up near the bottom of the block, leaving only a pitted surface.

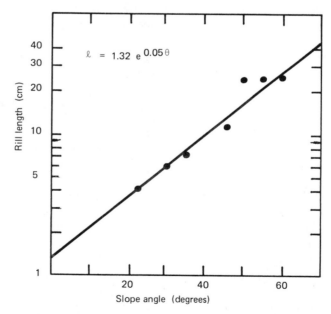

Figure 2.26 Relation of rill length to slope of the surface containing the rills. Data are from Glew (1977). The equation is least-squares fitted exponential function to Glew's data.

Surface landforms in karst regions / 53

Figure 2.27 Stepped surface (trittkarren) on massive limestone, in the French Pyrenees near Pierre San Martin Shaft.

entire rock face. Stepped karren (trittkarren) form on intermediate slopes.

Although it is perhaps stretching the concept of sheet flow, when water ponds on flat areas of limestone, it dissolves out solution basins known by the Slavic term *kaminitza* or the Spanish *tinajita* (Udden, 1925). Stagnant water in the basin accumulates carbon dioxide from moss, algae, and other plant debris and so is much more corrosive than the flowing sheets of water on bare limestone surfaces. Kaminitza are found on most bare rock pavements in many hydrogeologic environments. As they deepen some basins become bowl-shaped, wider at the bottom than at the top. Others develop overflow channels or inlet and outlet channels (Fig. 2.28).

2.4.2 *Hydraulic Forms Resulting from Channel Flow*

Of the channel features, the solution runnel (rinnenkarren) is ubiquitous. These are semicircular channels, usually rounded on the bottom and with a width ranging from a few centimeters to fractions of a meter. They are found on limestone surfaces of all slopes and on bare as well as soil-covered surfaces. On steep slopes, the runnels are nearly linear and are oriented along the slope of the bedrock. Branching or tributary patterns are rare. On more shallow slopes, tributaries are sometimes observed and the channels may meander

Figure 2.28 Solution pan with drainway (kaminitza or tinajita), near Lough Bunny, County Clare, in western Ireland.

(meänderkarren). Runnels cut in bare bedrock have sharp edges. When runnels form under a slight soil cover or merely become clogged with organic debris and vegetation, the extra CO_2 engendered by the organic material and the sponge action of the soils and vegetation enhance the solution of the bedrock in the bottom of the runnel, giving it a baglike cross section wider at the bottom than at the top (hohlkarren, in Bögli's terminology). Under deeper layers of soil, the flow is greatly inhibited, becoming a percolating rather than a channel flow, and the channel geometry is gradually lost. Instead of a cleanly defined runnel, the drainways under the soil mantle tend toward rounded forms, with only the vague outline of a runnel. These are called round karren (rundkarren).

2.4.3 Etched Forms Resulting from Solutional Attack on Structural Weaknesses

The solution of limestone in the absence of well-defined channel or sheet flow takes place most readily along joints, bedding planes, and other structural weaknesses. Smallest of these solution sculpturings, but common on exposed bedrock surfaces, are the solutional grooves in which the joints are etched into negative relief (Fig. 2.29). Solution grooves are common in both arctic and arid areas and are apparently better preserved where the precipitation rates are low. Davies

Surface landforms in karst regions / 55

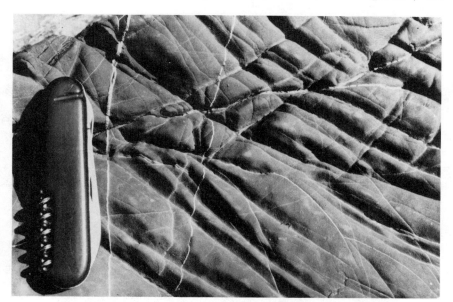

Figure 2.29 Solution grooving on limestone ledge. Wasatch Mountains at the head of Little Cottonwood Canyon, Utah.

(1957) found finely grooved limestone fragments, which he called rillenstein, to be the most common karst form in Greenland.

Most common of all karren features is the solutionally widened joint, cleft karren or kluftkarren (Fig. 2.30). They vary from a few centimeters in depth and a meter in length to meters in depth and many meters in length. Cleft karren in bare rock exposures have sharp edges and nearly vertical walls. The clefts easily become choked with soil, which retards the rapid vertical movement of water. The walls become more curved and there is solution into bedding planes along the sides. Pluhar and Ford (1970) distinguish between cleft karren as open joints extending indefinitely downward to terminate in narrow cracks, and trench karren, which terminate on resistant beds and therefore appear as flat-bottomed trenches. In their area on the dolomites of the Niagara escarpment in Ontario, the cleft karren were nearly linear while the trench karren often were sinuous. Cleft karren act as the primary paths for water to reach the subsurface. Runnels are often terminated by the clefts, and there is a transition from horizontal flow in the runnels to vertical flow down the clefts.

A horizontal equivalent to cleft karren is referred to as groove karren. Groove karren are horizontal grooves dissolved along bedding planes in horizontally bedded rocks. Downward percolating

Figure 2.30 Solution widening of joints (kluftkarren) on limestone pavement, in the Burren of western Ireland.

waters can move along either joints or bedding planes; they gradually dissolve away both of these surfaces, leaving isolated joint blocks behind (Fig. 2.31).

The solution of limestone along joints sometimes produces a mixed structure control–hydraulic control form, which may be called pit-and-tunnel karren (Fig. 2.32) (Pluhar and Ford, 1970). Instead of the joint being widened uniformly from the surface to depth, the flow is diverted into small vertical solution opening and then horizontally through what amounts to a tiny cave passage. Pit-and-tunnel karren seem to be common in both temperate climate

Surface landforms in karst regions / 57

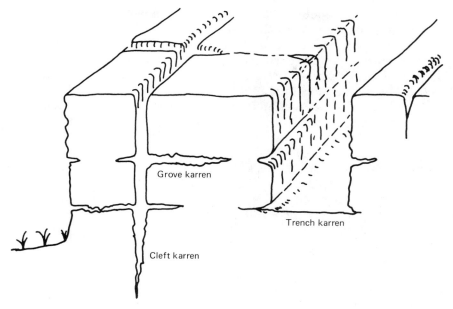

Figure 2.31 Joint blocks showing cleft, trench, and groove karren. The distinction between cleft and trench karren and cutters is simply one of scale.

Figure 2.32. Pit-and-tunnel karren in separated limestone boulder, in the French Pyrenees between Pic Casterau and Pic Paradis.

58 / *Geomorphology and hydrology of karst terrains*

karsts and alpine karsts. Either a thin soil cover or a snow pack may be useful in developing of this form.

Vertical solution along joints and fracture intersections or other locations that permit localized inputs of groundwater generate a larger vertical feature, which is not quite a doline and is too cylindrical and equidimensional to be considered a cutter. These are frequently seen in cross section in road cuts and may be called solution wells. Typical examples are a meter or so in diameter and a few meters deep. The walls are smooth-etched forms expected from solution in the presence of a soil plug that retards the flow velocity of infiltrating water and inhibits the development of hydraulic sculpturing such as fluting. These, like the pit-and-tunnel karren, are transitional between structural and hydraulic control. Since their size scale is continuous, from small pits a fraction of a meter in depth to large vertical features, they are also transitional with various forms of dolines and with other soil-filled solution features of negative relief.

2.4.4 *Etched Forms Resulting from Solutional Attack on Massive Bedrock*

Etching of the bedrock in the absence of obvious structural control also takes place. Smallest of these features is the rain pit, a few mil-

Figure 2.33 Pinnacle karren near sea level, southwest of Mayaguez, Puerto Rico. The height of the pinnacles is 2 to 3 m.

limeters to a few centimeters in diameter, roughly circular, symmetrical pits etched into the bare limestone surface. These occur where the bare bedrock is exposed in climates in which minor etching can be preserved. Rain pits are found on many limestones of the American Southwest. Davies (1957) has provided many examples on limestone fragments in Greenland.

One of the most variable and least understood of the karren forms is pinnacle karren (spitzkarren). Exposed limestone surfaces in temperate climates etch in such a way that isolated jagged spikes in the range of a few centimeters are left behind. In tropical climates, pinnacle weathering is the dominant karren form. These decorate the large residual hills, cones, and towers with chaotic mass of spiney, irregular pinnacles. Pinnacle forms seem to develop in the absence of extensive soil cover, but the role of plants, roots, and algae may be of considerable importance. The most jagged pinnacles (see, for example, the black phytokarst from Hell described by Folk et al., 1973) seem to be etched out primarily by the action of low-order plants. Tropical pinnacle karren (Fig. 2.33) are extremely irregular and jagged. In contrast, in alpine regions small, rounded pinnacles, often gently curved, sometimes flat-topped, develop. These seem to be associated with channel formation in the bedrock under a snow pack. The channels gradually widen and form an anastomosing pattern, and only the pinnacles are left as intermediate "islands."

3

Underground Landforms in Karst Regions

3.1 CAVES

3.1.1 *Definitions*

Caves are natural openings in the earth. They are characterized by their size and shape, that is, the diameter of the passages, their length, and the overall plan and layout of the openings. The concept of a cave is, however, essentially anthropomorphic: a void in the earth is interesting if it can be entered and explored by human beings. This defines a minimum diameter and length of a cave in terms of the size of the explorer, a notion that is built into the guidelines of several of the State Cave Surveys. The Alabama Cave Survey, for example, requires a cave to be at least 150 m long to be listed, whereas the Virginia Cave Survey (Holsinger, 1975) requires only a length of 6 meters. Clearly, what constitutes a cave depends largely on the definer. Some geomorphologists (e.g., Ford, 1977) would take issue with this statement and insist that "cave" is a strictly geologic concept. There is value in separating caves, as objects for exploration, from "conduits," which are the links in the underground drainage system. Conduits also have a minimum dimension (Chapter 9), on the order of 1 cm. Thus, one may say that "a cave is a natural opening in the earth, large enough to admit a human being, which some human beings choose to call a cave."

Curl (1964) wrote an equivalent definition in mathematical terms by defining an ideal spherical explorer of variable radius. By setting the lower limit of the size of the explorer to approximately the size of a human being, we can rigorously define a set of proper caves—those that can be explored by humans.

Our interest, in this and in most subsequent chapters, is in solution caves, that is, caves that have been formed by the dissolution of bedrock by circulating water. We may assume that all the following discussions of the sizes, shapes, and properties of caves refer only to solution caves. Caves formed by other processes are also of interest to explorers; they are discussed briefly in Chapter 12.

3.1.2 Entrances

Complex underground drainage systems are composed of continuous sequences of conduits and smaller openings in the bedrock. There is an interconnected set of channels with a characteristic morphology, pattern, and hydrologic function. As the landscape is cut down, progressive layers of conduit-containing bedrock are drained and become air-filled. Of these air-filled conduits, some reaches are sediment-choked, some are too small for human exploration, but most are inaccessible simply because there is no way to get there. Only where the underground channels have broken through to the surface with the formation of an entrance is a portion of the conduit accessible to human exploration. In most karst regions, the accessible caves represent a few percentage points or less of the total conduit system.

Cave entrance formation is a fortuitous event. Except for entrances in swallow holes or at spring mouths, the entrance need not be related to the sources and discharges of the water that formed the cave. Sometimes cave entrances are formed by the downcutting of surface valleys that intersect and truncate the underlying cave passages. However, truncation and dissection may not form entrances if thick soil covers or weathered bedrock slump into and obliterate the opening. Other entrances are formed by random processes of upward stoping of cave passages, sinkhole collapse, or intersection of vertical shafts with cave passages.

Curl (1958) found that most caves in Pennsylvania and West Virginia have only one entrance, a few have two, and the number of caves with three or more entrances falls off according to a Poisson distribution. Statistical distribution functions and geologic reasoning agree that a substantial class of caves with no entrances should exist. The number of caves with various numbers of entrances discovered in Alabama through April 1973 are shown in Fig. 3.1. Extrapolation predicts a very large number of zero-entrance caves in Alabama. Curl's calculations predict that the entranceless caves are generally smaller than those with one or more entrances. The undiscovered cave resources thus consist, on the average, of a very large number of rather small caves.

3.1.3 Terminations and Length Distributions

Caves have finite length. They are terminated by ceiling collapse (breakdown), clastic fills, the ceiling plunging below standing water

62 / *Geomorphology and hydrology of karst terrains*

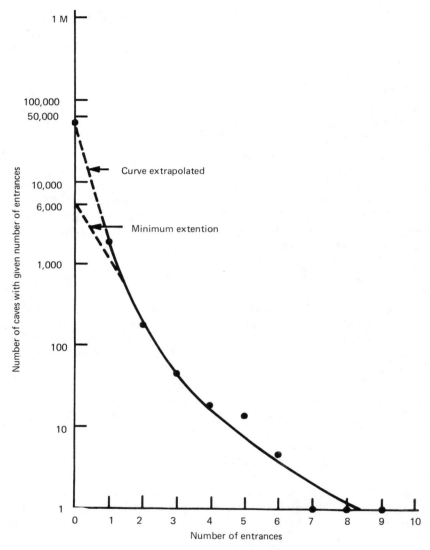

Figure 3.1 Distribution of caves in Alabama with a given number of entrances. [Adapted from Varnedoe (1973).]

levels (sumps), and pinching down to dimensions too small for human explorers. Only rarely do cave passages terminate in blind bedrock walls. The cave passages are dissolved out by flowing or percolating water that needed a place to enter and leave. The size of the cave and the configuration of passages are therefore determined by the arrangement of these terminations. Lengths of individual caves

Underground landforms in karst regions / 63

are determined by random events of breakdown and plugging rather than by the hydraulic requirements of a groundwater flow system.

Figure 3.2 shows the length distribution for several populations of caves. The cave lengths are grouped in half-log intervals. The larger groups of caves, such as those tabulated from the Alabama Cave Survey, produce smooth distribution functions. All of these are truncated populations, cut off at 30 m (100 ft) at the short end for

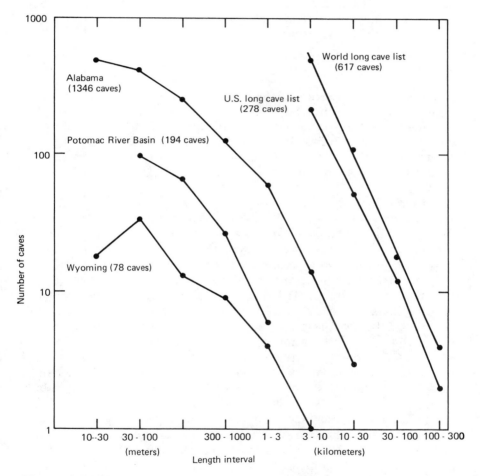

Figure 3.2 Cave length distributions for several populations of caves. Lengths of caves in Wyoming from Hill et al. (1976); cave lengths in the Potomac River Basin drawn from previous work of the author (White and White, 1974); data on Alabama caves from Varnedoe (1973). The U.S. Long Cave List includes all caves with lengths greater than 3 km, according to R. Gulden's November 1980 listing. The distribution of the world's long caves is from Chabert (1977).

most listings. It is unlikely that many of the existing surveys give comprehensive listings of very small caves. The length distributions of the longest known caves, taken from the U.S. Long Cave List and the World Long Cave List, are nearly linear in this plotting scheme. Extremely long caves are, in fact, quite uncommon. Table 3.1 shows the top 20, as the list stood in early 1987.

Cave length distributions can be analyzed by statistical functions (Curl, 1966a), leading to the concept of a "karst constant," a single parameter that characterizes the cave length distribution for a particular geologic region. The length distribution can also be described by the function (Curl, 1986)

$$N(l) = N(l_0)\left(\frac{l}{l_0}\right)^{-\nu} \tag{3.1}$$

where $N(l)$ is the fraction of the cave population with length greater than l, and l_0 is the shortest or cutoff length for the distribution. The exponent, ν, is the fractal dimension of the cave length population in the sense used by Mandelbrot (1983). Cave lengths and various elements that make up caves are similar in that the length and other parameters depend on the scale length used for their measurement.

3.1.4 Cave Maps

The primary descriptive unit of the underground conduit system is the cave map. Thousands of these have been prepared, almost entirely by the cave explorers themselves. Cave maps vary widely in accuracy, completeness, detail, and cartographic excellence, but all are constructed in essentially the same way. The pattern of the cave is measured by constructing a traverse line using tape and compass methods. The usual procedure is to traverse each passage in the cave, establishing a series of survey stations. The vector from the nth to the $(n + 1)$th station is determined by measuring the length with a tape, the angle with respect to horizontal with a clinometer, and the orientation with respect to north with a compass. The hand-held pocket transit is the compass of choice for most surveys, but other compasses are sometimes used. For greater accuracy, tripod-mounted pocket transits, or more rarely alidades or theodolites, are employed. The overall traverse accuracy is about 1 percent, with ordinary care, and can be improved to perhaps 0.1 percent by using pedestal-mounted compasses or other more accurate instruments.

Survey accuracy of a linear or branched traverse line is essentially uncontrolled. Surveys around looped passages that close on themselves provide a check on the overall precision and identification of unacceptably large errors, and they allow small random errors to be distributed over all traverse vectors during data reduc-

Table 3.1
A Short List of the World's Longest Caves[a]

Cave name	Length (m)	Country
Mammoth Cave—Flint Ridge System	500,506	U.S. (Kentucky)
Optimistceskaja	157,000	U.S.S.R. (Ukraine)
Hölloch	150,500	Switzerland
Jewel Cave	118,573	U.S. (South Dakota)
Ozernaja	107,300	U.S.S.R. (Ukraine)
Sistema Ojo Guareña	88,907	Spain (Burgos)
Système de la Coumo d'Hyouernedo	82,500	France (Haute Garonne)
Zolushka Cave	80,000	U.S.S.R. (Ukraine)
Siebenhengste—Hohgant System	80,000	Switzerland
Wind Cave	72,756	U.S. (South Dakota)
Friars Hole System	68,122	U.S. (West Virginia)
Fisher Ridge Cave System	63,795	U.S. (Kentucky)
Sistema Purificacion	60,848	Mexico (Tamaulipas)
Organ Cave System	60,510	U.S. (West Virginia)
Kananda Atea	54,800	Papua New Guinea
Reseau de la Dent de Crolles	53,800	France (Isère)
Red Del Rio Silencio	53,000	Spain (Cantabria)
Easegill—Lancaster Cave System	52,400	Great Britain
Reseau de l'Alpe	51,777	France (Isère/Savoie)
Gua Air Jernih (Clearwater Cave)	51,660	Maylaysia
Reseau de la Pierre St. Martin	51,200	France/Spain
Crevice Cave	45,384	U.S. (Missouri)
Fighiera–Corchia–Farolfi System	45,000	Italy (Tuscany)
Hirlatzhöhle	44,600	Austria
Cumberland Caverns	44,444	U.S. (Tennessee)

[a] Rankings are effective as of early 1987.
Source: R. Gulden, "List of the World's 100 Longest Caves," computer data base; see also Courbon and Chabert (1986).

tion. Magnetic induction or radiofrequency transmitters have been used to connect cave surveys to the surface, thus allowing closures to be made without the unpleasant task of surveying the cave twice (Charlton, 1966; Davis, 1970).

Data reduction is accomplished with a complete range of techniques from hand plotting of individual traverse vectors, to the use of large digital computers that store all the survey data, adjust random errors around loop closures, and plot the traverse lines (Schmidt and Schelleng, 1970; Rutherford and Amundson, 1974). Beyond the ease in data handling and improved error adjustment it offers, computer processing allows you to prepare stereographic representations of the map and has great potential, still largely unexploited, for interrogating the survey data for statistical properties of cave patterns and passage geometry.

The cave interior is a complex surface in three dimensions, and representing it two-dimensionally is a problem that has never been satisfactorily resolved. The usual procedure is to sketch a horizontal projection of the passage walls during the survey and to show the floors and ceilings with sketched cross sections. These sketches are then transferred to the traverse line when the map is plotted. Unfortunately, it is the passage walls and ceilings that are of geological interest, and the precision with which they are represented on the final map depends strongly on the care, skill, and artistic talent of the person who held the notebook on the primary survey.

The content of the cave, its sediments, speleothems, breakdown, channels, trenches, trails, and other features are represented on the map by means of special symbols (Hedges, 1979). The amount of detail surveyors choose to use in their maps varies from almost photographic completeness to none at all. The whole procedure, from underground surveying and note-taking, through data reduction, to final map preparation and drafting is discussed in such references as Brod (1962), Butcher and Railton (1966), and Hosley (1971). Cave surveying is perhaps one of the most tedious efforts at data collection in all of the earth sciences, and the cave geologist owes an immeasurable debt to the cave explorers for the tens of thousands of unpaid hours they have invested in this activity.

3.1.5 *Regional Cave Surveys and Cave Catalogs*

A regional cave survey is a systematic inventory of the known caves in a particular geographic area, regardless of their size or geologic interest. Such surveys have been carried out, usually on a state-by-state basis, in much of the eastern and central United States and in some states in the West. The usual procedure is to explore the karst regions within the state more or less systematically, with the objective of discovering all caves. The caves are explored, and frequently mapped, and descriptive texts written. The maps and descriptions are then compiled into volumes and published (Table 3.2).

Some surveys are the work of one person, but many are group efforts. Some groups, such as the Missouri Speleological Survey, are formal organizations that publish their data in journal form. Others work in cooperation with State Geological Surveys, which then produce the publications. Once very popular, cave surveys, or at least their publications, are falling into disrepute within the caving community. Cave surveys, their critics claim, make things too easy for the casual caver, and result in a great increase in traffic and overuse of cave resources.

Table 3.2
Regional Cave Surveys[a]

	Author	Date	Publisher[b]	Publication	Number of caves described
Alabama	Varnedoe	1980	NSS[e]	Compilation	2020
California	Halliday	1962	PP	Monograph	400
Colorado	Parris	1973	PP	Monograph	265
Idaho	Ross	1969	SGS	Monograph	24
Illinois	Bretz & Harris	1961	SGS	Monograph	63
Indiana	Powell	1961	SGS	Monograph	398
Kentucky[c]					
Maryland	Davies	1952	SGS	Monograph	57
	Franz & Slifer	1971	SGS	Monograph	104
Minnesota	Hogberg & Bayer	1967	SGS	Monograph	26
Mississippi	Knight et al.	1974	PP	Monograph	42
Missouri	Bretz	1956	SGS	Monograph	437
	MO Spel. Surv.	1958–present	NSS	Journal	
Montana	Campbell	1978	SGS	Monograph	302
New Jersey	Dalton	1976	SGS	Monograph	152
New York	NY Cave Surv.	1975–present	NSS	Bulletin series	
Ohio	White	1926		Journal article	21
Oregon	Greeley	1971a	SGS	Monograph	19
Pennsylvania	Reich	1974	SGS	Compilation	
	White	1976a			169
	MAR	1958–present	NSS	Bulletin series	
South Dakota[d]					
Tennessee	Barr	1961	SGS	Monograph	686
	Matthews	1971	SGS	Monograph	316
Texas	TX Cave Surv.	1958–1975	NSS	Journal	
Virginia	Douglas	1964	PP	Monograph	1790
	Holsinger	1975	SGS	Monograph	2319
Washington	Halliday	1963	SGS	Monograph	155
West Virginia	Davies	1958	SGS	Monograph	409
	WV Spel. Surv.	1971–present	NSS	Bulletin series	
Wyoming	Hill et al.	1976	SGS	Monograph	245

[a]Only major compilations are listed, and of these only the most recent are given. No attempt is made to follow the long history of some of these surveys. Much additional information on individual caves can be found in the literature.

[b]NSS = publication of the National Speleological Society or its Internal Organizations; PP = privately published (usually by the author); SGS = publication of a State Geological Survey.

[c]No regional survey has been published for Kentucky, in spite of the large number of caves. For a description of and other references on the Mammoth Cave are, see Palmer (1981a).

[d]No cave survey for South Dakota has been published. For descriptions of Jewel Cave, see Conn and Conn (1977); for Wind Cave see Palmer (1981b).

[e]The guidebooks published by the National Speleological Society for their annual conventions frequently contain additional descriptions of the caves in the convention area.

68 / *Geomorphology and hydrology of karst terrains*

3.2 THE SHAPES OF CAVES

3.2.1 *Passage Elements*

If the concept of "cave" itself is anthropomorphic, so are the elements that make it up. A passage is a segment of traversable cave, usually longer than it is wide or high. It is linked to other passages at intersections. Individual passage elements may be large or small, and may be only a few meters or many kilometers long. A cave is an assemblage of passages with various shapes and various relationships to each other, which may have formed at the same time or at different times, from a single water source or from different sources. Passages in the same cave may be genetically related or only fortuitously connected.

The total cave as seen in map projection is an assemblage of all accessible passage elements. The geometry of the passage elements is determined by structural and stratigraphic factors that control the route followed by the underground water.

Passage elements can be divided into two broad classes, those that consist of fragments of single conduit, and those that consist of fragments of interconnected maze passage. Within each of these classes, three distinct geometries can be recognized (Fig. 3.3) (Palmer, 1975; White, 1960):

A. Single conduit passages
   1. Linear passages
   2. Angulate passages
   3. Sinuous passages
B. Maze passages
   1. Network mazes
   2. Anastomotic mazes
   3. Spongework mazes

The labyrinthine character of the maze is the result of simultaneous enlargement of available joints, bedding planes, and other primary pathways. The very close-spaced components of the maze make up a single passage element.

A *linear passage* is simply a straight segment of passage without bends. Linear passages occur in regions of folded rocks, where they tend to follow the better-developed strike joints. Linear passages occur in other regions where a single major joint or fracture trace maintains control over the flow path.

Angulate passages are characterized by sharp bends—sometimes but not always at right angles—with intermediate straight sections. The passages often have the same arrangement as the regional joint set. In dipping limestones long reaches of passage may be oriented along the strike while the short segments cut across the bedding.

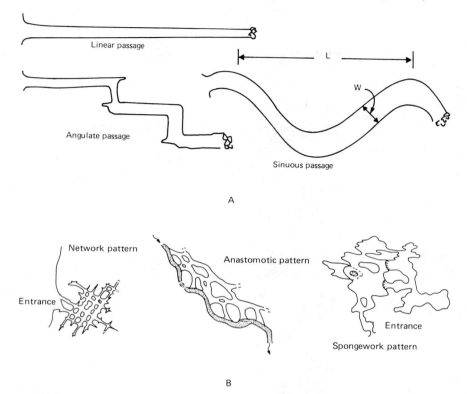

Figure 3.3 (A) Sketch maps showing the ground plans of single-conduit passages with linear, angulate, and sinuous patterns. (B) Sketch maps showing the ground plans of maze-type cave passages with network, anastomotic, and spongework patterns.

Angulate patterns also develop in flat-bedded limestones, where the hydraulic gradient is oriented diagonally to the regional joint set. The groundwater flow uses first one joint set and then the other to carve a zig-zag course through the joint blocks.

The orientation of an angulate cave can be displayed by dividing the azimuth into small segments (ten segments are commonly used) and accumulating the length of passage oriented along each segment. The cave can then be reduced to a rossette or bar graph, in which the dominant passage orientations can be clearly seen (Fig. 3.4). The orientations of joints in the vicinity of the cave can also be measured and displayed in similar fashion (Deike, 1969).

Sinuous passages usually occur in flat-bedded limestones. Instead of sharp-angle bends with straight reaches of passage between, sinuous passages have broad sweeping curves and relatively few really straight stretches. Frequently, the bends of sinuous passages are

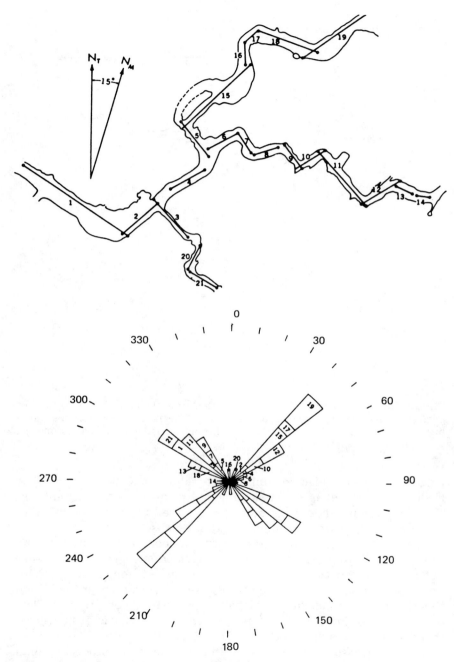

Figure 3.4 How to express cave passage orientation as a rossette. Adapted from Deike (1969).

uniformly spaced, so that the pattern has a regular "wavelength." The pattern is very similar to the meander patterns exhibited by surface rivers. Some passage meanders have small amplitudes, so that the deviation of the passage from the centerline is hardly more than the width of the passages. Others have wide, sweeping bends, of which a few examples such as Serpentine Cave of New South Wales (Ongley, 1968), almost double back on themselves to form incipient oxbows.

Network maze passages consist of a grid of intersecting passages, usually with high, narrow cross sections. The grid spacing is typically on the order of 10 m, and probably maps the regional joint spacing. The network pattern arises when all joints, regardless of their original permeability, are dissolved at a uniform rate, so that all possible mechanical openings enlarge into cave passages. The pattern typically occurs in rocks of low dip or with near-vertical joints.

Anastomotic maze passages are formed of curvilinear tubes of circular or elliptical cross section that intersect in a braided configuration. Anastomotic mazes are typically two-dimensional, following highly permeable bedding plane partings, but they can be a three-dimensional tangle of tubes that follow no particular geological horizon. Joint control is minimal. Small segments of individual tubes may follow joints, but the overall pattern is an arrangement of curvilinear forms.

Spongework maze passages consist of interconnected, nontubular solution cavities of varied size and irregular geometry arranged in an apparently random, three-dimensional pattern (Palmer, 1975). Spongework mazes are found in Carlsbad Caverns (the Boneyard) and Wind Cave in South Dakota. The spongework maze pattern appears in caves formed in young, porous, and poorly jointed limestones.

It is tempting to argue that network mazes are the result of strong structural control, anastomotic mazes the result of dominant hydraulic control, and spongework mazes the result of random solution of massive bedrock without either structural or hydraulic control. All three patterns appear to be the bedrock, closed-conduit, analogs of braided patterns in surface rivers.

3.2.2 Passage Cross Sections

The enlargement of the initial joint and bedding plane partings to form cave passages takes place by means of two distinctly different groundwater flow regimes. In the first case, groundwater circulates at depth under hydrostatic head through available secondary permeability. Gradual removal of the rock by solution allows the passage to grow in all directions around the flow line, until a mature passage is formed. In the second case, the water-filled passage, after a certain period of enlargement, develops a free-air surface. It then takes on the characteristics of a surface stream, flowing down gradient under the influence of gravity. Further solutional enlargement is mainly vertical, with relatively little widening. Once the stream begins to cut down, the ceiling and walls above the water level are not dissolved further. In both cases, the geometry of the conduit is strongly determined by the flow rate of the water, which may vary from slow percolation to velocities in the range of meters per second.

The cross section of cave passages represent a compromise between hydrodynamic forces that tend to shape the passage into smooth, streamlined forms and structural controls that give the passage the shape of the initial guiding joint or bedding plane parting (Figs. 3.5–3.7). High velocities and thick, uniform limestones tend to promote the hydraulic forms while low velocities and nonuniform bedding enhance the formation of conduits with complicated cross sections.

Passages formed under hydraulic control trend toward two end member types—the canyon and the elliptical tube.

Canyons are typically higher than they are wide, and have near-vertical walls, with little variation in passage width from top to bottom. They may show evidence of fast-moving water. Canyons may be formed by the gradual downcutting of a free-surface stream, but also by solutional widening of a very regular joint under completely water-filled conditions. The canyon shape itself is not proof of the type of flow. Canyon passages may be completely linear, following a straight guiding joint, or sinuous if the stream meandered as it downcut. Canyon passages with migrating meander bends can have quite complex cross sections, but the meandering is evidence for an origin from downcutting free-surface streams.

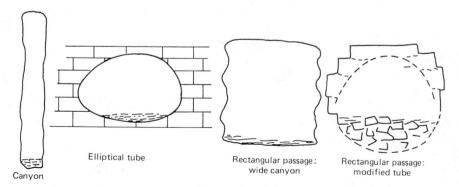

Figure 3.5 Cross sections of passages with shapes controlled by the hydraulics of flowing water.

Underground landforms in karst regions / 73

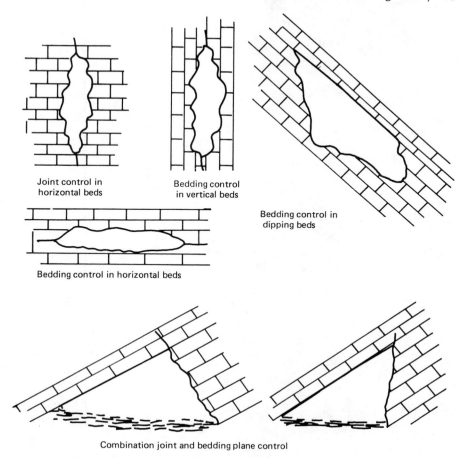

Figure 3.6 Cross sections of passages resulting from structural control of limestone dissolution.

Elliptical tubes form as groundwater flows under pipe-full conditions. They vary in size from openings too small to enter to tubes with diameters of tens of meters. Elliptical tube passages in flat-bedded limestones in low-relief terrains, such as in Missouri or Kentucky and in the Appalachian plateaus, tend to be wide, and the original mechanical opening is often a bedding plane parting. Tubes in higher-relief terrains become more circular. Circular borelike passages often contain evidence of high-velocity flow.

An intermediate geometry is the rectangular passage illustrated in Fig. 3.5. It can form as a wide canyon in a flat-bedded sequence where a joint of restricted vertical extent is present or the roof beds are relatively insoluble. Wide rectangular canyons also result from

Underground landforms in karst regions / 75

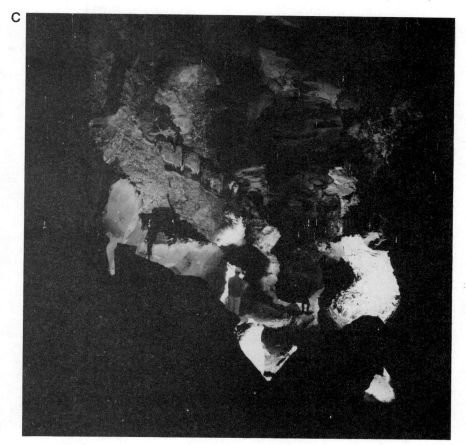

Figure 3.7 Photographs of some representative cross sections: (A) elliptical tube—Cleaveland Avenue, Mammoth Cave Kentucky—a sinuous tube developed mainly along one bedding plane; (B) rectangular canyon—Big Avenue in the New Discovery section of Mammoth Cave—the original tube can be seen at the top of the passage; (C) large irregular canyon passage—New Discovery section of Mammoth Cave.

the modification of an elliptical tube by breakdown, and perhaps by later solutional modifications that round off the fracture surfaces.

Passages controlled by rock lithology and structure can be extremely complicated, depending on whether joints or bedding plane partings or both were the primary groundwater routes. The dip of the beds has considerable influence.

3.2.3 *Composite Passage Cross Sections*

The channel of a surface stream usually represents a balance between water flow and sediment-carrying capacity. As the stream

76 / Geomorphology and hydrology of karst terrains

downcuts its channel and the discharge varies through piracies or climatic change, the channel shape adapts to the new conditions but destroys the old channel in the process. Immediate past history may be recorded in fragments of degraded channel on the flood plain, and fragmentary vestiges of earlier history may also be recorded in river terraces. However, most of the early history of the river channel and the valley in which it formed is irrevocably destroyed.

Cave passages, in contrast, preserve their earlier history in almost overwhelming detail. Cave passages differ from surface stream channels in that the processes of lateral erosion, slope retreat, and bank caving do not usually occur. As a result, much of the earlier history of the passage is preserved.

A common composite passage is one that has a T-cross section (Fig. 3.8), with an elliptical tube at the top and a canyon cut in its floor. Such cross sections can develop when the main recharge for the tube is pirated upstream and a residual flow, underfit to the tube, cuts a canyon in its floor as the base level is lowered. These passages can also form when a different water source, underfit to the tube, finds its way into the preexisting tube, using the larger passage as part of its route. Tubes are also transformed into canyons when the tube becomes too large to maintain pipe-full conditions.

Downcutting of an underground flow line may be continuous, giving a complex geometry, or disrupted, producing a sequence of passages. Transitions may occur from canyons to tubes and back again. These various components of the evolving passage sequence can take on independent routes; they may overlap in one section of passage and form separate passages in another.

3.2.4 Rooms

Nowhere is the anthropomorphic description of a cave more apparent than in the notion of "rooms." Rooms are marked on many cave maps but are features of diverse origin and size. To the explorer of

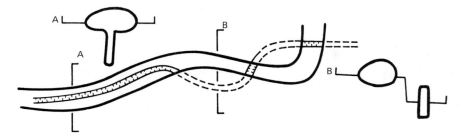

Figure 3.8 Sketch map showing characteristics of composite passage with canyon incised in the floor of a larger tube.

a tangle of crawlways, a place where he can stand up and stretch is a "room." Or a room may be a gigantic underground vault 100 m or more across. The usual room is a place in which the cave widens or heightens above the average passage through which the explorer has been traveling. Among the many diverse origins for rooms are the following:

1. Intersections of several passages.
2. A fragment of large conduit reached through a cave system consisting of much smaller conduit.
3. Places where breakdown fell while groundwater was actively circulating, removing the fallen blocks by solution. Block stoping, followed by solution, can lead to very large dome-shaped rooms. Some of the largest cave rooms known are of this origin.
4. Complexes of coalescing vertical shafts.
5. Genuine rooms, formed by more vigorous solution at one point, resulting in enlargement of the passage.

3.3 CAVE PATTERNS

Caves are made up of interconnected passage elements of various kinds. The three main variables are length, complexity of pattern, and vertical development. Caves range in length from the smallest opening that one chooses to call a cave up to the Flint Mammoth System, at present the longest known cave in the world. Length is generally taken to be simply the sum of the lengths of all component passages. Complexity arises from the diversity of passage elements that make up the cave and the number of distinct flow paths represented in the cave. Caves can enlarge vertically as a result of at least three mechanisms: (1) deep circulation of groundwater under hydrostatic head, (2) extensive downcutting and development of successively lower passages by water circulating near base level as the base level retreats, and (3) extensive vertical solution in limestone masses above the local base level.

3.3.1 *Single-Drainage-Line Caves*

The simplest caves, hydrologically, are those created by a single drainage line of groundwater flow. Just as single passage elements were separated into two broad classes—the single conduit passages and maze passages—the overall patterns of single-drainage-line caves can be subdivided into caves with closed loops and those without closed loops. The subdivision for entire caves is not so clear-cut, because different parts of the same cave may contain different passage elements. In general, however, one can subdivide cave patterns into the following classes:

1. Single-conduit caves
2. Branchwork caves
3. Tight maze caves
4. Loose maze caves
5. Three-dimensional mazes

Single-conduit caves are single passages with no side passages. Some are linear and some—for example, Fleming Cave (Fig. 3.9)—are angulate. Although Fleming Cave occurs in beds with near-vertical dip, the passage crosses from one bed to another and is not predominantly strike-oriented. Fletcher Cave (Fig. 3.10) is a single-conduit cave with mainly sinuous elements. It is a stream cave and serves as the underground route of a surface stream that sinks at the entrance and flows along the passage. Although the cave is essentially one conduit, it contains minor side passages and a number of closed loops. These loops arise from the superposition of earlier routes of the conduit on the present-day stream passage. The single-conduit cave is analogous to the main channel of a surface stream.

Branchwork caves are those with tributaries. The size can vary from meters to kilometers, and the branches can take on different geometries. Fisher Cave (Fig. 3.11) consists of branching tributaries of large and small tubes joining with each other to form a trunk that empties into the Meramec River Valley. Streams occupy some of the tributaries, but, because these have downcut more or less within the same channel, the minor anastomotic loops seen in Fletcher Cave are missing here. Fisher Cave consists of a single type of passage element: the sinuous elliptical tube. A constrasting type of branchwork cave can be seen in Skull Cave (Fig. 3.12). Although the cave follows rather strictly the branchwork geometry, the passages contain sinuous, angulate, and linear elements. The linear and angulate portions of the cave are strongly joint-controlled and the passages tend to be high, narrow fissures. The stream passage, although exhibiting a tendency to sinuosity, is more angulate than sinuous, suggesting a predominance of structure over hydraulic control. Unlike Fisher Cave, Skull Cave reveals no converging tributary pattern.

Tight maze caves consist entirely of maze passage elements, as previously defined. A good example is Brady's Bend Cave in western Pennsylvania (Fig. 3.13). Every joint in the closely jointed limestone has been opened by solution, resulting in a fairly large cave that consists of a network maze with a passage spacing of only a few meters. Mark Twain Cave (of Tom Sawyer fame) and nearby Cameron Cave, on the Mississippi River bluffs of northeastern Missouri, are examples of large tight network mazes. One of the most incredible exam-

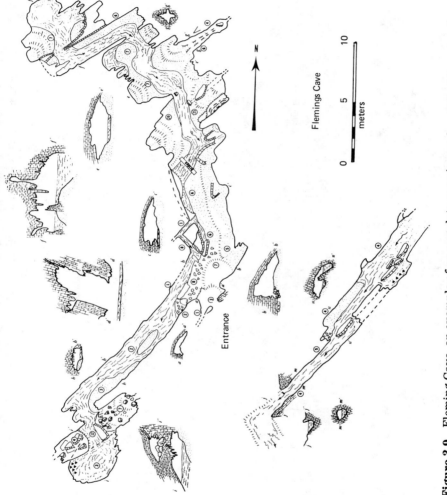

Figure 3.9 Fleming Cave, an example of an angulate cave in nearly vertical limestone. Fleming Cave is formed in Helderberg limestone in Huntingdon County, Pennsylvania. [Map prepared by B. L. Smeltzer.]

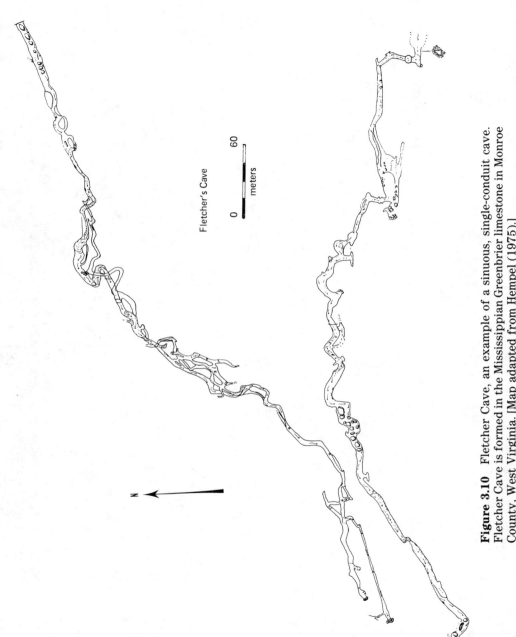

Figure 3.10 Fletcher Cave, an example of a sinuous, single-conduit cave. Fletcher Cave is formed in the Mississippian Greenbrier limestone in Monroe County, West Virginia. [Map adapted from Hempel (1975).]

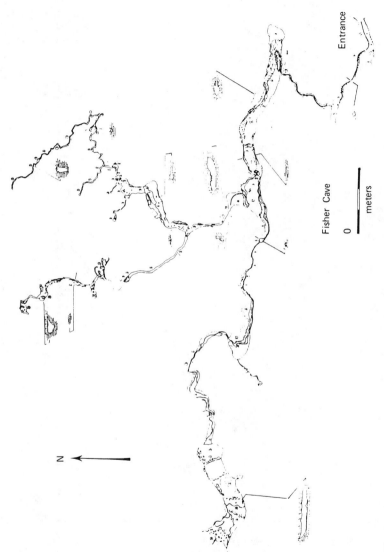

Figure 3.11 Fisher Cave, an example of a branchwork cave made of sinuous elliptical tubes. Fisher Cave is formed in the nearly horizontal Ordovician Gasconade dolomite in Franklin County, Missouri. [Map courtesy of the Missouri Speleological Survey.]

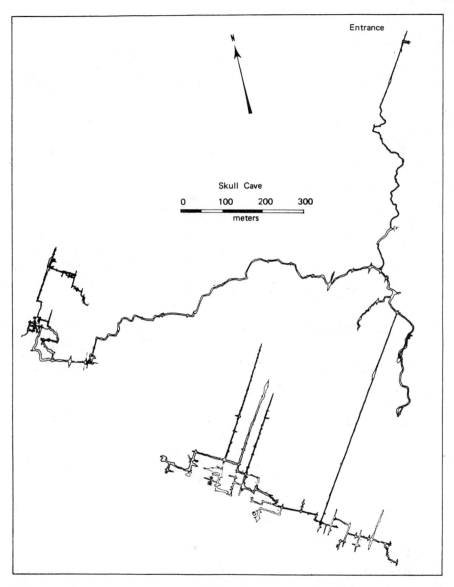

Figure 3.12 Skull Cave, an example of a branchwork cave made of predominantly angulate and linear passage elements. Skull Cave is formed in gently dipping Devonian Helderberg limestone, in Albany County, New York. [Map courtesy of Ernst Kastning.]

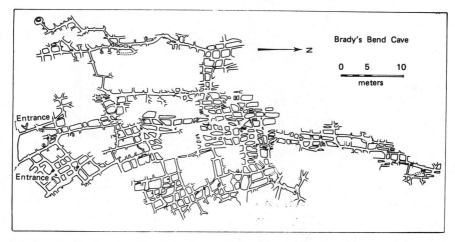

Figure 3.13 Bradys Bend Cave, an example of a tight network maze cave. Bradys Bend Cave is in the Pennsylvanian Vanport limestone in Armstrong County, Pennsylvania. [From White, 1976a. Map courtesy of the Pennsylvania Geological Survey.]

ples is Anvil Cave in northern Alabama, where some 20 km of network maze underlies a meadow no more than 500 m across. Nearly all the known tight maze caves are made up of network maze passage elements.

Loose maze caves are those with closed loops among more or less coplanar passages. The size of the loops varies and passage cross sections often vary as well, so that distinct flow paths through the loops are not readily discernible. A type example is Greenville Saltpetre Cave (Fig. 3.14). The passages are mainly elliptical tubes of varying size. The tubes interconnect in a random way, with little evidence of strong structural control. This may be considered an anastomotic maze expanded to the scale of a large cave. In caves of this type, evidence for directed flow is often absent. The large looped passages of the loose maze appear to carry water at the same time just as the tight network mazes carry water through all available joints at once.

There is more correspondence between the network and anastomotic maze caves and surface drainage systems than early theorists were willing to admit. If groundwater flow in the single-conduit and branchwork type of cave can be likened to flow in surface stream channels, flow in a maze cave can be likened to that in a swamp. Surface water in a swamp is not stagnant. The swamp is fed from streams entering at the perimeter, and it ultimately drains through some kind of channel. Within the swamp itself are many channels, wide pools, islands, and blind embayments. Gradients and flow velocities are extremely low, but because of the large cross section, an immense amount of water can be transmitted. Much the same situa-

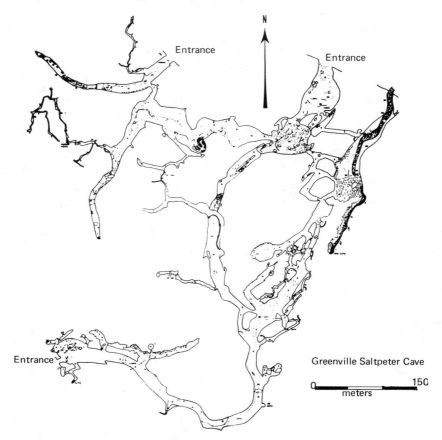

Figure 3.14 Greenville Saltpetre Cave, an example of a loose maze cave. Most cave development is along bedding planes, and most of the passages are elliptical tubes. Greenville Saltpetre Cave is formed in the Mississippian Greenbrier limestone in Monroe County, West Virginia. [From Hempel (1975).]

tion obtains in the maze cave. Maze caves tend to develop where gradients are very low or flow velocities are restricted through some geologic control (although some mazes are associated with extremely high flows; see Chap. 9).

A common source of closed loops in cave passages is the superposition of conduits cut by water from the same source at different times. Small-scale piracies are common in caves, and passages are easily bypassed and undercut. These passages may be referred to as three-dimensional mazes. The closed loops are clearly evident in ground plan, but the hydrology of their formation is somewhat different from that of the two-dimensional mazes described previously.

3.3.2 Tiered Caves

A concept that runs through the cave literature is that of the cave "level." Caves often consist of nearly horizontal passages that are stacked in tiers, with or without connecting passages. Many of these can be interpreted as the result of discontinuous downcutting by water moving from a catchment area to a discharge point concurrent with base level lowering.

Tiered caves may consist of a stacked sequence of individual, more-or-less horizontal caves in the same area, perhaps with some horizontal offset. Other tiers are zones within a single cave where there is a concentration of larger passages. These levels or tiers may be interconnected by canyons or shafts that formed contemporaneously with the development of the main levels, or they may be interconnected by canyons and shafts that developed much later, during the dissection of the topography overlying an older cave system. Tiered caves appear complex in map view because of the superposition of levels.

The Mammoth Cave–Flint Ridge Cave System is an exceptionally large and complex tiered cave. Miotke and Palmer (1972), among others, have used the existence of concentrations of passages at particular levels to interpret the geomorphic history of the cave system. Figure 3.15 shows some of these levels near the west end of Mammoth Cave. The main cave is a large trunk passage at 180-m (600-ft) elevation that appears to have carried underground drainage from a catchment of several hundred square kilometers to a paleospring on the ancestral Green River, perhaps in the middle to late Pliocene period. As the base level surface stream, the Green River, lowered its channel at varying rates during the Pleistocene, progressively lower levels developed in the caves, culminating with the present-day active trunks which are actually some 5 to 10 m below river levels because of in-filling of the river channel with Wisconsin gravel.

Tiered caves are the result of vertical piracy. New drainage routes are formed below the older ones, often following somewhat different paths, leaving the older routes to become dry upper-level cave passages. This is an alternative development to the continuously downcutting stream that leaves behind it a high canyon passage. Both types of passage development are common, and one of the tasks of cave origin models is to explain why this is so.

3.3.3 Multiple-Drainage-Line Caves

Multiple-drainage-line caves are those that represent connected fragments of two or more underground drainage lines. In a multiple-drainage line cave one could travel underground from one groundwater basin to another, as one would walk across a ridgetop divide

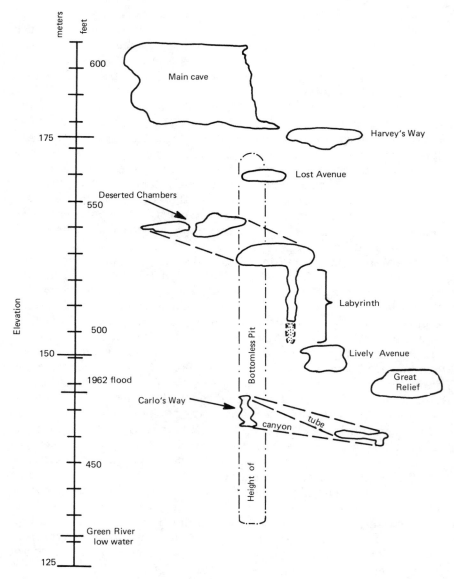

Figure 3.15 Relative sizes and vertical position of cave passages in the western end of Mammoth Cave. The National Park Service tours pass through or near many of these passages. See Palmer (1981a) for detailed descriptions. [Sketch adapted from Deike (1967).]

between two surface drainage basins. A traversable underground divide requires the shifting of drainage over time or a flood spillover route.

An example of this type of cave is the Mammoth Cave–Flint Ridge Cave System. These two caves were found to connect through several very tight passages. According to the anthropomorphic rules of cave length, the Mammoth Cave–Flint Ridge Cave System is a single cave—the largest in the world. It appears that the Flint Ridge System, in spite of its size, is a single-drainage-line cave. All of the passages appear to discharge toward a single spring location, which has shifted relatively little over time. Likewise, Mammoth Cave is a single-drainage-line cave, although it has the additional complication of draining at present through two springs about 1 km apart. The system as a whole, however, is a multiple-drainage-line cave. Some shaft drains near the drainage divide between the two underground catchments connect Flint Ridge with Mammoth Cave. Fortuitously, the drainage from the connection area at some time in the past flowed north through the Flint Ridge Cave System. Later, there was a small shift in the divide, and the flow changed to the south, through Mammoth Cave; thus, a just barely humanly traversable connection was formed. Hydrologically, the two caves are almost completely independent.

Many large caves are multiple-drainage caves. This is especially true for caves that occur in the headwaters of regional drainage systems. Indeed, a complex record is often preserved of piracy, shifts in drainage divides, reversals of flow directions, flood spillover routes, and other responses of the groundwater flow system to changing climate and surface drainage.

3.3.4 *High Gradient Caves*

Deep caves are restricted by geology to regions where thickness of the limestone and relief of the uplands above base level allow room for their development. These requirements make high gradient caves almost synonymous with alpine caves.

Most common of the high gradient caves are the vertical caves. If the entrance to a vertical cave is at the top, the distinction between a vertical cave as an underground feature and an open pit or shaft as a surface landform is a matter of semantics. Dolines, shafts, and other closed depression features are produced by water moving vertically downward through air-filled joints and fractures in the bedrock. Vertical caves are also produced mainly by vertical solution in the unsaturated zone. The elements of the vertical cave are the vertical shaft, the solution chimney, and larger vaultlike voids created by a combination of vertical solution and mechanical stoping. There is a strong tendency for vertical caves to be composed of nearly ver-

88 / *Geomorphology and hydrology of karst terrains*

tical components—shafts and chimneys of various kinds—interspersed with nearly horizontal passages. The result is a cave with a stair-step profile, in which the total vertical drop of the cave is achieved with a series of smaller drops offset by sections of more gently sloping passage.

The function of many high gradient caves is to carry runoff from high-altitude catchment areas to the valleys below. If one ignores the passage detail and smooths over the stair-step profile, many vertical caves are seen to be more or less parallel to the surface slope, but they provide a more direct, sometimes steeper, route for water to reach base level. This occurs, for example, in the Ragge Javre Raige Cave in Norway (Fig. 3.16). A catchment area at an altitude above 600 m drains by means of a stair-step vertical cave system inside the mountain, to discharge in springs near sea level in the fjord below.

If the high-altitude catchment is large, and the region is one of high precipitation, very large high gradient caves can develop. The Sistema Huautla in Oaxaca, Mexico, is the largest in the Western Hemisphere, with a surveyed depth of 1225 m (Stone, 1983; Minton, 1984). Various subbasins on the Huautla Plateau drain through a

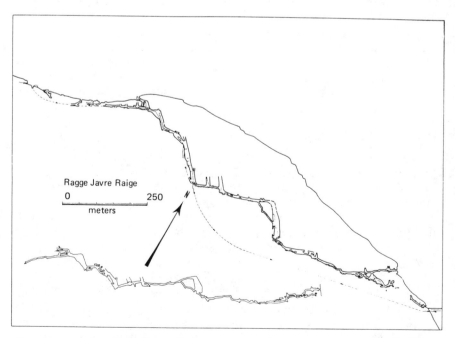

Figure 3.16 Profile of Ragge Javre Raige Cave in Norway, showing relation of stair-step profile of the cave to the slope of the land surface. [Adapted from Courbon (1972).]

Underground landforms in karst regions / 89

complicated series of vertical caves, apparently following major fracture systems (Fig. 3.17). Although there is a base level series of passages that are partially flooded conduits, the bulk of the system developed completely in the vadose zone. This system also has considerably less of the stair-step character.

The stair-step profile is not restricted to alpine caves. Many caves in moderate-relief terrains exhibit stair-step profiles. In the Appalachian Mountains, the cavernous limestone often crops out along the flanks of dissected plateaus. Mountain streams rising on the pla-

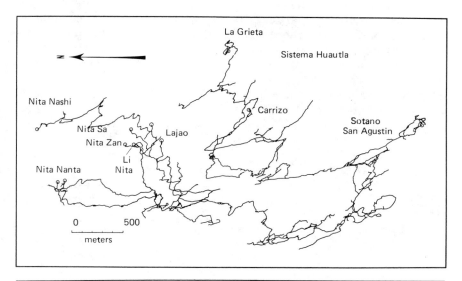

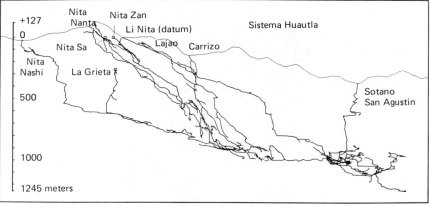

Figure 3.17 Plan and profile of the Sistema Huautla, Oxaca, Mexico. The names on the map are those of the various entrances to the system. [Courtesy of the Association for Mexican Cave Studies.]

90 / *Geomorphology and hydrology of karst terrains*

teau uplands flow down the mountain sides and sink at the limestone contact. These streams descend by steep gradient routes, which are often stair-stepped, to base level master drains. Swago Pit in West Virginia (Fig. 3.18) provides an example. The steep stream gradient is provided by a combination of vertical shafts with intervening nearly horizontal stream passage.

Also found mainly in alpine areas are long, single-conduit caves containing near-vertical passages through which water must be driven under hydrostatic head from bottom to top. Two examples are Castlegard Cave in British Columbia and the Tantalhöhle in Austria. Castlegard Cave extends beneath the Columbia ice field, deriving its water from melt water around the glacial margin. The overall slope of the cave is toward the discharge point on the Castlegard River, but the profile of the cave shows several pits that the water must fill to reach the mouth of the cave. The Tantalhöhle is a similar drain that crosses a ridge of the Austrian Alps. The profile of the Tantalhöhle shows a series of up and down sections that require water traversing the cave to be driven by hydrostatic pressure.

Most high gradient caves are single-conduit caves or branchwork

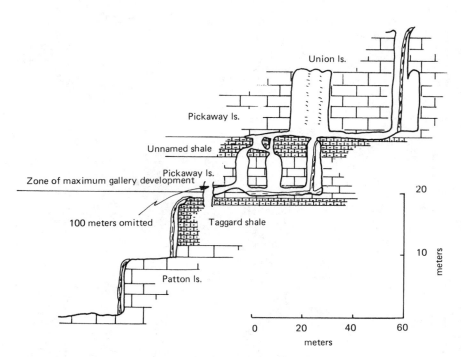

Figure 3.18 Profile of the Swago Pit section of the Swago-Carpenters Cave System, Pocohantas County, West Virginia, showing stair-step profile.

caves. It is possible, however, to find maze caves with considerable overall relief. These are planar or two-dimensional caves, except that the plane along which the cave is developed is tilted rather than horizontal. One such example is Breathing Cave in the Butler Cave–Sinking Creek System in Virginia (Deike, 1960a; White and Hess, 1982). The main axis of the Butler Cave–Sinking Creek System is a trunk conduit developed along the axis of a syncline. Breathing Cave and some of the tributary sections of Butler Cave are network mazes developed along the flanks of the syncline. They follow the dip of the beds, giving a relief on the order of 100 m although the limestone units in which the caves are formed are only 20 m thick. Another example is Big Brush Creek Cave in Utah, an intermediate case between a branchwork and a loose maze cave with many superimposed spongework elements (Palmer, 1975). The internal relief of the cave is nearly 250 m. Other examples are the network caves that follow the dip of the Pahasapa limestone where it is lapped on the flanks of the Black Hills Dome in South Dakota (Howard, 1964).

Large caves with uniform three-dimensional development and a considerable vertical extent are not common—the Cave of the Winding Stair in California (Halliday, 1962) comes close. It has a mixture of horizontal passages, sloping passages, pits, and large rooms, but it lacks distinct levels. Nor are the pits the distinctly vertical elements found in many other high gradient caves.

3.4 SOLUTIONAL SCULPTURING

Solutional attack removes bedrock in a nonuniform way, creating a complex suite of solutional forms etched or sculptured on the walls and ceilings of cave passages. These small sculpturings, the underground equivalent of the karren forms that appear on surface karst, are called speleogens (Halliday, 1955) in some U.S. literature.

Bretz (1942) attached special significance to solutional sculpturing. He named many of the features and used them as evidence for determining the vadose or phreatic origin of caves. The Bretz binary classification of solutional sculpturing into phreatic features and vadose features is no longer satisfactory. There is an implied genetic significance with which not all later workers agree. Also, the Bretz list includes such items as domepits (vertical shafts), which are passage elements on a scale larger than the solution features. Other features have since been identified that do not appear on the Bretz list.

There are four main categories of solutional features: (1) Channel features have a characteristic channel geometry, but are smaller than the cave conduit in which they are formed. (2) Hydraulic features are sculptured forms whose shape is determined by the flow dynamics of moving water. They therefore record information about the behavior of past flow systems in now-abandoned conduits. (3) Etched features are forms sculptured by chemical attack on the bedrock along lines of structural or lithologic weakness. (4) Residual features [called petromorphs by Halliday (1955)] are residual insoluble portions of the bedrock left projecting in relief from the cave wall as the surrounding bedrock dissolved.

3.4.1 Channel Features

Ceiling channels appear as upside-down stream beds in the roofs of cave passages. Typically, the channel width is less than 1 m and can be traced for distances of tens of meters. Some ceiling channels are half tubes; others are deeply incised with depths many times their widths. Some are single channels that are frequently sinuous. Others have a complex system of tributaries. The type locality is Seawra Cave in Mifflin County, Pennsylvania, where there is one of the most complex arrays of ceiling channels known, with multiple tributaries, sinuous bends, and undercut banks downstream from meander bends. Bretz (1942) interpreted these features as true stream channels cut upward into the ceiling when in-filling of sediment on the floor forced free-surface streams against the bedrock roof of the passage.

Many of the ceiling channels in the Mammoth Cave area are small canyons that extend distances of many meters upward above the ceiling of the passage, although they are only a decimeter wide. These channels are cut by descending vadose waters that accidentally cut through the ceilings of older horizontal tubes. Many of these carry water after storms, and some are related to vertical shafts. The Mammoth Cave type of ceiling channels have an entirely different origin from that of the Seawra ceiling channels.

Floor slots are narrow channels cut in the bedrock floors of cave passages. They are transitional to the residual canyons cut in floors of tubular passages by residual free-surface streams, distinguished from them mainly by size. Lange (1954) has argued that such features can also be produced by flow under pipe-full conditions.

Incised meanders are formed when small free-surface streams set up a meandering pattern in the unconsolidated fill on the floor of a large passage, cutting against the bedrock walls. As the meandering channel cuts deeper, its outer bends extend farther back into the wall of the larger passage, resulting in a half cone of residual bedrock. Incised meanders can evolve into independent cave passages. Incised meanders that begin high on the wall of a large passage could be taken as evidence of a previous sediment filling in the passage, because a previous level of fill was necessary to initiate the meandering free-surface stream.

Bedding plane and *joint plane anastomoses* are networks of tiny tubes occurring mainly along bedding planes. The tubes are usually circular or elliptical in cross section and range in size from a few centimeters to meters. At the upper end of the size scale, they blend continuously into the anastomotic maze form of passage element. Most anastomosis tubes are small and unenterable and can be seen only where they intersect large passages or have been exposed by breakdown falling away along the plane of weakness created by the anastomosis. Ewers (1966), from a study of bedding plane anastomoses in the central Kentucky karst and from modeling the features in salt blocks, concluded that these braided nets of tubes are among the earliest solutional openings in rocks where bedding planes provide the main routes of groundwater flow. They develop while flow rates are very small and the hydraulic head is large; they cease to grow significantly when the system resistance becomes low. Efficient flow paths through anastomoses may develop into cave passages. Bretz (1942) assigned an important role to anastomosis channels as the precursor pathways that formed deep in the phreatic zone before cave passages developed. Ewers' conclusions bear out their antecedent role, although not necessarily the depth of origin. There is also some evidence that anastomosis channels are formed contemporaneously with cave passages by water driven back into the bedding planes by local damming of the main flow line or hydrostatic pressure of seasonal fluctuations in groundwater inputs.

3.4.2 *Hydraulic Features*

Vertical rills (Fig. 3.19) are narrow grooves cut vertically in bedrock walls, especially over ledges. They appear to have much in common with the rillenkarren observed on the surface. The grooves are uniformly spaced and only one to a few centimeters wide; there are usually many of them in a neatly parallel array. Vertical rills in the central Kentucky caves are frequently associated with the mouths of anastomosis channels. It is possible that they form when the passage is in the floodwater zone. During floods, the passage fills with water, which is stored in the anastomosis channels. When the flood recedes and the main passage drains, the water drains out of the anastomosis system rather slowly, creating the rills as it flows down the walls in sheets.

Horizontal grooves are cut in cave walls, sometimes cutting the bedding at a low angle. When bedding is exactly horizontal, the hydraulically controlled grooves are difficult to distinguish from grooves originating with differential solubility of the beds. The horizontal groove is strictly horizontal, regardless of the dip of the bedding. Some are shallow; others are cut many meters into the wall. It has been argued that they form by lateral solution of shallow free-surface streams. Evidence for the role of free-surface streams is

Figure 3.19 Vertical rills in relation to small anastomosis channels. The photograph was taken in Fossil Avenue in the New Discovery section of Mammoth Cave, Kentucky.

found in many Missouri caves, where the stream flows in a narrow groove far off to one side of the main passage. In time, the new stream channel is downcut and a composite passage with offset components connected by a narrow slot results.

Pendants are vestiges of bedrock hanging below the ceiling or flat-bottomed ledge. They usually occur in groups and grade laterally into a smooth ceiling. Pendants were believed by Bretz to be remnants of an older ceiling that was carved and dissected into the pendant shapes when the passages were flooded by free-surface streams. Other instances of pendants are the residual "islands" found among highly developed anastomosis channels that are exposed when the lower beds fall away (Fig. 3.20). Pendants occur near the mouths of many caves in tropical karst. Lehmann (1954a) described giant pendants around the base of the karst towers using the term *Deckenkarren* ("ceiling karren").

High gradient stream beds in caves are often eroded into complex forms, of which circular basins called potholes are common. These vary in size from a few centimeters to several meters. Most potholes are shallow—less deep than they are wide. The bottom is usually a smoothly curved bedrock basin, sometimes containing rounded peb-

Underground landforms in karst regions / 95

Figure 3.20 Pendants resulting from exposure of closely spaced anastomosis channels by breakdown along the bedding planes containing the channels. Rickwood Caverns, Blount County, Alabama. [Courtesy of Barry F. Beck.]

bles and sand, which may have played a role in scouring out the basin. Ford (1965a) shows that potholes are both created and destroyed in high gradient cave streams and that abandoned or "hanging" potholes on cave walls are useful indicators of past flow conditions.

Sharp-edged pinnacle-like spires of residual country rock have been found in the beds of tropical cave streams. These were termed echinoliths by Aley (1964). Echinoliths resemble the pinnacle weathering or spitzkarren form that occurs on exposed limestone surfaces in tropical areas. The cave form may be related to the flooding in cave streams that carry only a small clastic load.

Vertical grooves are cut in the walls of vertical shafts and occasionally on vertical cave walls, where water streams down in thin sheets. They lend a serrate pattern to the cross-sectional view of vertical shafts. Some writers call these vertical grooves vertical fluting.

The term *flute* was used in an entirely different sense by Curl (1966b) and others to describe asymmetric cuspate grooves that circle some cave passages. Flutes, in this usage, are special cases of the cuspate oyster-shell-shaped solution depressions in cave walls known as scallops. Scallops and flutes occur in large groups, sometimes covering all exposed surfaces of cave passages. Figure 3.21

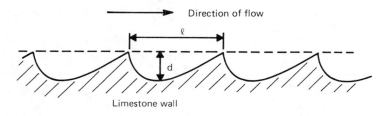

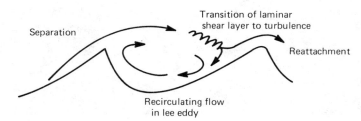

Figure 3.21 *Top:* Cross section of scalloped wall showing characteristic length. *Bottom:* Flow pattern over the scallop surface, as described by Curl (1974).

illustrates the characteristic cross section. Scallops and flutes are asymmetric with sharp cusplike edges. They can be described by a characteristic length, l, the distance from cusp to cusp between adjacent scallops. The characteristic length varies from a few centimeters to a meter or more, but the lengths of all scallops or flutes within any one group are very similar.

It has long been recognized (Coleman, 1949) that scallops and flutes are indicators of the direction of water flow. The smooth slope is on the downstream side of the scallop and the steep cusp is on the upstream side. The size of scallops and flutes is inversely related to the flow velocity of the water that sculptured them. Small scallops are characteristic of cave passages with high-velocity flow (Fig. 3.22), whereas larger scallops form where flow velocities are lower. At the large scallop–low-velocity end of the scale, the sizes of the scallops become comparable to the size of the passage, and scallops are difficult to distinguish from bends in the passage wall.

By analyzing the process of flute and scallop formation, Curl (1966b) was able to show that they are strictly a hydraulic phenomenon. The characteristic flute and scallop geometry depends on the properties of the fluid and not, to a first approximation, on the characteristics of the bedrock. Scallops cut in ice by wind action have the

Figure 3.22 Small scallops on walls, floor, and ceiling of a well-scoured flood overflow passage in Horn Hollow Cave, in Carter Cave State Park, Kentucky.

same hydraulic geometry as scallops cut in limestones by groundwater solution. The parameters are related by a dimensionless Reynolds number:

$$\frac{\bar{v}\bar{L}\rho}{\eta} = N_R \tag{3.2}$$

where $\bar{v}$ is the velocity of fluid flowing past the scallop, $\bar{L}$ is the mean scallop or flute length, ρ is the density of the fluid, and η is the fluid viscosity. Flutes were shown to be a special case of scallops for which the flow conditions were constant over long periods of time. Flutes propagate into the wall and downstream at the same angle—about 71° to the horizontal—and have rounded crests. They are usually observed to have short (2–10-cm) crest-to-crest distances, which are associated with a high-velocity flow. Under such conditions the channel flow is turbulent and the velocity near the wall is not too different from the mean velocity in the channel. Curl's estimated Reynold's number is 22,500 for flutes.

Scallops are much more common than flutes, but using scallops to measure paleoflow velocities has proved a difficult problem. Goodchild and Ford (1971) investigated scallop patterns by using laboratory models consisting of plaster blocks. The expected scallop patterns developed, but it became apparent that the material on which the scallops formed did exert some influence on the final distribution of scallop sizes. The main difficulty, however, was in providing a practical definition for $\bar{v}$. The developing scallop responds to the flow velocity in the fluid layer immediately above it, but the velocity profile is determined by both the flow velocity in the channel and the hydraulic radius of the channel.

A more elaborate analysis of scallop formation (Blumberg and Curl, 1974; Curl, 1974) provides the needed relationships. Rather than use the arithmetic means that were used previously, Curl argues that the Sauter mean

$$\bar{L}_{32} = \frac{\Sigma l_i^3}{\Sigma l_i^2} \tag{3.3}$$

provides the best measure of scallop size. The l_i values are lengths of individual scallops in a homogeneous population. By dimensional analysis it was shown that

$$\frac{\bar{L}_{32}v^*\rho}{\eta} = N_R^* \tag{3.4}$$

v^*, a friction velocity, equals $\sqrt{\tau\rho}$, where τ is the average shear stress at the wall and ρ is the fluid density. The scallop Reynold's number, N_R^*, based on friction velocity, is a universal constant for scallop formation and was found from model experiments (Blumberg and Curl, 1974) to have the numerical value $N_R^* = 2200$. A scallop Reynold's number that applies to actual channel velocity can be calculated from the friction velocity Reynold's number and the velocity distribution in the channel. Curl (1974) assumed two limiting geometries for cave passages—a circular cross section and a rectangular cross section:

$$N_R = N_R^* \left[2.5 \left(\ln \frac{D}{2L_{32}} - \frac{3}{2} \right) + B_L \right] \tag{3.5}$$
(circular conduit)

$$N_R = N_R^* \left[2.5 \left(\ln \frac{D}{2L_{32}} - 1 \right) + B_L \right] \tag{3.6}$$
(rectangular conduit)

D is the diameter of a circular conduit or the distance between the parallel walls of a rectangular conduit. The quantity B_L is determined only by the wall roughness and was found from model studies to be equal to 9.4. The scallop Reynold's number can thus be determined for a particular channel geometry, and from Eq. 3.2. the mean channel velocity, $\bar{v}$, can be calculated.

Given a now-dry cave passage with scalloped walls, one need

only assume a temperature for the water that flowed through the conduit in the past, measure the passage size, and enough scallop lengths to calculate the Sauter mean and the paleovelocity and paleodischarge of the passage can be calculated (Fig. 3.23). Discharge through caves varies with the time of the year, and it is not yet clear what discharge the scallop records. If the rate of attack of groundwater on the cave wall is not strongly velocity-dependent, then the scallop pattern should record the most-probable, flow which is often close to the mean flow.

3.4.3 Etched Features

Etched features are those that appear to have been dissolved out of the bedrock by more or less stagnant water. Their form, therefore, depends on the vagaries of relative rates of solution of different parts of the rock.

Etchpits are the smallest of these features. Etchpits consist of small, usually circular, holes in the cave walls. They range in size from a few millimeters to several centimeters. They usually occur in groups, so that large expanses of the wall are pitted. They differ from scallops in having symmetric profiles. The bedrock near the etchpits is often spongy or has a rotted look.

Spongework is a highly complicated system of holes, tubes, and interconnecting cavities of various sizes and shapes. It gives the cave wall a "Swiss-cheese" appearance. Bretz (1942) argued that spongework was the result of differential solution by the etching action of nearly stagnant water. It is not a common solution feature in caves. The Boneyard in Carlsbad Caverns is the type locality, and large areas of spongework occur in Wind Cave in South Dakota. A spongework-like feature occurs in Big Brush Creek Cave in Utah, which Palmer (1975) argues is a result of high-velocity flow under high hydrostatic head. The high head during periods of flood runoff forces water through all possible openings in the rock, thus creating the spongework pattern.

Wall and ceiling pockets are larger than spongework and are usually separated by stretches of unaffected rock. Wall and ceiling pockets often resemble kettle holes and are usually circular or elliptical in outline. Some are hemispherical and others, although circular, extend some meters into the rock as cylindrical openings. Ceiling pockets seem to be more common in tropical caves (Fig. 3.24). Most of these are smooth openings in the bedrock; there is no joint to provide a line of weakness for solutional attack or to transport additional water. Although Bretz proposed differential solution of the limestone as a cause, there are no sedimentary structures in limestone that would etch out into these smooth cylindrical shapes. Ceiling pockets occur in great profusion in caves of Sarawak (Wilford,

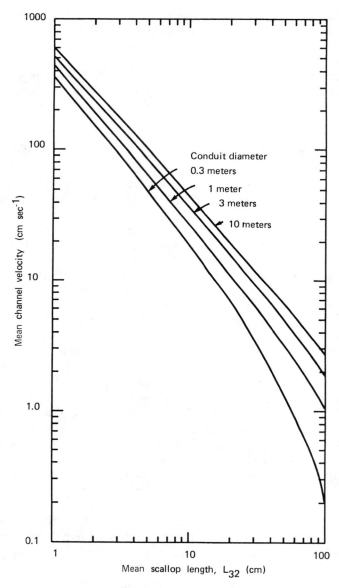

Figure 3.23 Relation between the mean channel velocity and mean scallop length determined by Eqs. 3.1 and 3.5. The curves were calculated using a calculator program kindly provided by Dr. Curl.

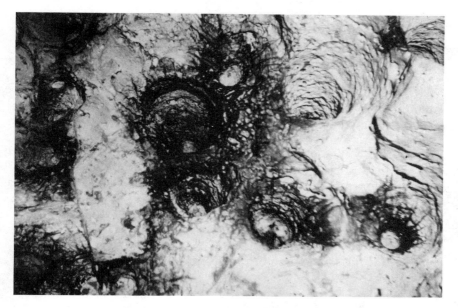

Figure 3.24 Cylindrical ceiling pockets in Runaway Bay Cave, Jamaica. The pockets are circular cylinders of varying height in a massive limestone unit with no obvious joint or bedding plane control.

1966), inspiring the bizarre hypothesis that the holes were formed by the solvent action of bat urine.

Joint-determined wall and ceiling grooves and pockets are common in caves in all climates. They typically occur as deep, narrow slots extending along joints. Some are nearly circular, while others are elongated along the joint. The sizes range from a few centimeters to many meters. When the joint-controlled groove forms in the ceiling parallel to the passage, it produces a gothic-arch cross section. When joints are oblique to the passages, deep canoe-shaped channels form in the ceiling. These features appear to form in a zone of active solution caused by the mixing of waters of different chemistry. As will be shown later, two waters, both saturated with calcium carbonate, can become undersaturated when mixed. If one kind of water is flowing along the cave passage under pipe-full conditions, and another kind enters the passage through the joint, the mixing zone will be a zone of active solution that dissolves out the joint-controlled groove or pocket.

3.4.4 Residual Features

Parts of the bedrock that dissolve less readily are left behind in relief as the passage enlarges. Continuous rock spans, natural bridges and pillars are obvious examples of the massive limestone itself having been left as the passage around it enlarged.

Other residual features are merely insoluble constituents of the limestone. Thin shale partings in the limestone are sometimes decomposed, resulting in grooves in the wall parallel to bedding planes. In other situations, the shale partings stand out in relief as clay fins, sometimes very thin and delicate. Chert nodules remain standing in relief as, of course, do smaller silicified fossils. Veins of crystalline calcite tend to dissolve more slowly than the fine-grained limestone around it, and they also tend to stand in relief. Much attention has been given to a residual calcite vein form in a reticulated pattern known as boxwork. Thin boxwork veins of calcite or quartz occur in many caves. However, the best-known locality is the caves of the Black Hills of South Dakota, where there is a complex, multilayered boxwork. The well-developed boxwork of the Black Hills is a result of more complicated mineralization processes (White and Deike, 1962) and is not common in other caves.

4

Karst Landscapes

4.1 CONTROLLING VARIABLES FOR KARST LANDSCAPES

4.1.1 *Boundaries on Karst Landscapes*

Karst landscapes, made up of assemblages of karst landforms, are many and varied over the surface of the earth. The same kind of landforms, however, tend to dominate individual karst regions. Thus, the rolling limestone plains of the Kentucky Pennyroyal and the Tennessee Highland Rim are called doline karsts because dolines are the most obvious and universal landform in the areas. Caves, sinking streams, and other karst features also occur, but in the field, on topographic maps, or on air photographs, dolines are the most immediately eye-catching landform. Other karst areas are given climatic labels. One speaks of the karst belt of northern Puerto Rico as tropical karst, although the dominant landforms are cones and towers. The karst areas of the Alps, Pyrenees, or Rocky Mountains are often referred to as alpine karst, a term that describes the setting but says little about either the climate or the dominant landforms. The boundaries of a particular style of karst may be defined along the outcrop line of the karstic rocks, by physiographic boundaries, or by drainage basin.

Much of the research on karst is concerned with water budgets and chemical balances of various types. The drainage basin is a convenient unit for such work, although many karst drainage basins contain areas not on karstic rocks.

4.1.2 *Variables of the Karst*

Figure 4.1 shows a highly schematic drawing of the essential components of karst development. The physical object is a block of the

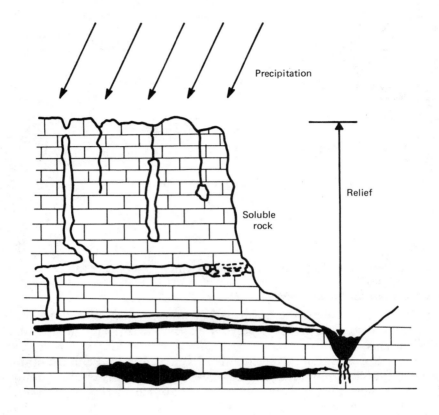
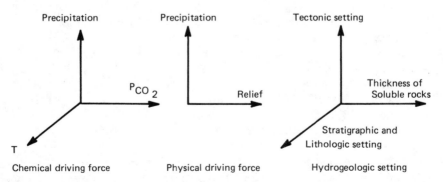

Figure 4.1 The essential features of the karst process. The "coordinate systems" illustrate in a broad way the independent variables that guide the development of karst landscapes.

earth's crust containing rocks that are capable of being dissolved by surface or groundwater. The karst process, by definition, is chemical solution. The observed suite of karst landforms is the result of operating on the block of rock by the chemical processes. The rocks and the process may be examined individually to identify, again in a systematic way, the controlling variables listed in Fig. 4.1 as seven independent coordinates.

We can think of Fig. 4.1 as describing a rate process. All rate processes represent a balance between forward reactions and back reactions. Chemical solution is the forward driving process. The resistance of the bedrock to solution can be thought of as a sort of back reaction. If the bedrock is highly resistant to solution, karst landforms are sparse, subdued, and form slowly even when the chemical and physical driving forces are optimum. On the other hand, highly soluble rocks such as pure limestones or gypsum do not develop karst surfaces when the chemical and physical driving forces are weak. There are balances and trade-offs among all the variables, and much of the remainder of this book is devoted to delineating these.

For the exposed bedrock to be dissolved and karst landforms to develop, a source of energy is required. Energy gradients provide the driving force for the karst processes. Soluble limestone and gypsum are often out of equilibrium with CO_2-rich aqueous solutions. The chemical reactions that result in the dissolution of the rocks are driven by the excess free energy in the chemical reactions. In broad terms, three chemical variables control the rate of karst development: the amount of water available (equated with precipitation), the amount of carbon dioxide available (which is related to plant and soil cover), and the temperature. These are exactly the variables that are often lumped together as "climate"; this shows us both why such terms as tropical karst are used and why they are indeterminate. Hot, wet climates are chemically quite different from hot, dry climates.

Left to themselves, the chemical reactions drive toward equilibrium; solutions become saturated and the dissolution of the bedrock ceases. For the process to continue, the saturated solutions must be removed and replaced with fresh solutions. Thus, precipitation is also a physical driving force. The energy source for the circulation of groundwater is the earth's gravitational field. The action of gravity on water in the atmosphere and in the uplands drives water over and through the rock mass, always toward base level streams at lower elevations and ultimately to the ocean. The hydraulic gradient provides the primary physical driving force, and its geomorphic representation is the relief of the landscape.

The geologic setting of the karstic rocks divides broadly into three master variables: the thickness of the soluble rock sequence, their stratigraphic and lithologic characteristics, and their structural or tectonic setting. These are independent coordinates, in a conceptual sense, although only the rock thickness is readily fitted with a numerical scale.

The thickness of the soluble rock sequence places limits on the extent of karst development. The Loyalhanna limestone of western Pennsylvania is the only soluble rock in a thick sequence of shales, sandstones, and coals. There is extensive cave development, but since the Loyalhanna is only 10 to 20 m thick, the surface expression of karst is limited to a few dolines and small springs. This contrasts with the classic karst of the Adriatic Coast, which has developed on roughly 9000 m of carbonate rock.

The stratigraphic and lithologic characteristics are a family of related variables that include rock type, lithologic characteristics, grain size, accessory minerals, and the stratigraphic mix of various lithologies. There are four important karst-forming rocks: limestone, dolomite, gypsum, and salt. Most of what is discussed in this book is limestone karst. For reasons discussed later, karst landforms do not develop easily on dolomite. Although dolomite karsts are common, they tend to contain subdued surface forms and only minor cave development. Gypsum and salt karsts are discussed separately in Chapter 11. To this list should be added some oddities such as thermokarst and karst developed on slightly soluble rocks. Some discussion of these topics is given in Chapter 12.

The structural setting is also a family of related variables. On a large scale, the rocks may be flat-lying, folded, block-faulted, or overthrusted to various degrees, and many complex patterns are possible. The structural setting may have originated from old tectonic movements, and thus the structural environment is static, on the time scale of karst processes. The area may be one of young tectonics, presently active, with faults still in motion, in which the structural setting is still mobile on the time scale of the karst processes. The Adriatic karst appears to be of this type. On a smaller scale, the density of joints and fractures, their linear extent, and their orientation are important for defining the primary pathways of groundwater movement and for setting the pattern of cave development and underground drainage. Structural features such as insoluble rocks in anticlinal cores, or blocks of insoluble rock raised by normal faulting, can act as groundwater barriers and so control or modify drainage patterns.

All of the geologic factors taken together constitute the hydrogeologic setting. The hydrogeologic setting is a framework within which the karst processes are constrained to operate, and to a large degree, it dictates the types and extents of karst landscapes that are possible. Any given karst landscape must be analyzed in terms of the process variables and the hydrogeologic variables. Arguing, as some writers have done, that one or more of the master variables, such as climate, is all-important is not a constructive approach to understanding karst.

4.2 SOME COMMON KARST LANDSCAPES

The principle seems to hold generally that any karst landscape is constructed from a stochastic but repetitive arrangement of a limited set of karst landforms. A useful description can be made by simply naming the dominant landforms:

  Doline karst
  Cockpit karst
  Cone and tower karst
  Fluviokarst
  Pavement karst
  Polje karst
  Labyrinth karst
  Cave karst

A much more detailed analysis of various approaches to the classification of karst is given by Quinlan (1978).

4.2.1 *Doline Karst*

Doline karsts are the most widely distributed of any karst landscape type. They are landscapes dotted with sinkholes (Fig. 4.2), although doline sizes and densities vary by orders of magnitude from one region to another. Table 4.1 lists depression densities for some doline karsts in various parts of the world. The depression density is determined by details of local geology, particularly the spacing of fractures and lithology of the bedrock, rather than by regional or climatic parameters. This can be seen in the Appalachian karst, where there is a log-normal distribution of depression densities within a population of drainage basins, all in roughly similar geologic and climatic settings (Fig. 4.3).

Doline karsts in the United States include the Mitchell Plain of southern Indiana (Palmer and Palmer, 1975), and southcentral Kentucky (White et al. 1970; Quinlan, 1970; Miotke, 1975a), and the Highland Rim of central Tennessee, all of which are developed on a low dip sequence of Mississippian limestones that facilitate the development of sinkhole plains. Other doline karsts occur in northern Florida, the Lexington Plain of Kentucky, southwestern Missouri, the Greenbrier karst of West Virginia, and the Appalachian Great Valley, which extends from Pennsylvania, through the Shen-

108 / *Geomorphology and hydrology of karst terrains*

Figure 4.2 Doline karst. Sinkholes developed on the Mississippian Greenbrier limestone, in Monroe County, West Virginia. [Courtesy of William K. Jones.]

andoah Valley of Virginia (Hack, 1960a; Hubbard, 1984), to east Tennessee. Many other doline karst areas exist. Regions with sparse karst such as Kansas (Merriam and Mann, 1957) could be described as "doline karst."

4.2.2 *Cockpit Karst*

Workers on tropical karst seem to agree that cockpit karst is different from doline karst, but they do not agree just how it is different. Cockpits are larger depressions than the dolines, and to maintain a uniform circular or star shape, the initial points of development must be more widely spaced than the spacing of dolines. Well-developed cockpit karsts are not common; the Cockpit Country of Jamaica is the type example (Sweeting, 1958; Smith et al., 1972).

The doline area ratio varies from near zero in landscapes where only a few scattered dolines are superimposed on a fluvial topography to near unity in the doline karsts of Kentucky and Indiana, where all runoff is internal and there is no surface drainage. Cockpit karsts also have doline area ratios near unity, but they have fewer and larger depressions. If the mean size of the depression is more or

Table 4.1
Characteristics of Some Doline Karsts

Karst region	Area km²	Depression density, (km⁻²)	Doline area ratio	Reference
South Harz	0.42	80.95	0.0134	
Blaubeurer Alb	137.67	2.38	0.00032	
Gräfenberger Alb	41.25	0.87	0.00097	Cramer (1941)
Wiesentalb	10.58	6.04	0.00116	
Altmühlalb	202.42	0.81	0.00060	
French Jura	6.63	31.60	0.043	
Velebit Mountains	60.81	1.15	0.045	Cramer (1941)
Kanzian, Istria	2.98	15.40	0.030	
Doberdo, Istria	8.35	49.32	0.049	
Mount Kaijende, New Guinea	13.18	13.05	1.00	Williams (1971)
Darai Hills, New Guinea	13.84	13.58	1.00	Williams (1971)
Brownstown, Jamaica	13.0	2.85	0.42	Day (1976)
Shenandoah Valley, VA Ordovician rocks	—	1.66	—	Hack (1960)
Cambrian rocks	—	0.59	—	Hack (1960)
Sinkhole Plain, Kentucky	153	5.41	—	Troester et al. (1984)
North Florida	427	7.94	—	
Puerto Rico	799	5.39	—	
Dominican Republic	1262	5.71	—	

less the same, the depression density must increase with the doline area ratio; however, the cockpit karsts have high doline area ratios but lower depression densities. More quantitative data on closed depressions in various karst areas would be helpful. It may well be that cockpit karst is not a distinguishable landscape.

4.2.3 Cone and Tower Karst

The requirements for cone and tower karst seem to be a thick massive limestone and a well-developed fracture system. Dissolution along the fractures separates limestone uplands into isolated blocks. The solution corridors that separate the blocks deepen and widen, and the tops of the blocks are rounded and dissected into secondary karst surfaces with shafts and various forms of karren. The dissolution along the fractures is eventually terminated by change in lith-

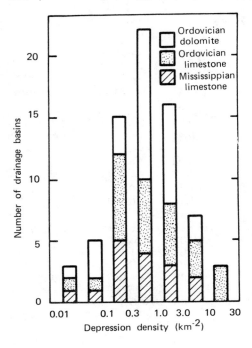

Figure 4.3 Distribution of drainage basins with depression density for 62 Appalachian basins. Data are from White and White (1979).

ology, by base level, or by accumulated residual clastic material. The trenches then widen and transform into an alluvial plain separating the increasingly isolated cones and towers.

Cone and tower karst and cockpit karst are both found in the thick massive limestones of tropical regions. Figure 4.4 presents a rather idealized distinction. Deep solution along fractures produces trenchlike depressions and convex slopes on the residual hills of the cone and tower karst. More extensive solution and internal transport of water and clastic residues produces the broader cockpit depressions and a concave slope on the intermediate residual hills. In cockpit karst there is a continuous network of pyramidal hills with intermediate saddles. The cockpits are star-shaped depressions located on fracture intersections. In cone and tower karst the solution is more uniformly distributed along the fractures that form a continuous network of trenchlike depressions with isolated intermediate rock masses. In a certain geometric sense, cockpit karst and cone and tower karst are topologic inverses of each other. Gradations between these landforms are common.

Cone and tower karst is found mainly in Central America and the South Pacific. In Central America, cone and tower karst from Tabasco, Mexico (Gerstenhauer, 1960) and Belize (McDonald, 1976a) have been briefly described. The Mogote karst of Cuba consists of

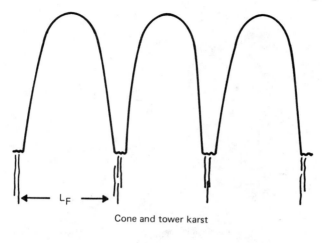

Cone and tower karst

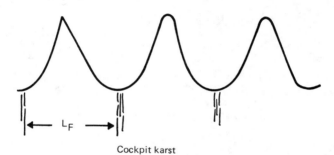

Cockpit karst

Figure 4.4 An idealized view of the distinction between cone and tower karst and cockpit karst, based on the curvature of the slopes. L_F is the spacing between fractures.

squat limestone towers with nearly vertical sides, with a complex of foot caves and interior cave passages. Some are isolated towers standing above an alluviated plain; others appear to be separated from a partially intact upland (Sierra de los Organos) (Lehmann, 1954a,b; Núñez-Jiménez, 1959; Panoš and Štelcl, 1968). The northern limestone region of Puerto Rico is mixed cone and tower karst with some very steep towers and some conical hills. Some of the cones and towers are nestled tightly together with only rough gorges between them, and others occur as isolated hills with an intervening plain underlain by blanket sands (Fig. 4.5) (Monroe, 1976). Similar cone and tower karst apparently occurs in the Dominican Republic.

The classic tower karst occurs widely in south China in Yunnan, Guangxi, and Guizhou Provinces, with some of the most spectacular topography near Guilin. Both nested and isolated towers occur. The

112 / Geomorphology and hydrology of karst terrains

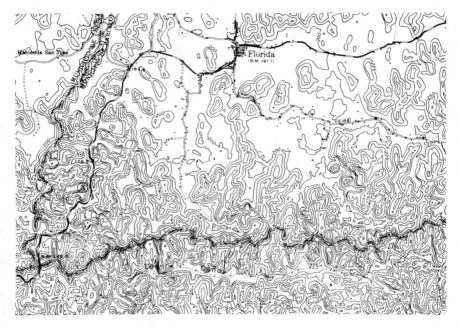

Figure 4.5 Cone and tower karst from the northwestern karst belt of Puerto Rico; a section from the U.S. Geological Survey Florida quadrangle. The contour interval is 10 m.

isolated towers are often relatively narrow with vertical sides—bare limestone monoliths jutting from the plain (Von Wissmann, 1954; Chinese Academy, 1976). The cone and tower karst of China extends southward into Vietnam. Post-Pleistocene sea level rise has innundated part of the karst, so that limestone towers emerge from the Gulf of Tongkin (Silar, 1965).

Herbert Lehmann (1936) first described cone and tower karst from his studies in Java. Similar landforms occur in the Celebes (Sunartadirdja and Lehmann, 1960; McDonald, 1976b), in other parts of Malasia (Tjia, 1969; Verstappen, 1960), and particularly in Borneo and Sarawak. The mountains of New Guinea provide several styles of cone and tower karst (Verstappen, 1964; Williams, 1971, 1972a).

4.2.4 *Fluviokarst*

Much of the karst of the eastern United States is fluviokarst. In regions where the limestone thickness is less than the relief, both soluble and nonsoluble rocks crop out in the same drainage basins. Fluviokarst is a landscape of deranged drainage, blind valleys, swallow holes, large springs, closed depressions, and caves. In many fluviokarst regions, the larger rivers maintain their surface courses, often fed by underground tributaries.

There is a variation of fluviokarst that may be termed plateau margin karst. Some of the best examples in the United States occur along the dissected margin of the Cumberland Plateau in Tennessee and northern Alabama. The Cumberland Mountains are underlain by several hundred meters of Mississippian carbonate rocks capped with resistant sandstones and underlain with shale and shaly limestone. The rocks are nearly horizontal, so that all of the deep valleys cut into the plateau margin are underlain with the carbonates. Surface streams with catchments on the plateau flow down the escarpment and cut deep gorges in the limestone. Much of this drainage moves in subsurface routes, forming caves at various levels in the escarpment where vertical movement of the streams is interrupted by resistant sandstone or shale layers. Overall, the drainage profile is a stair step, often with both surface and underground components (Fig. 4.6).

If boundaries are drawn around geological formations, or around local physiographic settings, then fluviokarst with its arrangements of karst valleys is only one of a set of karst landscapes. If boundaries are drawn around drainage basins, particularly large drainage basins, then nearly all karst is fluviokarst, as Roglić (1964) has argued.

4.2.5 Pavement Karst

Pavement karsts are areas of bare limestone, usually sculptured into karren of various types (Williams, 1966b). Solution along joints laces the bedrock surface with a network of kluftkarren. Pavement karsts occur in alpine terrains where soils are thin or stripped by alpine glaciation (Fig. 4.7). Pavement karsts occur in northern regions where Pleistocene continental glaciation has stripped the soils and left the underlying carbonate rocks open to solutional attack. Once solution has opened the joints to permit internal drainage, and the soils have been stripped, it is extremely difficult to reestablish the soil cover. Rainstorms flush incipient soils quickly to the subsurface and the bare bedrock pavement is preserved. The result can be a barren rocky plain with no vegetation cover at all (Fig. 4.8).

Much of what has been described as alpine karst is actually pavement karst. Bare rock surfaces are exposed to continuous solutional attack, resulting in the development of many karren forms, because severe climate at high altitudes, steep slopes, and thick snow pack prevent development of soils. Many karst areas in the Rocky Mountains of the United States and Canada contain segments of pavement karst (Campbell, 1979; Ford, 1979; Medville et al., 1979; White, 1979; Wilson, 1979). Pavement karst is also the common karst land-

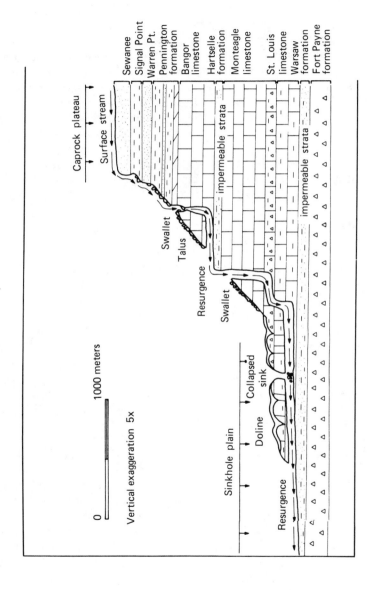

Figure 4.6 Profile for representative plateau margin karst, in Cumberland Mountains, Tennessee. [Adapted from Crawford (1979).]

Figure 4.7 Joint-controlled kluftkarren and limestone pavement, in South Darby Canyon, Teton Mountains, Wyoming. [Courtesy of J. A. Stellmack.]

Figure 4.8 Pavement karst of the Gort Plain, in County Clare, western Ireland.

scape at high altitudes, for example, on Spitzbergen Island (Corbel, 1957) or Greenland (Davies, 1957).

Pavement karst is a characteristic landscape in arid and semiarid regions. Human influences, particularly overgrazing, can destroy vegetative cover. Rapid loss of soil through solution openings results in a jumbled chaos of loose blocks and ledges. In arid climates the soil, which may have been relict from Pleistocene pluvial periods, cannot be reestablished and the bare pavement karst may become a permanent feature. Much of the karst of the countries around the Mediterranean is this type.

4.2.6 *Polje Karst*

Because poljes as individual landforms are very large closed depressions, a polje karst covers a lot of territory. This, in turn, requires a great thickness of carbonate rocks. Although closed depressions that meet the definition of polje are scattered through the karst areas of the world, the major polje karst is the Dinaric karst of Yugoslavia. Only here is the entire landscape contructed of poljes alternating with intermediate mountain ranges (Gams, 1969).

4.2.7 *Labyrinth Karst*

Labyrinth karst (Sweeting's term) is a landscape dominated by intersecting solution corridors and solution canyons. The landscape was originally described from the Fitzroy Basin of western Australia (Jennings and Sweeting, 1963), where the margin of a limestone plateau is dissected into a network of solution corridors and box canyons. Jennings and Sweeting referred to it as a "giant grike-land." The karst of the Nahanni River Basin in the Northwest Territories (Brook and Ford, 1973) contains a similar rectilinear canyon pattern, although in a quite different geological setting.

4.2.8 *Cave Karst*

There are regions where limestones or other soluble rocks crop out at the surface, where there are caves and a well-developed underground drainage, with little surface expression in the form of closed depressions or other karst landforms. The best examples come from arid climates such as the American Southwest. In the Guadalupe Mountains, New Mexico, is the massive El Capitan reef limestone. Surface expressions of karst are very subdued; some bedrock pavement is sculptured with rain pit and rillenkarren forms. Dolines and other closed depressions are rare. In the subsurface are Carlsbad Caverns and the other major caves of the Guadalupes. Part of the Edwards Plateau of Texas likewise displays few surface karst landforms, despite the occurrence of many caves and an important limestone aquifer at depth. Although those who demand dramatic expression of solution processes might not call these landscapes karst at all, here they are termed cave karsts.

4.3 CLASSIFICATION BY COVER

The master variables described previously allow most contemporary karst landscapes to be interpreted, and individual karst regions to be described, in terms of their dominant landforms. This leaves, however, some residua of ancient landscapes and other karst phenomena outside the conceptual framework. An alternative is to classify karst phenomena in terms of their cover and their relationships to the contemporary landscape.

The classification that follows comes primarily from Quinlan (1967, 1978). Quinlan uses as his basic criterion for classification the occurrence and nature of cover over the karst surface. These types are then modified by the lithology of the rock, the climate, the geologic structure, the physiography, and any modifications during or after karstification.

A. *Covered Karst.* Dissolved bedrock surface is covered with some sort of material, soil, or rock.
   1. *Subsoil karst* (bodenbedeckter karst). Karst surface is covered with soil either transported or its own residuum.
   2. *Mantled karst.* Covered with allochthonous rock or sediments. It is part of the contemporary landscape, and older than its cover.
   3. *Buried karst* (paleokarst). Covered with allochthonous rock or sediment. It is not part of present landscape and older than its cover.
   4. *Interstratal karst.* Covered with autochthonous rock or sediment. It may or may not be part of the contemporary landscape, and is younger than its cover. The term refers to solution of soluble rock beneath relatively insoluble rocks. This type does not include caves developed beneath caprock.
   5. *Subaqueous karst.* Including drowned karst (e.g., by sea level rise), subfluvial karst, developed beneath a river, and submarine karst, developed beneath the tidal zone.
B. *Exposed Karst.* Bare rock surface is exposed.
   1. *Naked karst* (nacktkarst). Developed and maintained without any cover, or beneath a temporary cover of snow or water.
   2. *Denuded karst.* Subsoil karst or interstratal karst that has been exposed by the erosion of its cover.

3. *Exhumed karst.* Mantled karst or buried karst that has been divested of its cover by erosion. It is the reexposed portion of a former karst landscape.
4. *Relict karst.* The topographic or physical remains of a karst that has not been covered and in which all or most of the karsted rock has been removed by subsequent erosion.

Most of the karst landscapes of the temperate United States are subsoil karst. The rolling karstlands of Indiana, Kentucky, Tennessee, Virginia, West Virginia, and Missouri have a sculptured bedrock surface that lies below the soil. The landforms that one sees in the field or on topographic maps are smoothed and subdued by the soil cover. Only in a few places—quarries, roadcuts, and areas of extreme erosion—can the bedrock surface be seen.

Buried karst or paleokarst is an ancient karst landscape that has been buried under later sedimentation. Perhaps the most outstanding example in the United States occurs at the top of the Mississippian limestones in the Midwest and West. In the Black Hills of South Dakota, the top of the Pahasapa limestone was pitted with sinkholes, shafts, and caves at the close of Mississippian time. The Pennsylvanian Minnelusa sandstone filled the surface features, and the karst remains to the present day.

Interstratal karsts are solution features that develop as part of the present-day groundwater circulation patterns beneath an insoluble rock cover. Quinlan, in his original definition, excluded caves from this category, but there is no good boundary line between the subsurface aspect of other types of karst and interstratal karst. For example, cave systems occur in many parts of the Appalachian plateaus and in the interior lowlands under clastic rock-capped plateau fragments or ridges. Such caves are not interstratal karst. However, small network caves form at shallow depth in limestones, sometimes only a few feet thick and sandwiched between insoluble rocks. These solution features are considered a form of interstratal karst.

Naked karst occurs primarily in alpine regions, where soils are poorly developed. Such karst occurs in the Rockies and other mountain ranges of the American West, covering a considerable area. Naked karst is characterized by sharp, spikey sculpturing of the bedrock, whereas bedrock of subsoil karsts is more rounded. Snow melt and glacial action play a role in many naked karsts and may be as important in determining the final shape of the landforms as is the absence of soil. Many naked karsts are the same as the pavement karsts described earlier.

5
The Chemistry of Carbonate Dissolution

This chapter summarizes the physical chemistry of carbonate dissolution. Those with greater interest in the chemistry of groundwater or chemical processes in karst are urged to seek more information in Garrels and Christ (1965), Stumm and Morgan (1981), and, on a more practical level, Loewenthal and Marais (1976).

5.1 CONSTITUENTS OF KARST WATER

Landscapes evolve in carbonate terrains through the process of chemical dissolution and transport of calcite and dolomite. Most simply, then, we are concerned with the heterogeneous equilibria (or disequilibria) between a small set of solid phases (the minerals in the limestone or dolomite bedrock), a liquid phase (an aqueous solution containing a variety of dissolved species), and a gas phase (usually, air containing carbon dioxide as the only active ingredient). Accessory species are sulfate ions from gypsum beds or weathered sulfide minerals, chlorides from deeper circulating groundwaters or from seawater mixing, and sodium and potassium from similar sources.

If underground waters in karst regions are the habitat of aquatic organisms or a source of water supply for people, then the number of important chemical parameters increases. To the inorganic constituents is added dissolved oxygen, the key component in waters that support life. There are also pollutants that dissolve in groundwater from agricultural activities, sewage, industrial wastes, and other sources of chemical contamination. Table 5.1 lists some of the

Table 5.1
Some Relevant Inorganic Species for Chemical Reactions in Karst

	Solid phase		Liquid phase	Gas phase
Essential species	$CaCO_3$ $CaMg(CO_3)_2$	Ca^{2+} Mg^{2+} H^+	H_2O $(CO_2)_{aq}$ H_2CO_3 HCO_3^- CO_3^{2-} OH^-	CO_2
Accessory species	$CaSO_4 \cdot 2H_2O$ SiO_2 FeS_2	Na^+ K^+	SO_4^{2-} Cl^- $(O_2)_{aq}$ H_4SiO_4	H_2S O_2
Contaminant species		NH_4^+ Heavy metals	NO_3^- NO_2^- Phosphates	NH_3 CH_4

important species. Discussion of the contaminating species is deferred to Chapter 14.

5.1.1 *A Note on Units*

Concentrations of dissolved species are reported in the following units of measure:

- Parts per million (ppm) = weight of solute per million weight units of solution.
- Milligrams per liter (mg/L) = weight of solute per liter of solution.
- Molal concentration, m_i = moles of solute per kilogram of solvent.
- Molar concentration, $[i]$ = moles of solute per liter of solution.
- Thermodynamic activity, a_i = activity of species i. Must be defined with respect to a chosen standard state. (See works such as Stumm and Morgan (1981) for in-depth discussion.)

The aqueous solutions encountered in karst terrains are usually very dilute, and their densities differ from that of pure water by an amount less than the experimental error. For this reason, ppm and mg/L concentrations are numerically equal and molal and molar concentrations are essentially equal. It is not usually valid, however, to set thermodynamic activities equal to concentrations.

5.2 CHEMICAL EQUILIBRIA

5.2.1 *Dissolution Reactions for Carbonate Minerals*

Calcite and dolomite are ionic salts. In pure water they dissociate into their constituent ions

$$CaCO_3 \rightleftharpoons Ca^{2+} + CO_3^{2-} \tag{5.1}$$

$$CaMg(CO_3)_2 \rightleftharpoons Ca^{2+} + Mg^{2+} + 2\,CO_3^{2-} \tag{5.2}$$

These reactions are described by solubility product constants

$$K_c = \frac{a_{Ca^{2+}} a_{CO_3^{2-}}}{a_{CaCO_3}} = a_{Ca^{2+}} a_{CO_3^{2-}} \tag{5.3}$$

$$K_d = \frac{a_{Ca^{2+}} a_{Mg^{2+}} a_{CO_3^{2-}}^2}{a_{CaMg(CO_3)_2}} = a_{Ca^{2+}} a_{Mg^{2+}} a_{CO_3^{2-}}^2 \tag{5.4}$$

where a is the activity of the dissolved species and is closely related to concentration. The solubility product constants are functions of temperature (Table 5.2). The solubility of calcite in pure water is very small—about 6 ppm at 10°C—and is actually less than the solubility of quartz.

The carbonate ions that form by the dissociation of alkaline earth carbonates hydrate when in contact with water

$$CO_3^{2-} + H_2O \rightleftharpoons HCO_3^- + OH^- \tag{5.5}$$

forming a mildly alkaline solution. Increasing the hydroxyl concentration by raising the pH decreases the carbonate solubility. However, introducing hydrogen ions by lowering the pH drives the reac-

Table 5.2
Equilibrium Constants for Carbonate Reactions[a]

$T(°C)$	pK_w	pK_{CO_2}	pK_1	pK_2	pK_c	pK_a	pK_d
0.0	14.94	1.11	6.58	10.63	8.38	8.22	16.56
5.0	14.73	1.19	6.52	10.55	8.39	8.24	16.63
10.0	14.53	1.27	6.46	10.49	8.41	8.26	16.71
15.0	14.35	1.34	6.42	10.43	8.43	8.28	16.79
20.0	14.17	1.41	6.38	10.38	8.45	8.31	16.89
25.0	14.00	1.47	6.35	10.33	8.48	8.34	17.00
30.0	13.83	1.52	6.33	10.29	8.51	8.37	17.12
35.0	13.68	1.58	6.31	10.25	8.54	8.41	17.25
40.0	13.53	1.68	6.30	10.22	8.58	8.45	17.39

[a]All values are expressed as negative logarithms (pKs). K_w, K_{CO_2}, K_1, and K_2 are from Harned and Owen (1958); K_c and K_a are from Plummer and Busenberg (1982); K_d is from Langmuir (1971).

tion to the right, greatly increasing carbonate solubility. Most carbonate minerals are readily soluble in acid, and the acid most important to karst processes is carbonic acid, formed by the dissolution of gaseous CO_2.

The solution of carbon dioxide from the gas phase takes place in two steps. First CO_2 gas is transported across the gas–liquid interface to form CO_2(aqueous) in solution. The dissolved CO_2 then reacts with water to form neutral carbonic acid. Since there is no convenient experimental way of separating the two reactions, they are usually considered together.

$$CO_2(\text{gas}) \rightleftharpoons CO_2(\text{aqueous}) \tag{5.6}$$

$$CO_2(\text{aqueous}) + H_2O \rightleftharpoons H_2CO_3 \tag{5.7}$$

We use a bulk equilibrium constant that describes all neutral carbon-bearing species (see Table 5.2)

$$K_{CO_2} = \frac{a_{H_2CO_3}}{P_{CO_2}} \tag{5.8}$$

where P_{CO_2} is the carbon dioxide partial pressure expressed in atmospheres. The concentration of dissolved CO_2 increases with increasing carbon dioxide pressure in the gas phase that coexists with the aqueous solution. Dissolved CO_2, however, decreases with increasing temperature (Fig. 5.1).

Neutral carbonic acid dissociates in solution to form the bicarbonate ion, which in turn dissociates to form the carbonate ion. At the pH and ionic strength of most carbonate-bearing waters, the bicarbonate ion is the dominant species.

$$H_2CO_3 \rightleftharpoons HCO_3^- + H^+ \tag{5.9}$$

$$HCO_3^- \rightleftharpoons CO_3^{2-} + H^+ \tag{5.10}$$

The equilibrium constants are

$$K_1 = \frac{a_{HCO_3^-} a_{H^+}}{a_{H_2CO_3}} \tag{5.11}$$

$$K_2 = \frac{a_{CO_3^{2-}} a_{H^+}}{a_{HCO_3^-}} \tag{5.12}$$

Numerical values at various temperatures are listed in Table 5.2.

The ionization of carbonic acid releases hydrogen ions, forming a mildly acidic solution. The connection between these reactions and

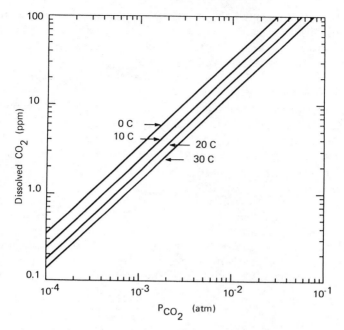

Figure 5.1 Solubility of carbon dioxide as a function of the CO_2 partial pressure in the coexisting gas phase. Calculated from the equilibrium constants in Table 5.2 using Eq. 5.8.

the hydration of the carbonate ion formed by dissociation of carbonate minerals is the dissociation of water:

$$H_2O \rightleftharpoons H^+ + OH^- \tag{5.13}$$

$$K_w = \frac{a_{H^+} a_{OH^-}}{a_{H_2O}} = a_{H^+} a_{OH^-} \tag{5.14}$$

These are homogeneous reactions in solution, and each species has a unique activity common to all reactions in which it appears. The activity of the carbonate ion links these reactions to the solubility of calcite and dolomite. The activity of H^+, measured by pH, may be controlled by other reactions in the system. The activity of carbonic acid ties the system to the external carbon dioxide pressure.

The net reactions for dissolution of calcite and dolomite by carbonic acid are

$$CaCO_3 + H_2O + CO_2 \rightleftharpoons Ca^{2+} + 2\,HCO_3^- \tag{5.15}$$

$$CaMg(CO_3)_2 + 2\,H_2O + 2\,CO_2 \rightleftharpoons Ca^{2+} + Mg^{2+} + 4\,HCO_3^- \tag{5.16}$$

The net reactions accurately portray the mass balance and identify the species in solution but tend to obscure many of the details. For this reason calculations are usually made with individual reaction steps rather than with the net reactions.

5.2.2 Activity Coefficients

The equilibrium constants for these various reactions are written in terms of activities of the constituent species. Only the H$^+$ activity is determined experimentally by measuring pH. The other ions are determined experimentally as concentrations. Concentration is related to activity by the expression

$$a_i = \gamma_i m_i \quad (5.17)$$

where m_i is molal concentration.

The activity coefficient, γ_i, connects the activity, a thermodynamically idealized concentration, with the real concentration. It is unity for ideal solutions. Activity coefficients can be calculated for each ion in dilute solutions by means of the Debye-Hückel equation

$$-\log \gamma_i = \frac{A z_i^2 \sqrt{I}}{1 + \mathring{a}_i B \sqrt{I}} \quad (5.18)$$

The parameters A and B are constants for a given temperature and for a given solvent. Values for aqueous solutions are listed in Table 5.3. z_i is the formal charge on the ion and $\mathring{a}_i$ is a parameter specific to each ion that effectively measures the ionic diameter. Values for the species of most importance to karst waters are listed in Table 5.4. Note that the numerical values are of the right order of magnitude to be ionic diameters but some values, particularly those of small cations, tend to be too large. This is because the cation carries with it a

Table 5.3
Values for the Constants of the Debye-Hückel Equation

T(°C)	A	B
0	0.4883	0.3241 × 10^8
5	0.4921	0.3249
10	0.4960	0.3258
15	0.5000	0.3262
20	0.5042	0.3273
25	0.5085	0.3281
30	0.5130	0.3290
35	0.5175	0.3297
40	0.5221	0.3305

Source: Manov et al. (1943), for aqueous solutions.

Table 5.4
Values for the $å_i$ Parameter in the Debye-Hückel Equation

Cation	$å_i$	Anion	$å_i$
Ca^{2+}	6×10^{-8}	CO_3^{2-}	4.5×10^{-8}
Mg^{2+}	8×10^{-8}	HCO_3^-	4×10^{-8}
Na^+	4×10^{-8}	Cl^-	3×10^{-8}
H^+	9×10^{-8}	SO_4^{2-}	4×10^{-8}

Source: Selected values adapted from Garrels and Christ (1965).

coordination sphere of water molecules, and $å_i$ measures the effective diameter of the whole complex.

The quantity, I, is the ionic strength and is a measure of the total concentration of charged species in solution, whether or not these species take part in the reactions under consideration. Ionic strength is defined as

$$I = \tfrac{1}{2}\Sigma m_i z_i^2 \tag{5.19}$$

All terms in the Debye-Hückel equation are available in tables, except for the ionic strength. The equation is valid up to ionic strengths of about 0.1. Other methods must be used to calculate activity coefficients in strong electrolytes, but the Debye-Hückel equation is generally adequate for karst waters.

5.2.3 Equilibrium Constants

Equilibrium constants such as those listed in Table 5.2 can be derived from thermodynamic quantities. Consider as an example the overall reaction for calcite dissolution:

$$CaCO_3 + CO_2 + H_2O \rightleftharpoons Ca^{2+} + 2\,HCO_3^-$$

If all species in the reaction are in their standard states, the standard Gibbs free energy for the reaction can be written as

$$\Delta G_R^0 = \Sigma\,\Delta G_f^0\,(\text{products}) - \Sigma\,\Delta G_f^0\,(\text{reactants}) \tag{5.20}$$

The usual standard states are solids as the pure crystal (no solid solution) at atmospheric pressure, gas at a pressure of 1 atm, and dissolved species as ideal solutions at one molal concentration (unit activity).

The ΔG_f^0 are standard free energies of formation, and specify the energy change when the compounds or species are created from their constituent elements. These have been compiled in Table 5.5 for compounds and species of interest to karst water chemistry. From these,

Table 5.5
Selected Thermodynamic Quantities at 298.15 K[a]

Species		ΔH_f^0 (kJ mol^{-1})	ΔG_f^0 (kJ mol^{-1})
Calcite	CaCO$_3$	−1207.370	−1128.842
Aragonite	CaCO$_3$	−1207.430	−1127.793
Dolomite	CaMg(CO$_3$)$_2$	−2324.480	−2161.672
Gypsum	CaSO$_4 \cdot$2H$_2$O	−2022.628	−1797.197
CO$_2$ (gas)		−393.510	−394.375
H$_2$O (gas)		−241.814	−228.569
H$_2$O (liquid)		−285.830	−237.141
H$^+$ (aqueous)		0.0	0.0
Ca^{2+} (aqueous)		−542.830	−553.540
Mg^{2+} (aqueous)		−466.850	−454.800
Na$^+$ (aqueous)		−240.300	−261.900
OH$^-$ (aqueous)		−230.025	−157.328
CO$_3^{2-}$ (aqueous)		−677.140	−527.900
HCO$_3^-$ (aqueous)		−691.990	−586.850
H$_2$CO$_3$ (aqueous)		−699.650	−623.170
SO$_4^{2-}$ (aqueous)		−909.270	−744.630
Cl$^-$ (aqueous)		−167.080	−131.270

[a]Standard states are $P = 1$ bar; ideal solutions at unit molality.
Source: All data are from Robie, Hemingway, and Fisher (1978).

free energies of reaction can be calculated for any reaction of interest.

The free energy of reaction when the species are not in their standard states is

$$\Delta G_R = \Delta G_R^0 + RT \ln K \qquad (5.21)$$

If the reaction is at equilibrium, $\Delta G_R = 0$, thus

$$\Delta G_R^0 = -RT \ln K \qquad (5.22)$$

R is the gas constant, which has a numerical value of 1.9872 cal mol^{-1} K^{-1} or 8.3143 J mol^{-1} K^{-1} (CODATA Task Group values of 1976). By setting the two expressions for the free energy of reaction equal, we can include the activities of species not in their standard states in the calculations.

Reference thermodynamic data are usually given only at the standard temperature of 25°C. Estimates for the equilibrium constants at other temperatures in the groundwater range of 0 to 40°C may be made with the Van't Hoff equation

$$\frac{d \ln K}{dT} = \frac{\Delta H_R^0}{RT^2} \quad (5.23)$$

The Van't Hoff equation can be integrated, remembering that $\left(\frac{\partial \Delta H}{\partial T}\right)_p = \Delta C_p$

$$\log K(T) = \log K(T_{std}) - \frac{\Delta H_R^0}{2.303R}\left(\frac{1}{T} - \frac{1}{T_{std}}\right)$$
$$+ \frac{\Delta C_p}{R}\left[\frac{1}{2.303}\left(\frac{T_{std}}{T} - 1\right) - \log \frac{T_{std}}{T}\right] \quad (5.24)$$

Where ΔH_R^0 is the enthalpy of reaction and may be calculated from the enthalpies of formation given in Table 5.5, by assuming that $\Delta C_p = 0$ over the temperature range of calculation.

Equilibrium constants for many complexation reactions are tabulated by Smith and Martell (1976).

5.2.4 Complexes and Ion Pairs

Calculation of the equilibria in karst groundwaters would be relatively straightforward if the dissolved species occurred as bare ions. The activities would be determined by the ionic strength, which could be calculated from measured ion concentrations. As concentrations increase, the bare ions tend to associate with each other, either as molecular units called complexes or as transient associations called ion pairs. Complexing and ion pair formation take place either between the species of interest or between these species and other dissolved materials. Those of most importance to carbonate groundwaters are the following:

$CaHCO_3^+$ $MgHCO_3^+$

$CaCO_3^0$ $MgCO_3^0$

$CaSO_4^0$ $MgSO_4^0$

Equilibrium constants for the dissociation of these pairs to the bare ions are given in Table 5.6.

Ion pairs and complexes influence the carbonate equilibria through their effect on the ionic strength. Since the pairs are either neutral or of lower charge than the bare ions, their presence effectively reduces the ionic strength, thus changing the activity coefficients and therefore the activities of ions in solution. Neglect of ion

128 / *Geomorphology and hydrology of karst terrains*

Table 5.6
Dissociation Constants for Ion Pairs

T(°C)	pK CaHCO$_3^+$	pK MgHCO$_3^+$	pK CaCO$_3^0$	pK MgCO$_3^0$	pK CaSO$_4^0$	pK MgSO$_4^0$
0.0	0.90	0.79	3.01	2.76	2.20	1.97
5.0	0.96	0.84	3.02	2.77	2.22	2.03
10.0	1.02	0.88	3.04	2.79	2.25	2.09
15.0	1.06	0.91	3.07	2.81	2.27	2.15
20.0	1.11	0.94	3.11	2.84	2.29	2.21
25.0	1.14	0.97	3.15	2.88	2.31	2.27
30.0	1.17	0.99	3.21	2.92	2.33	2.32
35.0	1.19	1.01	3.27	2.97	2.36	2.38
40.0	1.21	1.02	3.33	3.02	2.38	2.44

Sources: pK (CaHCO$_3^+$), pK (MgHCO$_3^+$) from Reardon (1974); pK (CaCO$_3^0$), pK (MgCO$_3^0$) from Reardon and Langmuir (1974); pK (CaSO$_4^0$) from Reardon and Langmuir (1976); pK (MgSO$_4^0$) from Nair and Nancollas (1958).

pairs for dilute and relatively pure carbonate waters does not introduce large errors into the equilibrium calculations. The importance of ion pairs increases with ionic strength and the concentration of other species in solution. Ion pairs with the sulfate ion are particularly important.

5.2.5 *Solubility of Carbonate Minerals*

Groundwater in contact with limestone contains at least the following species: Ca^{2+}, H_2CO_3, HCO_3^-, CO_3^{2-}, H^+, and OH^-. The water may also be in equilibrium with crystalline calcite and with a gas phase characterized by a partial pressure of CO_2. Because there are seven composition variables, the complete determination of the system requires specification of seven independent relationships between the variables. The equilibrium constant expressions of Eqs. 5.3, 5.8, 5.11, 5.12, and 5.14 provide five of these; the general requirement for charge balance provides the sixth. For the seventh constraint, three possibilities are of interest:

1. *The closed system.* A certain volume of groundwater contains a known concentration of dissolved CO_2 before the reactions are "turned on." Carbonic acid dissolves calcite, and the dissolved CO_2 is consumed in the process as the system comes to final equilibrium. There must be an overall mass balance in a closed system, and this provides the seventh constraint. This case describes the limestone–soil contact when overlying soils are impermeable and inhibit free exchange of CO_2 with the upper organic-rich part of the soil and the atmosphere.

2. *Constant CO_2 reservoir.* The water is assumed to exchange CO_2 freely with a large gaseous reservoir of known P_{CO_2}. Setting P_{CO_2} as a constant provides the seventh constraint. This case is applicable to the limestone–soil interface when soils are permeable and to the deposition of carbonate minerals in caves.
3. *System with fixed pH.* The hydrogen ion activity is assumed to be buffered by reactions other than the carbonate reactions. Dissolution of calcite proceeds until the concentration of Ca^{2+} reaches equilibrium with the fixed pH. This case is applicable when other sources of acid are present.

Case 2, the most relevant to karst processes, is discussed in some detail. Taking the ratio of the equilibrium constants for the first and second ionization reactions of carbonic acid removes the pH dependence and yields

$$\frac{K_1}{K_2} = \frac{a_{HCO_3^-}^2}{a_{CO_3^{2-}}\, a_{H_2CO_3}} \tag{5.25}$$

The activities of carbonic acid and the carbonate ion can be expressed in terms of the CO_2 pressure and the solubility product constant of calcite, respectively.

$$a_{H_2CO_3} = K_{CO_2} P_{CO_2} \tag{5.26}$$

$$a_{CO_3^{2-}} = \frac{K_c}{a_{Ca^{2+}}} \tag{5.27}$$

$$\frac{K_1}{K_2} = \frac{a_{HCO_3^-}^2 \, a_{Ca^{2+}}}{K_c K_{CO_2} P_{CO_2}} \tag{5.28}$$

Now we need to invoke the charge balance condition, which in general form is

$$m_{H^+} + 2m_{Ca^{2+}} = m_{HCO_3^-} + 2m_{CO_3^{2-}} + m_{OH^-} \tag{5.29}$$

If the full statement of charge balance is included in the calculation, the resulting solubility equation is a fourth-order polynomial that must be solved reiteratively. Alternatively, the pH range can be restricted to $6 < pH < 9$ where H^+, OH^-, and CO_3^{2-} are all very small compared with Ca^{2+} and HCO_3^- concentrations. The charge balance then reduces to

$$2m_{Ca^{2+}} = m_{HCO_3^-} \tag{5.30}$$

Placing this statement in Eq. 5.28, replacing activities with concentrations, and rearranging terms gives the concentration of dissolved calcite as a function of the CO_2 partial pressure:

$$m^3_{Ca2+} = P_{CO_2} \frac{K_1 K_c K_{CO_2}}{4K_2 \gamma_{Ca2+} \gamma^2_{HCO_3^-}} \quad (5.31)$$

The dissolved calcite concentration varies with the cube root of the CO_2 partial pressure.

An alternative calculation can be made in which the activity of carbonate species are expressed through Eqs. 5.8, 5.11, and 5.12 as functions of P_{CO_2} and a_{H+}. The charge balance condition, using Eq. 5.30, relates calcium ion activity to bicarbonate ion activity. Substitution of these into the expression for the solubility product constant allows the equilibrium hydrogen ion activity to be written as a function of P_{CO_2}:

$$a^3_{H+} = P^2_{CO_2} \frac{K_1^2 K_2 K^2_{CO_2} \gamma_{Ca2+}}{2K_c \gamma_{HCO_3^-}} \quad (5.32)$$

The solutions to Eqs. 5.31 and 5.32 are plotted in Fig. 5.2 for several temperatures. Calcite is more soluble in cold water than in warm water, but the temperature effect is small compared with the effect of CO_2 pressure over the range of temperature and P_{CO_2} expected in karst environments.

Case 1 follows a similar line of development. An initial concentration of dissolved CO_2 is described by

$$P^i_{CO_2} = \text{constant} \quad (5.33)$$

Again the charge balance is as given in Eq. 5.30. In this case the seventh constraint is provided by the mass balance in a closed system, which takes the form

$$m^i_{H2CO3} = K_{CO_2} P^i_{CO_2} = m_{H2CO3} + m_{HCO_3^-} - m_{Ca2+} \quad (5.34)$$

because the dissolved carbonate species must come from either the initial charge of CO_2 or the dissolution of calcite. Equation 5.25 is used as a starting point. The activity of bicarbonate ion is taken from the statement of charge balance, the activity of carbonate ion from the solubility product constant, and the activity of carbonic acid from the statement of mass balance (Eq. 5.34). Replacing activities by concentrations and rearranging terms leads to the cubic equation

$$m^3_{Ca2+} + \frac{K_1 K_c \gamma^2_{HCO_3^-}}{4K_2 \gamma^3_{Ca2+}} m_{Ca2+} - \frac{K_1 K_c K_{CO_2} P^i_{CO_2} \gamma^2_{HCO_3^-}}{4K_2 \gamma^3_{Ca2+}} = 0 \quad (5.35)$$

This equation can be solved by reiteration methods to give equilibrium calcium concentrations for any initial charge of CO_2.

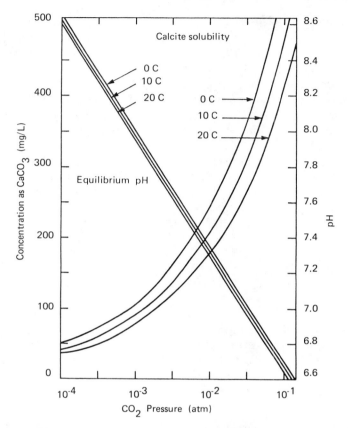

Figure 5.2 Solubility curves for calcite as a function of carbon dioxide partial pressure. Solubility curves were calculated from Eq. 5.31 using the equilibrium constants given in Table 5.2. At equilibrium, the saturated solutions will have the pH values shown as calculated from Eq. 5.32.

5.3 MEASUREMENTS

The characterization of karst waters requires certain chemical analyses and other measurements. Those that must be measured in the field are temperature, conductivity, and pH. Cations, alkalinity, and other anions can be determined later in the laboratory. What constitutes a "laboratory" depends on the field area. Samples collected for alkalinity measurement should be analyzed within 24 hours. In remote areas many investigators find it convenient to set up a field laboratory for analyses of alkalinity and preliminary analyses of Ca^{2+} and Mg^{2+}.

5.3.1 CO_2 in the Gas Phase

Gas phases of interest include the surface atmosphere (in which the CO_2 content is nearly constant), soil atmospheres, and cave atmospheres. Some measurements have been made by an adsorption/titration apparatus that required, in effect, carrying a small chemistry laboratory into the cave (Delecour et al., 1968; Ek et al., 1968), but this technique has been largely superseded by the Dräger apparatus. The Dräger pump is a commercially available device designed for the rapid analysis of atmospheric gases in mines and industrial plants. The active elements are glass tubes containing chemicals specific to the gas to be analyzed. A hand pump is used to draw a fixed volume of gas through the glass tube. The active ingredient for carbon dioxide is hydrazine hydrochloride, which turns purple by reaction with CO_2. The front edge of the advancing color change is read directly as volume percent of CO_2 on the calibrated glass tube. Although not very precise, the measurement can be performed in a few minutes. The apparatus is compact, robust, and easy to carry into difficult caves. A metal sampling tube can be driven into the soil for direct measurement of the soil atmosphere at different depths (Miotke, 1972).

Carbon dioxide concentrations are reported in volume percent, as concentration in milligrams per liter, and as the partial pressure of CO_2 in the total gas phase. The relationships between these are

$$P_{CO_2} = \frac{V(CO_2)}{V(gas)} P_{gas} \tag{5.36}$$

$$P_{CO_2} = \frac{C_{CO_2} (mg/L) \times 22.4}{1000 \times 44} = 5.092 \times 10^{-4} C_{CO_2} (mg/L) \tag{5.37}$$

5.3.2 Temperature and Conductivity

Temperature of surface waters, underground waters, cave air, or soil air can be measured directly with liquid-in-glass thermometers or the thermistor probe usually attached to other instruments, such as a conductivity meter. Measurement to the nearest degree is needed for later thermodynamic calculations. Temperature variations in karst waters and in the underground environment are subtle, and careful measurements to 0.1°C may reveal meaningful fluctuations.

Specific conductance (Spc), the electrical conductivity of the aqueous solution, is directly related to the concentration of ionic species. Field conductivity meters actually measure the electrical resistance of the solution. The resistivity is the resistance of a quantity of material 1 cm thick and 1 cm² in cross section. The unit is the ohm-centimeter. Conductivity is the reciprocal of the resistivity with units of $ohm^{-1} cm^{-1}$. The SI unit is the siemens per meter (1 siemens

$m^{-1} = 100$ ohm^{-1} cm^{-1}). An obsolete unit for conductance is the mho and many measurements in the literature are reported in "micromhos" although this must mean micromhos cm^{-1}. The thickness and area parameters vary with individual conductance cell design and are usually combined into a cell constant that, in turn, is incorporated into the calibration of the instrument set up to read conductance directly from a meter or digital display. In practice one simply immerses the cell in the water and reads the meter. Repeated measurements on muddy water, sewage, or algae-clogged surface streams may result in contamination of the electrodes, which must then be cleaned. The calibration of the instrument can be checked by measuring the conductivity of a 0.00702 N KCl solution with a conductivity of 0.1 siemens m^{-1} at 25°C.

Conductivity is a sensitive function of temperature. All measurements must be corrected to the standard reference temperature of 25°C. A set of corrections specifically designed for the bicarbonate ion (Jacobson and Langmuir, 1974a) can be fitted to the function

$$\text{Spc}(25°) = 1.81 \text{ Spc}(T)e^{-0.023T} \tag{5.38}$$

where T is in degrees Celsius and the equation ranges from 0 to 40°C.

Conductivity provides a direct measure of ionic strength through the empirical relation of Jacobson and Langmuir (1970)

$$I = 1.88 \times 10^{-5} \text{ Spc} \tag{5.39}$$

A field measurement of Spc provides an easy source of ionic strength, particularly if the water contains ions not included in the chemical analysis.

5.3.3 pH

The hydrogen ion activity is expressed as pH (pH = $-\log a_{H+}$) and can be measured directly with a pH meter. The pH probe is a glass membrane electrode, permeable to H$^+$, and the measurement is of the junction potential across the membrane (see Bates, 1964, for detailed discussion). The electrical connections are with a reference electrode constructed inside the glass electrode and a second reference electrode immersed in the water sample being measured. Equilibrium must be attained between the H$^+$ in the water and the liquid medium inside the electrode. Because the electrical resistance of the glass membrane is on the order of 10^8 ohms, and the membrane potential is on the order of 50 mV per pH unit, very precise and high-stability electronics are necessary for precise measurement.

The geochemistry of karst waters became an active area of research in the early 1960s. The first investigators had trouble with their pH measurements, and some went so far as to argue that accurate field pHs could not be measured. They turned to other methods for determining the saturation (or aggressivity) of the waters. Much of the early difficulty stemmed from trying to use low-cost pH meters, which simply did not have the necessary stability and precision. Geochemical investigations of karst waters require good middle to top-of-the-line meters. To determine $[H^+]$ to the same accuracy as other chemical analyses, ±5 percent, requires that the pH be measured to ±0.02 pH units.

pH is a relative measurement and must be standardized against a buffer. Good-quality meters have adjustments for both slope and intercept and so require two buffers—usually pH = 4 and pH = 7—for standardization. The potential measured across the electrodes as well as the pH of the reference buffers change with temperature. The first effect can be compensated electrically by the meter. The second requires that one know the pH/temperature characteristics of the buffers and standardize the meter to the appropriate buffer values for the measurement temperature.

There are three main sources of drift and instability in the field measurement of pH: (1) Some surface waters, particularly streams flowing from sandstones or quartzites, have very low ionic strength and therefore little buffer capacity. It may take some time for the H^+ equilibrium to establish itself across the glass electrode. (2) The glass electrode is an extremely high impedance circuit and is therefore very sensitive to moisture condensation on cables and plugs, and within the instrument itself. The moisture problem is most troublesome in caves, where relative humidities are high. (3) Temperature imbalances between sample, buffers, and electrodes are the source of much of the difficulties researchers have had with pH measurements. It is extremely important that solution, buffers, and electrodes all be at the *same* temperature; otherwise, the pH reading continuously drifts. A useful procedure for measuring cave streams, surface streams, and springs where there is plenty of "sample" is to immerse buffers and electrodes in the water and wait until the temperatures have equilibrated. However, pH should not be measured in flowing water because of streaming potentials, which produce systematic errors in the readings.

5.3.4 *Alkalinity*

Alkalinity is the acid-neutralizing capacity of water and reflects the presence of carbonate, bicarbonate, hydroxyl, and other basic constituents. In unpolluted carbonate groundwaters, the alkalinity is

$$\text{Alk} = [HCO_3^-] + 2[CO_3^{2-}] + [OH^-] - [H^+] \qquad (5.40)$$

In the usual pH range of karst groundwaters, only the bicarbonate ion makes a significant contribution to the alkalinity, and thus an acid titration is essentially a determination of HCO_3^-. A convenient laboratory method is to titrate a 50-mL sample with standard acid using the pH meter. pH is monitored as a function of the volume of added acid. The pH begins to fall abruptly between pH 4 and 5, and the inflection point on the plot marks the endpoint of the titration. Potassium biiodate ($KH(IO_3)_2$) is a convenient alkalinity titrant because it can be weighed accurately to make up a stable 0.01 normal solution.

Alkalinity and pH are both strongly dependent on CO_2 concentration. For this reason, it is important to measure pH on site and to perform the alkalinity analysis as quickly as possible. If the alkalinity samples are collected in tightly capped bottles and stored in an ice chest, CO_2 losses can be retarded sufficiently to permit titration delays of a day or two.

5.3.5 Cations

Analysis for Ca^{2+} and Mg^{2+} is required for any geochemical interpretation of karst waters. Other ions such as Sr^{2+}, Na^+, K^+, NH_4^+, and heavy metals may also be required for trace element or pollution studies. Calcium and magnesium ions can be titrated with ethylenediaminetetraacetic acid (EDTA) as a complexing agent using murexide as an indicator for Ca^{2+} endpoint and calmagite as an indicator for $Ca^{2+} + Mg^{2+}$ titration. Magnesium is then determined by difference. Premixed chemicals for EDTA titration are available commercially.

Atomic absorption spectroscopy (AA) has been the method of choice for most cations. The method has a high sensitivity for alkali and alkaline earth ions and can measure these in the parts per billion level. It requires expensive equipment and the preparation of standards for calibration. The main drawback of AA is that it analyzes only one element at a time. In recent years, AA analysis has been suplanted by atomic emission (AE) analysis. Atomic absorption draws the water sample into a flame and measures the absorption of the characteristic atomic lines of a special lamp that contains a cathode of the element being analyzed. Atomic emission draws the water sample into a direct current arc (DCP-AE) or inductively coupled plasma (ICP-AE) and directly measures the light emitted by the characteristic atomic lines of the sample. Current atomic emission spectrometers contain echelle gratings and multiple detectors that permit simultaneous analysis of up to 20 elements. Most have on-board computers for calibrations, data processing, and data readout.

Titrametric methods for Ca^{2+} and Mg^{2+} analysis are convenient for field determinations, but the laboratory methods are more precise, allow determination of more cations, and are much less time-consuming for large numbers of samples. However, one can do karst geochemistry with a pH meter and a box of glassware in the back of

136 / Geomorphology and hydrology of karst terrains

a truck parked in the Mexican jungle rather than in an air-conditioned laboratory with a quarter of a million dollars worth of fancy equipment.

5.3.6 *Other Anions*

Anions other than bicarbonate, particularly SO_4^{2-} and Cl^{-1} give useful information. For analytical methods, see the American Public Health Association publication (1985). During the early 1980s an anion chromatograph appeared that permits the direct determination of mixtures of anions much as atomic emission spectroscopy permits determination of mixtures of cations.

5.4 DERIVED PARAMETERS

The measurable parameters described in the previous section are not themselves particularly useful for interpreting the chemistry of karst waters. Over the years it has proved more useful to combine the raw data into karst water parameters that relate more directly to the hydrology and geomorphology of the karst system. The more useful parameters include the hardness that measures the amount of carbonate rock taken into solution, the CO_2 partial pressure, several saturation indices, which describe whether the water is at equilibrium with a given mineral, and the Ca/Mg ratio, which relates to the dolomitic character of the bedrock.

5.4.1 *Hardness*

Water hardness refers to the concentration of ions that prevent soaps from lathering. Of these, Ca^{2+} and Mg^{2+} are the most important in natural waters. The original test for hardness was a soap test; now hardness is usually defined as the sum of the calcium and magnesium concentrations expressed as parts per million of $CaCO_3$

$$Hd = 1000 \ MW(CaCO_3)([Ca^{2+}] + [Mg^{2+}]) \tag{5.41}$$

The hardness has no theoretical significance but is a useful measure of the amount of dissolved carbonate rock in a groundwater.

If a karst water contains only Ca^{2+}, Mg^{2+}, and HCO_3^- without large quantities of other ions (such as Na^+, Cl^-, and SO_4^{2-}), the hardness can be estimated directly from the conductivity. Figure 5.3 shows some data from central Pennsylvania. The linear relationship can be expressed by the regression equation

$$Hd = A + B \ Spc \tag{5.42}$$

Comparison of the regression coefficients (Table 5.7) of natural waters with those from laboratory investigations of pure calcite and pure dolomite show that Spc tends to be higher in natural waters for

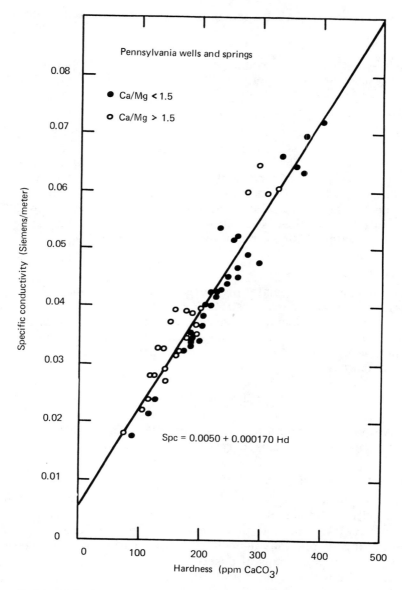

Figure 5.3 Relation between specific conductance and hardness determined from a set of wells and springs in Pennsylvania. Ca/Mg < 1.5 are dolomite waters; Ca/Mg > 1.5 are mainly limestone waters. Data are from Langmuir (1971).

Table 5.7
Regression Coefficients for Hardness/Specific Conductivity Relation[a]

Source of data	a	b
Single crystal calcite in laboratory	−21.24	7725
Massive dolomite in laboratory	−12.63	6500
Pennsylvania wells and springs (Fig. 5.3)	−29.6	5870

[a]Hd (mg/L as $CaCO_3$) = $a + b$ Spc (siemens m^{-1} at 25°C).

a given hardness. The extra contribution to the conductance is due to small quantities of other ions that are always present. Thus, the portable conductance meter provides a quick and easy means of tracing the uptake of $CaCO_3$ along an underground stream or through a drainage system. It is best calibrated, however, by using analyses from the water of the system under investigation.

5.4.2 CO_2 Partial Pressure

The theoretical CO_2 partial pressure is the CO_2 pressure of a hypothetical coexistent gas phase that is exactly in equilibrium with the analyzed water. It is not necessarily the actual CO_2 pressure present in the atmosphere where the sample was collected. P_{CO_2} is calculated from measured bicarbonate ion concentration and pH by combining Eqs. 5.8 and 5.11.

$$P_{CO_2} = \frac{a_{HCO_3^-} a_{H^+}}{K_1 K_{CO_2}} \tag{5.43}$$

The values of the equilibrium constants listed in Table 5.2 will give the P_{CO_2} in units of atmospheres.

5.4.3 Saturation Indices (SI)

It has long been recognized that most karst waters are not in equilibrium with solid calcite or dolomite. Many are undersaturated (aggressive) and capable of dissolving more carbonate; a few are supersaturated and may deposit speleothems or travertine. The saturation index is a means of describing quantitatively the deviation of carbonate waters from equilibrium. It is defined as either

$$\Omega = \frac{K_{iap}}{K_{sp}} \quad \text{or} \quad SI = \log \frac{K_{iap}}{K_{sp}} \tag{5.44}$$

where K_{iap} is the ion activity product for the dissociation of the mineral and K_{sp} is the solubility product constant of the mineral. The ion activity product is determined from the actual concentrations of cations and carbonate ions. The ratio describes a scale, as follows:

```
       Aggressive            Supersaturated
    _____|_____
       Negative        0         Positive
```

If the water is exactly saturated with the dissolving carbonate mineral, $K_{iap} = K_{sp}$ and SI = 0. Some authors (e.g., Berner and Morse, 1974; Thrailkill, 1972) define the saturation indices without the logarithm. However, the logarithmic scale provides equal intervals between equal degrees of either supersaturation or undersaturation.

Specific definitions for calcite and dolomite saturation indices are

$$SI_c = \log \frac{a_{Ca^{2+}} a_{CO_3^{2-}}}{K_c} \tag{5.45}$$

$$SI_d = \log \left[\frac{a_{Ca^{2+}} a_{Mg^{2+}} a_{CO_3^{2-}}^2}{K_d} \right]^{1/2} \tag{5.46}$$

The activities of calcium and magnesium are determined directly from the measured concentrations. The activity of CO_3^{2-}, a minor species in most karst groundwater, is calculated from the pH and alkalinity:

$$a_{CO_3^{2-}} = \frac{a_{HCO_3^-} K_2}{a_{H^+}} \tag{5.47}$$

This assumes that the dissociation reaction of bicarbonate is very fast compared with the rate of reaction of the water with the wall rock.

In terms of measured quantities, the calcite saturation index is

$$SI_c = \log \frac{\gamma_{Ca^{2+}}[Ca^{2+}] \gamma_{HCO_3^-}[HCO_3^-] K_2}{a_{H^+} K_c} \tag{5.48}$$

Trombe (1952) published a series of saturation curves for calcite plotted as a function of pH and $CaCO_3$ concentration, based on the equilibrium data of Tillmann. The Trombe curves were widely reproduced in the karst literature, often with pH and hardness for water samples plotted on them. The Trombe curves are now known to be somewhat inaccurate and tend to overestimate the undersaturation by values on the order of 0.25 pH units (Jacobson and Langmuir, 1972). The main drawback of the Trombe curves and related methods for expressing saturation is that they provide no way to account for varying ionic strength or other sources of protons. The saturation index allows the effects of temperature, other ions, and CO_2 pressures to be explicitly included. The departure from equilibrium is represented by a single number, which can be used in other calculations and interpretations.

5.4.4 Ca/Mg Ratio

The atomic ratio of calcium to magnesium can be immediately derived from the measured concentrations

$$Ca/Mg = [Ca^{2+}]/[Mg^{2+}] \qquad (5.49)$$

Ca/Mg = 1 for waters in contact with dolomite. Nearly all limestones contain some magnesium, and the observed values of Ca/Mg vary from 2 to 10, with 6 as a common value. The parameter provides information on the type of rock that a water sample has contacted.

5.5 CHEMICAL KINETICS

If a quantity of calcite is added to a beaker of water containing carbon dioxide, the calcite will begin to dissolve. The pH and calcium ion concentration of the solution increase toward their saturation values, as calculated in the previous section. However, several days are required for the reaction to approach within 90 percent of equilibrium, and much longer is needed to reach complete saturation. Runoff streaming down the sides of sinkholes or conical hills, or underground streams flowing through cave systems, can go a long distance in a few days. Thus, in many, perhaps most, situations in the karst, the residence time for water is substantially less than the time needed for complete chemical reaction. Karst waters are often out of equilibrium with the limestone or dolomite rock, and the rates of chemical reactions must be taken into account.

5.5.1 Principles

Figure 5.4 shows the dissolution of calcite on an atomic scale. The model assumes the exterior surface of the calcite crystal to be separated from the bulk solution by a thin stationary fluid boundary layer. The main steps in the dissolution process are as follows:

1. Hydration of aqueous CO_2 to form neutral carbonic acid.
2. Ionization of the carbonic acid to form a bicarbonate ion and a proton.
3. Transport of the proton through the boundary layer to the solid surface, where it is absorbed.
4. Reaction of the proton with a carbonate ion on the crystal surface to form a second bicarbonate ion.
5. Desorption of the second bicarbonate ion and its diffusion across the boundary layer into the bulk solution.
6. Release of the calcium ion from the crystal.
7. Diffusion of the calcium ion across the boundary layer into the bulk solution.

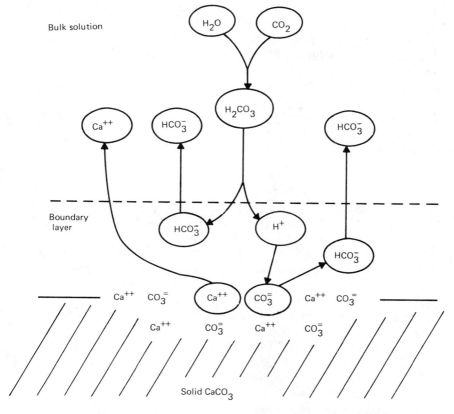

Figure 5.4 Schematic diagram of some of the important atomic steps in the dissolution of calcite in CO_2-containing aqueous solutions.

Steps 1 and 2 are homogeneous reactions that take place in the solution. Steps 3, 5, and 7 are transport steps involving the diffusion of ions across the static liquid layer. Steps 4 and 6 are chemical reactions that take place on the solid surface.

Carbonic acid is ionized in a matter of milliseconds, which can be considered instantaneous for most geological applications. The dissolved carbonate species H_2CO_3, HCO_3^-, and CO_3^{2-} are always in equilibrium with each other in solution.

Carbon dioxide exists in solution mainly as dissolved CO_2. At equilibrium only 0.17 percent of the dissolved CO_2 is hydrated to H_2CO_3, as described by Eqs. 5.6 and 5.7. Further, the rate of hydration of CO_2 is very sluggish. The rate equation can be written as

$$\frac{d[CO_2]}{dt} = k_{CO_2}[CO_2] \tag{5.50}$$

The rate constant, $k_{CO_2} = 0.03$ sec^{-1} (Kern, 1960), means that the time scale for the hydration of dissolved CO_2 is on the order of 30 seconds. Many occurrences, such as flow over bare bedrock surfaces or down vertical shafts, take place fast enough that the solution is depleted in carbonic acid before the CO_2 in solution has time to hydrolyze. Thus, in some situations step 1 may be rate controlling.

If the transport steps (3, 5, and 7) are rate controlling, the rate of solution depends on fluid flow velocity, because the thickness of the boundary layer changes with velocity. The mass transport rate equation has the form

$$\frac{dC}{dt} = \frac{A}{V}\frac{D}{\delta}(C - C_s) \tag{5.51}$$

where A is the surface area of the solid, V is the solution volume, D is the diffusivity of ions in the boundary layer, δ is the thickness of the boundary layer, C is the concentration of the ion of interest in the bulk solution, and C_s is the concentration of the same ion at the surface. The reaction is driven by the concentration gradients across the boundary layer, and D/δ becomes a mass transfer coefficient for the process.

If the surface reactions are rate controlling, other rate equations must be written, depending on which species controls the rate. Earlier models of carbonate dissolution kinetics assumed transport control (Curl, 1965; Weyl, 1958). More recent interpretations, discussed in the next sections, argue for surface reaction control. In fact, the issue is probably not settled.

Reaction rates increase exponentially with temperature, as described by the Arrhenius equation

$$k(T) = A_0 e^{-E_a/RT} \tag{5.52}$$

where E_a is an activation energy with values on the order of 8 to 20 kJ/mole for mass transport control and 40 to 100 kJ/mole for surface reaction control. Although the temperature differences between arctic and tropical climates have only a small effect on equilibrium concentrations of dissolved carbonates, they are sufficient to produce large changes in reaction rates.

5.5.2 Calcite Dissolution Kinetics

As calcite reacts with an undersaturated carbonic acid solution, a number of things go on simultaneously. The concentration of calcium ions in solution increases, the pH increases, and the saturation index approaches the equilibrium value. Berner and Morse (1974) determined the rate of calcite dissolution as a function of the undersaturation. Their results (Fig. 5.5) show that the rate is determined in part by the CO_2 partial pressure and is strongly dependent on the degree

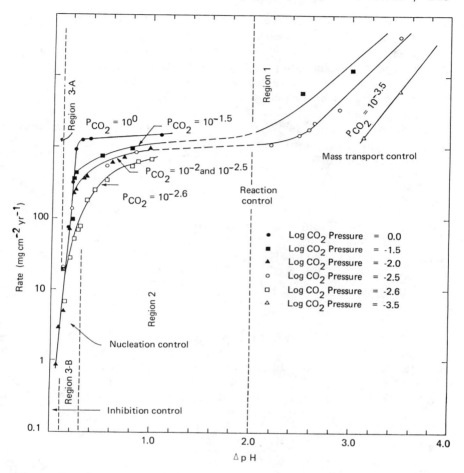

Figure 5.5 Rate of dissolution of calcite in synthetic seawater at various carbon dioxide partial pressures. Data are from Berner and Morse (1974). The rates were determined at constant undersaturation by the pH-stat methods. ΔpH = pH (equilibrium) − pH (experimental).

of saturation. The Berner and Morse data are given in terms of ΔpH, the difference between the pH at equilibrium and the pH of the solution. ΔpH is directly related to the saturation index

$$\Delta pH = -\tfrac{1}{2}SI_c \qquad (5.53)$$

There are three clearly defined rate regimes. At large undersaturations, Region 1, the rate increases proportionally to ΔpH—that is, with increasing hydrogen ion activity. In Region 2, the rate is nearly independent of the undersaturation but is partly a function of the CO_2 pressure. As the solution approaches saturation in Region

3, the rate begins to decrease rapidly when ΔpH falls below 0.3 (SI_c = −0.6). When ΔpH reached 0.1, the rate had fallen by 3 orders of magnitude and was too slow for accurate measurement under laboratory conditions. The complex nonlinear relationships between dissolution rate and undersaturation have much to do with the development of cave passages and karst landforms.

The decrease in rate as the system approaches equilibrium is a result of the effect of back reactions—in this case, the precipitation of calcium carbonate. Plummer and Wigley (1976) first described the approach to equilibrium by a fairly standard rate equation

$$\frac{dC}{dt} = \frac{A}{V} k_T (C_0 - C)^n \tag{5.54}$$

where k_T is the reaction rate constant and C_0 is the concentration at equilibrium. The reaction order, n, was found to be equal to 2 over much of the experimental range. In a similar approach, Sjöberg (1976) and Rickard and Sjöberg (1983) proposed the rate equation

$$\frac{dC}{dt} = kA(1 - \Omega^{1/2}) \tag{5.55}$$

where Ω is the saturation ratio defined by Eq. 5.44.

The most comprehensive investigation of calcite dissolution kinetics reveals at least three competing rate processes driving the forward reactions (Plummer et al., 1978a; Plummer et al., 1979). Each of the species present in the boundary layer shown in Fig. 5.4 can attack the calcite surface.

$$CaCO_3 + H^+ \rightleftharpoons Ca^{2+} + HCO_3^- \tag{5.56}$$

$$CaCO_3 + H_2CO_3 \rightleftharpoons Ca^{2+} + 2\,HCO_3^- \tag{5.57}$$

$$CaCO_3 + H_2O \rightleftharpoons Ca^{2+} + HCO_3^- + OH^- \tag{5.58}$$

The observed rate is the sum of the rates of these three processes minus the decrease in rate due to back reactions:

$$\text{Rate} = k_1 a_{H^+} + k_2 a_{H_2CO_3} + k_3 a_{H_2O} - k_4 a_{Ca^{2+}} a_{HCO_3^-} \tag{5.59}$$

The rate is given in units of millimoles per centimeter square per second. H_2CO_3 is used in the usual sense of the sum of dissolved CO_2 plus neutral carbonic acid.

The first term describes the rate of reaction of calcite with protons and is dominant under conditions of high acidity corresponding to Region 1 in Fig. 5.5. In solutions with pH of less than 4, this term accounts for most of the dissolution, but natural systems dominated by carbonic acid seldom get into this regime. Experiments with the dissolution of calcite under controlled flow conditions show that this term is transport controlled. The diffusion of H^+ across the boundary layer is the rate-controlling process (Rickard and Sjöberg, 1983; Herman, 1982; Sjöberg and Rickard, 1984; Compton and Daly, 1984).

The second term describes the reaction of calcite with carbonic acid and contains the dependence of the reaction rate on carbon dioxide pressure. The second term is dominant in undersaturated CO_2-bearing waters and corresponds to Region 2 of Fig. 5.5. The carbonic acid appears to be mainly surface-reaction controlled; there is little dependence of the rate on flow velocity when the undersaturation is in Region 2.

The third term describes the dissolution of calcite in water, the kinetic counterpart to the dissociation reaction listed as Eq. 5.1. Because the activity of water is near unity in most karst waters, this term acts as a constant background to the other dissolution processes.

The back-reaction term has a complicated dependence on all of the other variables

$$k_4 = \frac{K_2}{K_c} \left\{ k_1' + \frac{1}{a_{H+(s)}} [k_2 a_{H2CO3(s)} + k_3 a_{H2O(s)}] \right\} \quad (5.60)$$

Here K_2 and K_c are the equilibrium constants previously described, and k_1' is the effective rate constant for H^+ ions at the surface. The difficulty here, as in many rate calculations, is that one does not know the concentrations of species at the surface. Experimental comparisons were made by equating k_1' with k_1 and making surface concentrations equal to bulk concentrations. A major success of the Plummer-Wigley-Parkhurst model is that it could also be used to calculate precipitation rates for highly supersaturated solutions (Reddy et al., 1981).

None of the kinetic models described here work well near equilibrium. According to the Berner and Morse data, the rates fall by 3 or 4 orders of magnitude by the time the solutions become 90 percent saturated. The details of the solution rate in this region are sensitive to the presence of many impurities that can inhibit calcite dissolution. Magnesium, phosphates, and various organics, among others can poison nucleation sites or sites of initial solution attack. It appears that dissolution by nearly saturated solutions (or precipitation from slightly supersaturated solutions) is controlled by complex surface processes not readily amenable to description by bulk rate equations. Present kinetic theory, therefore, describes dissolution processes far from equilibrium, but it is less accurate for near equilibrium karst waters.

The temperature dependence of the forward reaction rate constants of the Plummer-Wigley-Parkhurst model were determined as follows:

$$\log k_1 = 0.198 - \frac{444}{T} \tag{5.61}$$

$$\log k_2 = 2.84 - \frac{2177}{T} \tag{5.62}$$

$$\log k_3 = -5.86 - \frac{317}{T} \quad (T < 298 \text{ K}) \tag{5.63}$$

$$\log k_3 = -1.10 - \frac{1737}{T} \quad (T > 298 \text{ K}) \tag{5.64}$$

It is therefore possible to include a temperature dependence in kinetic models of karst processes.

5.5.3 Dolomite Dissolution Kinetics

The problem of dissolution kinetics of dolomite is only partially solved. It is apparent from a comparison of karst development on dolomite with karst development on limestone that dolomite dissolves more slowly than calcite. Indeed, the freshman "acid test" to distinguish limestone (which fizzes in dilute hydrochloric acid) from dolomite (which does not) is further evidence that the two chemically similar rocks do not respond in the same way to solutional attack.

Busenberg and Plummer (1982) examined the dissolution kinetics of dolomite and constructed a rate equation analogous to Eq. 5.59:

$$\text{Rate} = k_1 a_{\text{H}^+}^n + k_2 a_{\text{H}_2\text{CO}_3}^n + k_3 a_{\text{H}_2\text{O}}^n - k_4 a_{\text{HCO}_3^-} \tag{5.65}$$

Measurements on spinning disks of dolomite single crystal (Herman and White, 1985) generally agreed with the rate equation; further, they showed that, as in the case of calcite, the first term in the equation is transport controlled and its contribution to the rate depends on the velocity of fluid flow. The exponent, n, in Eq. 5.65 is 0.5 at temperatures below 45°C, but the exponent for the hydrogen ion term increases with increasing temperature.

The dolomite rate equation applies only when the system is far from equilibrium. Figure 5.6 illustrates the problem. When dolomite is dissolved in water at a CO_2 pressure of 1 atm, the dissolution proceeds rapidly at first but then flattens out as $SI_d \rightarrow -2$. At 1 percent saturation, a major change in the dissolution mechanism occurs. Dolomite waters do come to equilibrium eventually because wells drilled in dolomite aquifers produce water that is nearly saturated with dolomite. Extrapolation of the nearly flat part of the curve in Fig. 5.6 to $SI_d = 0$ suggests that waters take years to reach equilibrium. Because dolomite and calcite are similar in both chemical com-

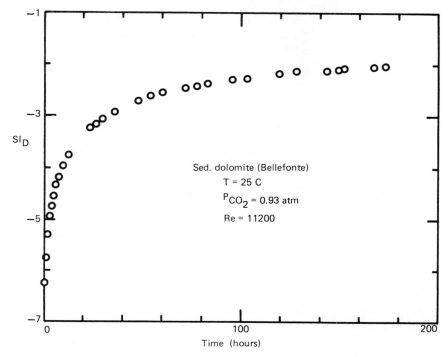

Figure 5.6 Dissolution rate of the Ordovician Bellefonte dolomite expressed in terms of the saturation index. Data are from Herman (1982).

position and crystal structure, we deduce that the problem has to do with the role of magnesium in modifying the surface reactions that are rate controlling near equilibrium. There is no information at present on the mechanism of dolomite dissolution under near-saturation conditions.

5.6 DATA PROCESSING

With the ready availability of programmable calculators, there is really no excuse for not making fairly precise calculations. The specific conductance can be used to estimate the ionic strength from which the activity coefficients are obtained by means of the Debye-Hückel equation. This allows one to enter the measured concentrations of cations and anions as activities and calculate the saturation indices and carbon dioxide pressures according to Eqs. 5.48 and 5.43. A single pass through the sequence of calculations can be written into one calculator program.

At least three fairly elaborate programs for karst groundwater data processing were written in the early 1970s: one by Roger Jacobson at The Pennsylvania State University, one by J. J. Drake at McMaster University, and one by John Thrailkill at the University of Kentucky. The widely circulated Jacobson program formed the basis for many interpretations of groundwater chemistry. Another program was WATSPEC, which could deal with most of the common minerals in carbonate aquifers (Wigley, 1977). One of the most widely used codes in the United States is WATEQF, a descendant of the general water speciation program WATEQ, written by A. H. Truesdell and B. F. Jones of the U.S. Geological Survey in 1973. Updated versions of WATEQF (e.g., Plummer et al., 1978b) take into account many more species and more recent refinements of the equilibrium constants. For other computer models of aqueous systems, see Jenne (1979).

The modeling of equilibrium processes in carbonate waters has reached a mature stage of development, but kinetic modeling is in its beginnings. Some tentative steps toward modeling karst processes have been taken using rate equations described in the previous section. Much more can be done. The microcomputer revolution, and the transformation of computer programming from an arcane art to a popular pastime, suggests that karst geochemistry of the future will depend on speciality programs or computer models written especially for the problems being investigated.

6
Karst Hydrology

6.1 DRAINAGE BASINS AND AQUIFERS

6.1.1 *Aquifers*

All rocks are permeable to some degree and can retain and transmit water through the pore spaces between the mineral grains. Rock formations capable of retaining large quantities of water are termed aquifers. Rocks with low permeability, called aquicludes, can transmit little or no water. Aquifers are characterized by thickness, lateral extent, rock type, and water-transmitting properties.

Water infiltrating from the surface into the underlying aquifer moves vertically downward under the influence of gravity, until it reaches the level at which all the pores are water-filled. The surface that separates the water-saturated zone from the zone with air-containing pores is the water table. The water-saturated zone below the water table is the phreatic zone; the unsaturated region above the water table is the vadose zone (Fig. 6.1). If the vadose zone communicates freely with the land surface, so that the position of the water table can rise and fall as the amount of infiltrating water changes, the aquifer is said to be unconfined. A contour map of the water table can be constructed by collecting water-level data from a large number of wells that penetrate into the phreatic zone.

If a permeable rock zone allows water to move beneath impermeable confining beds, the aquifer is said to be confined and may be under pressure. There is no water table surface free to adjust to variations in recharge. Instead, there is a piezometric surface, which describes the amount of pressure head developed beneath the confining layer. If a well is drilled into a confined aquifer (and if it is cased off from hydraulic communication with any unconfined aquifer

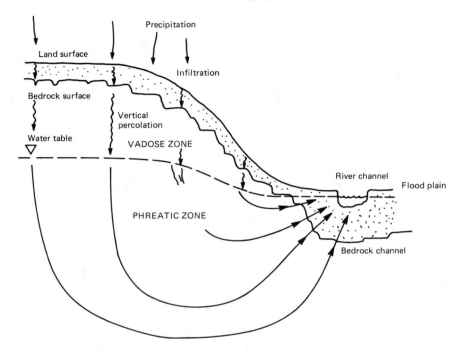

Figure 6.1 Water table and vadose and phreatic zones of an unconfined aquifer.

above, as is shown in Fig. 6.2), then water rises in the well bore to the level of the piezometric surface. The piezometric surface may lie far above the confining layer; it may even lie above the land surface. In the latter case, water spontaneously flows from the well and the aquifer is said to be artesian.

Water moves in aquifers under the influence of hydraulic gradients from recharge regions, where the water table stands highest, to discharge regions, where the water table is at its lowest. Flow patterns become more complex in artesian aquifers and where several aquifers are hydraulically connected. The water is ultimately lost from the groundwater system by surface discharge through springs and seeps into surface streams, lakes, or the sea.

Carbonate aquifers are those in which the main water-bearing formations are carbonate rocks. These, as will be seen, vary widely in their properties. Some behave much like aquifers in sandstones, some are more like fractured granites or basalts, and still others are unique karstic aquifers. Karstic aquifers differ from all other (with the exception of some basalts) in that they contain integrated sys-

Karst hydrology / 151

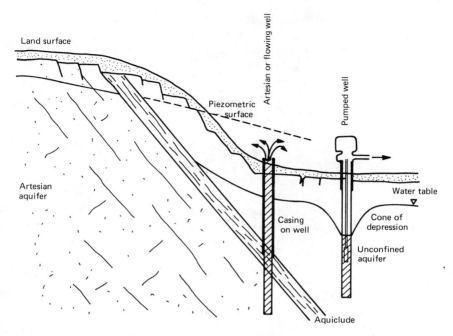

Figure 6.2 Confined (artesian) and unconfined aquifers with the expected response of water wells to each type. The artesian well is assumed to be cased through the unconfined aquifer down to the aquiclude.

tems of pipelike conduits that act as underground drains for the highly localized transport of water.

For background information, the reader is referred to such standard textbooks as Davis and DeWiest (1966) and Freeze and Cherry (1979); Zötl (1974), Smith, Atkinson, and Drew (1976), and Milanović (1981) focus specifically on karst hydrology.

6.1.2 *Porosity and Permeability*

The porosity of a rock is defined as the volume fraction occupied by empty space. Porosity is the fraction not occupied by mineral grains, which can be defined as

$$\theta = \frac{V_{pores}}{V} = \frac{V - V_{min}}{V} = 1 - \frac{V_{min}}{V} \tag{6.1}$$

where V is the bulk volume of the rock and V_{min} is the net volume of the mineral grains. Porosity is a dimensionless quantity.

The permeability of a rock is its ability to transmit fluid. To be permeable a rock must be porous, but its pores must also interconnect so that the fluid can move between them.

The permeability enters in a fundamental way into D'Arcy's law, which states that the fluid flow through a porous medium is proportional to the hydraulic gradient:

$$q = \frac{Q}{A} = -K\frac{dh}{dl} \qquad (6.2)$$

where Q is the flow (m^3 sec^{-1}) across an area, A (m^2), of the medium. Since the hydraulic gradient, dh/dl, is dimensionless, both K and q have units of velocity (m sec^{-1}); q is the unit flow in terms of volume per unit area per unit time. The hydraulic conductivity, K, depends on the properties of both the fluid and the aquifer bedrock. These can be separated, and D'Arcy's law restated as

$$q = -\frac{Nd^2\rho g}{\eta}\frac{dh}{dl} \qquad (6.3)$$

where g = gravitational acceleration, ρ = density of the fluid, d = grain diameter, η = viscosity of the fluid, and N = dimensionless shape factor. The quantity Nd^2 is the effective permeability of the bedrock. The shape factor, N, is difficult to determine, and the product is either calculated from laboratory measurements or determined from aquifer behavior. Permeability has units of area. It is often given in square centimeters, square meters, or as a poorly defined unit called the darcy. In dimensionally consistent units, 1 darcy = 10^{-8} cm^2 or 1 μm^2, but see Lohman (1972) for a discussion of alternative (and dimensionally inconsistent) definitions. Figure 6.3 shows the permeability ranges for some common aquifers and aquicludes.

Pore spaces are formed in rocks by the packing of mineral grains. Other voids may be formed by solution during or after diagenesis. All openings of this type are termed primary porosity. Joints, fractures, and bedding plane partings also produce openings through which groundwater can move. Joints and fractures are produced by orogenic processes, by the jostling of continental plates, and perhaps by earth tides. They form in all rocks, but massive brittle rocks such as sandstones, granites, limestones, and dolomites are better able to maintain the mechanical openings. Joints in softer rocks such as shales tend to be self-sealing because of flow of the rock under overburden pressure. The combination of joints, fractures, and bedding plane partings is the secondary or fracture porosity. Karstic aquifers are distinguished from all others by the presence of large solution cavities and integrated conduits. These make up the conduit porosity (Fig. 6.4; Table 6.1).

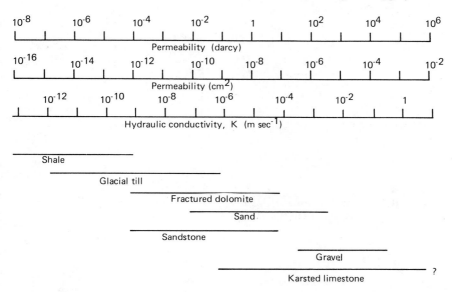

Figure 6.3 Ranges of permeability found in various types of aquifer materials. The hydraulic conductivity is for water at 25°C.

6.1.3 Surface and Subsurface Basins

In nonkarstic terrains there is a clean distinction between surface water and groundwater, and between surface drainage basins and groundwater aquifers. Drainage basins, which determine the catchment and runoff characteristics of surface water are defined by drainage divides set by local topography. Drainage basin boundaries remain fixed in time, at least on the time scale of surface water runoff. Surface water is often only loosely coupled to the groundwater system because of the slow rate of infiltration. Aquifers are usually continuous across surface divides, and may span many surface drainage basins. In unconfined aquifers, water table highs usually correspond to topographic highs and thus surface water divides, but in structurally complex regions, groundwater gradients and resulting flow directions may be quite different from the direction of surface runoff.

The conceptual distinctions between basins and aquifers are blurred in karst regions because of the integrated system of conduits that carry water through the subsurface. There are three components to the karst hydrologic systems:

1. The aquifer.
2. The patchwork of surface basins.
3. The patchwork of groundwater basins.

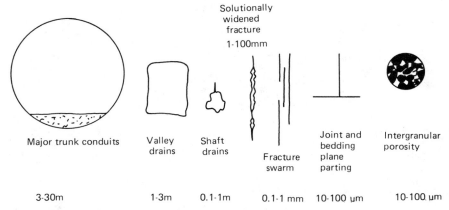

Figure 6.4 Types of porosity in karsted carbonate aquifers, with the size scale for the various openings.

**Table 6.1
Aquifer Properties for Types of Porosity**

	Primary porosity	*Fracture porosity*	*Conduit porosity*
Physical situation	Intergranular pores Vugs Isolated joint and bedding plane partings	Concentrations of joints and fractures Bedding plane partings (may be enlarged by solution)	Open channels and pipes of various sizes and shapes
Homogeneity	Usually isotropic	Usually anisotropic because of fracture spacing and preferred orientations May be statistically isotropic over large volumes	Usually highly anisotropic
Flow regime	Laminar D'Arcy flow	Laminar May deviate from D'Arcy flow	Turbulent Non-D'Arcy flow
Water table	Well-defined water table surface	Irregular surface	Behaves as subsurface drains, which may be at, above, or below adjacent water table
Response to short-term events	Slow	Moderate	Rapid

The groundwater basins are defined by the relationship between swallow holes and discharge points (Fig. 6.5). The groundwater basins are closely related to the overlying surface basins because the flow paths through the conduit system are alternative routes to the flow down the system of surface channels. In certain cases, the boundaries of the surface and subsurface basins are identical and their discharge routes are identical, the conduit merely serving as an underground bypass to the surface stream. In general, however, subsurface basins are not precisely congruent with the surface basins.

Underground conduit systems, particularly low-gradient ones, may have more than one channel carrying water, and there may be downstream distributaries so that water from a single-drainage basin discharges at more than one spring. The number of conduits carrying water and the details of the flow path change depending on groundwater levels. Rising water levels flood conduits that may be dry cave passages when the water level is low. The Hidden River system (Fig. 6.6) is one of the better mapped examples. Detailed mapping in the Adriatic karst in connection with hydroelectric power projects has revealed even more complex distributary systems (Milanović, 1984).

6.1.4 Water Balance and the Hydrologic Cycle

The hydrologic cycle is a fundamental concept of hydrology. It is a statement of mass conservation for the waters of the earth. Water that falls as rain supplies both surface streams and groundwater. Groundwater returns to surface streams, which aggregate to large rivers, which drain to the ocean. Evaporation from the ocean provides the moisture that drifts over the continents to fall as rain, thus completing the cycle. No water is gained; no water is lost.

The hydrologic cycle and water balance concept can be applied to individual drainage basins. Precipitation is considered as input; the master surface stream draining the basin is considered output. Within the basin, the water budget should balance.

Precipitation falling onto the land surface divides into three parts. One part infiltrates the plant cover and soil, penetrating the underlying bedrock, where it continues downward through the vadose zone to the water table. A second part flows overland into small gullies and rivulets and ultimately into the master base level stream. The third component returns to the atmosphere by evaporation and transpiration. For any rainstorm the water budget can be balanced at the land surface

$$P - E = I + R \tag{6.4}$$

where P = precipitation, I = infiltration, R = runoff, and E = evapotranspiration. Water balances are often expressed in units of milli-

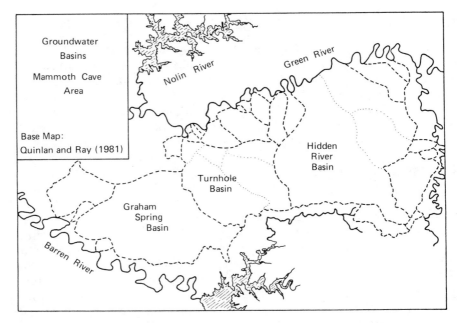

Figure 6.5 Karstic drainage basins in the Mammoth Cave region of Kentucky. The basin boundaries were determined by tracing sinking streams, cave streams and sinkholes to their outlet points in springs along Green River. [Adapted from Quinlan and Ray (1981).]

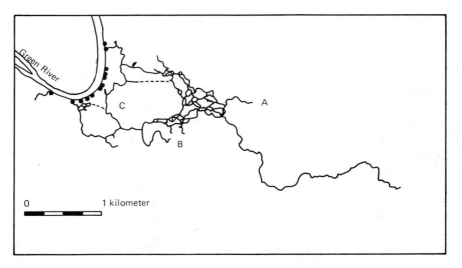

Figure 6.6 Map showing underground distributaries in the Hidden River drainage basin of the Mammoth Cave area in Kentucky. Water enters the horizontal cave at point A by rising 17 m up a breakout dome. The route from B to C is a river 12 m wide, 2.5 m deep, and 2.4 km long. During high flood, the water flows from A to 33 springs at 10 locations along the river. The entire cave is water-filled. [Adapted from Quinlan and Rowe (1977).]

Karst hydrology / 157

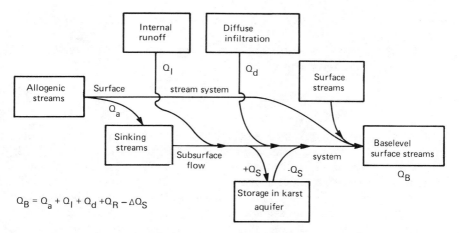

Figure 6.7 Water balance for a fluviokarst drainage basin.

meters (or inches), which would represent the water depth on the land surface, for example, if the precipitation were allowed to accumulate.

A somewhat more elaborate hydrologic cycle can be set up for karst drainage basins (Fig. 6.7). The karst aquifer and associated groundwater basins are assumed to be embedded in a larger basin containing nonkarstic rocks. Runoff from the nonkarstic borderland appears as streams flowing onto the karst (here assumed to be downstream, although other arrangements could be sketched). Some streams may remain on the surface across the karst (Q_R). Other streams from the borderlands sink into swallow holes at the margin of the karst and become part of the subsurface drainage system (Q_a). Rainwater may penetrate the soils on the karst surface as a diffuse infiltration through joints and fractures in the limestone (Q_d). There is little runoff on karst surfaces. Overland flow disappears into sinkholes to enter the groundwater system as internal runoff (Q_I). Surface and subsurface flow rejoin in the baselevel surface streams draining the region with total discharge (Q_B).

The water budget for the basin can be written as

$$Q_{in} - Q_{out} = \Delta Q_S \qquad (6.5)$$

where Q_{in} is the total input from all sources into the karst system and Q_{out} is the discharge from the system during the same period. ΔQ_S is the change in storage. Storms raise the water table and increase the amount of water stored in the aquifer. During dry periods, inputs fall below outputs, and the water in storage gradually decreases. ΔQ_S may be either positive or negative, depending on whether the aquifer has a net gain or net loss during the time period considered.

The instantaneous water balance at the land surface is written in somewhat expanded form as

$$q_p dA = \int_A q_E dA + \sum_i q_{di} A_i + \sum_j q_{Ij} A_j + \int_A q_R dA \qquad (6.6)$$

where the integrations and summations are over the area of the drainage basin. The q values represent instantaneous flow rates into or out of each area element. Precipitation, q_P, evapotranspiration, q_E, and surface runoff, q_R, are written as integrals because of their continuous variation over the basin. Diffuse infiltration, q_d, and internal runoff, q_I, are written as summations because these parameters are determined by the bedrock. The summations are over the outcrop areas of each rock type in the basin.

Water budgets are usually written with the instantaneous flows integrated over a specified period of time, which can be a full water year, the duration of a single storm, or any other period. Each of the components of the flow system is related to the instantaneous flow by

$$Q_P = \int\int q_P dA dt \qquad Q_I = \int \sum q_{Ij} A_j \, dt \qquad \text{etc.} \qquad (6.7)$$

Rewriting the water balance at the land surface, we obtain

$$Q_P = Q_E + Q_d + Q_I + Q_R \qquad (6.8)$$

which is exactly the same as Eq. 6.4, except for the separation of the diffuse infiltration and the internal runoff.

A detailed input–output budget for the basin can also be written. The karst aquifer is considered to have a single output—the base level stream with flow, Q_B, leaving the karst area downstream from the springs.

$$Q_B = Q_a + Q_I + Q_d + Q_R - \Delta Q_S \qquad (6.9)$$

Equation 6.9 defines a conservative basin. If the water budget balances over a sufficiently long time for the changes in storage to average to zero, the basin must be conservative and the conduit flow system must be well-behaved. If the budget does not balance, water must be entering or leaving the system through unknown routes.

6.1.5 Sediment Balance

All surface streams carry sediment loads. Nonkarstic aquifers carry no sediment load because pore sizes are too small and flow velocities are too low for sediment transport. The porous medium acts as an effective filter, allowing infiltrating groundwater to pass but rejecting solid material, which is left on the surface as a weathered residuum.

Water moving at high velocities through conduit systems can and does carry a sediment load. It is therefore necessary to consider the sediment budget as well as the water budget in karstic aquifers. Sediment budgets cannot be written in terms of a closed cycle. The source for the sediments is the weathering of karstic and nonkarstic rocks; the ultimate sink is the delta, where the river flows into the sea. The sediment budget can be written only in input–output form with source and sink terms (Fig. 6.8).

Weathering products from the erosion of the borderlands must move through the stream system that drains the terrain; if these streams flow underground through the karst, so do the sediments. Carbonate rocks always contain a certain percentage of insoluble material, and these materials accumulate as soils on the karst surface. A third source of sediments, unique to karstic aquifers, is the weathering residuum created in the subsurface by the solutional enlargement of the conduit system.

Sediment transport is very sensitive to flow regime and the rate of sediment transport can vary greatly over the year, depending on storm intensity and overall flow rates. At low flow times, the sediments tend to accumulate in the conduit system, and at high flow they are flushed out. A storage term accounts for the placement or removal of sediments in the conduit system.

The overall sediment budget is thus written as

$$S_B = S(\text{surface routes}) + S(\text{underground routes})$$
$$S(\text{surface}) = S_{as} + S_{Rs} \qquad (6.10)$$
$$S(\text{underground}) = S_{au} + S_{Ru} + S_c - \Delta S_{sto}$$

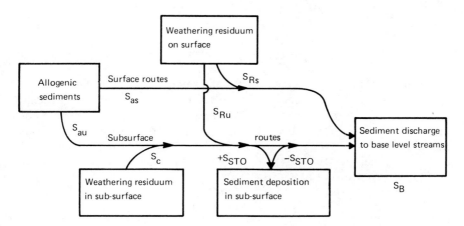

Figure 6.8 Sediment balance for a fluviokarst drainage basin.

160 / *Geomorphology and hydrology of karst terrains*

The sediment balance is important in the overall development of the conduit system because the sediment transport efficiency varies with climatic regime, particularly rainfall. Conduit systems that are competent to carry the sediment load when rainfall is high may not be competent to carry it in periods with less rainfall. The continued deposition of sediments could fill up the conduit system, forcing the drainage back to surface routes.

6.2 MECHANICS OF GROUNDWATER FLOW

Karst hydrology is a tale of flowing water. Unlike groundwater flow in most other aquifers, water in karstic aquifers flows in pipelike conduits and open cave stream channels as well as through fractures and pores. Thus, we must draw on the theories of flow in porous media and flow in pipes and channels. Porous media flow is discussed in texts on hydrology (e.g., Bear, 1972), and the fluid mechanics of pipes and channels has been the province of the civil engineers (e.g., Vennard and Street, 1982).

6.2.1 *Conduit Flow*

The conduit permeability of karst aquifers consists of integrated systems of openings ranging from solutionally widened joints and bedding plane partings to pipelike passages many meters in diameter. Consider, for the present, flow in large openings that may be approximated as irregular pipes.

The flow of water in natural systems of pipes and channels uses the earth's gravitational field as an energy source. If we imagine steady, frictionless flow in a closed system, the total energy must be conserved and equal to the sum of the kinetic and potential energy of the system. The expression of energy conservation usually takes the form of the Bernoulli equation:

$$gz + \frac{P}{\rho} + \frac{\bar{v}^2}{2} = \text{Constant} \qquad (6.11)$$

The potential energy has two parts, one resulting from the elevation, z, and the other from the pressure on the system, P. The kinetic energy of the system is contained in the motion of the water, as measured by its velocity, $\bar{v}$. It is the custom in both hydrology and fluid mechanics to use the hydraulic head, h, rather than the potential energy, for calculation. Dividing by the gravitational acceleration, g, sets the Bernoulli equation in the form

$$h + \frac{P}{\rho g} + \frac{\bar{v}^2}{2g} = h_T \qquad (6.12)$$

where the sum of the elevation head, the pressure head, and the velocity head equals the total head, h_T, of the system.

Real conduits, of course, are not frictionless. Much of the effort in fluid mechanics calculations goes into estimating the friction head, h_f, which measures the energy loss from friction between the fluid and the walls of the conduit. Balancing the energies between the inlet and outlet of the conduit gives the practical form of the Bernoulli equation:

$$h_T = \frac{P_1}{\rho g} + \frac{\bar{v}_1^2}{2g} + h_1 = \frac{P_2}{\rho g} + \frac{\bar{v}_2^2}{2g} + h_2 + h_f \qquad (6.13)$$

Most of the physical sciences use mass, length, and time as the fundamental units of measurement from which energy, velocity, force, and other units are derived. It has been the custom in fluid mechanics literature to use force, length, and time as the fundamental units. The distinction between force and mass changes the scale of fluid flow equations by a factor of the gravitational acceleration, g. These equations are written in many fluid mechanics texts in terms of the specific weight, γ:

$$\gamma = \rho g \qquad (6.14)$$

The sections that follow have been adapted to the mass, length, and time units to be consistent with the remainder of the book. The equations look slightly different from those in standard textbooks.

Water flowing in closed conduits is subject mainly to two forces: inertial forces associated with the momentum of the mass of water in motion, and viscous forces generated by the layers of fluid sliding past each other. The relative importance of inertial and viscous forces is described by their ratio, a dimensionless quantity called the Reynolds number

$$\frac{F_i}{F_v} = \frac{\rho \bar{v}^2 / R}{\eta \bar{v} / R^2} = \frac{\rho \bar{v} R}{\eta} = N_R \qquad (6.15)$$

Two flow regimes are similar if they have the same Reynolds number, regardless of the values of the other parameters taken individually. The quantity R is the hydraulic radius, which for a circular pipe, is taken as the pipe diameter, $R = 2r$. The flow behavior depends on the density, ρ, and viscosity, η, of the fluid, both of which are functions of the temperature (Table 6.2).

At low velocities, water moves through a smooth pipe in streamlines, with no mixing across these streamlines except by molecular diffusion. Viscous forces are dominant. As velocity increases, the streamlines become unstable. Irregularities in the walls of the pipe introduce disturbances that propagate and enlarge. Over a transition

Table 6.2
Physical Properties of Water[a]

T (°C)	Viscosity (poise)	Density (g mL^{-1})
0.0	0.01787	0.99987
5.0	0.01519	0.99999
10.0	0.01307	0.99973
15.0	0.01139	0.99913
20.0	0.01002	0.99823
25.0	0.008904	0.99707
30.0	0.007975	0.99567
35.0	0.007194	0.99406
40.0	0.006529	0.99224

[a] 1 poise = 1 g cm^{-1} sec^{-1} = 1 dyne sec cm^{-2} = 0.1 N sec m^{-2} = 2.089 × 10^{-3} lb(force) sec ft^{-2}.
Source: Data from *Handbook of Chemistry and Physics*, 59th ed. (CRC Press, Boca Raton, FL).

region, as velocity continues to increase, the flow changes from a laminar to a turbulent flow regime. In smooth pipes the transition begins at about $N_R = 2100$, but in rough pipes, such as cave passages, turbulence begins at lower Reynolds numbers.

The quantities of interest are the flow velocities and the head loss along a reach of conduit. For laminar flow, the energy equations can be solved exactly for an incompressible fluid. The laminar flow head loss for three common conduit geometries is given by the following:

$$h_f = \frac{12\eta \bar{v} L}{\rho g B^2} \quad \text{Parallel walled fracture} \tag{6.16}$$

$$h_f = \frac{8\eta \bar{v} L}{\rho g r^2} \quad \text{Circular conduit} \tag{6.17}$$

$$h_f = \frac{\eta \bar{v} L}{\rho g (Nd^2)} \quad \text{Porous medium} \tag{6.18}$$

where B is the spacing between the walls of the fracture (the aperture) and N is the permeability shape factor. The equation for the circular conduit is the Hagen-Poiseuille equation, often given as an expression for total discharge:

$$Q = \frac{\pi \rho g r^4}{8\eta} \frac{dh}{dL} \tag{6.19}$$

The head loss equation for the porous medium is simply a restatement of D'Arcy's law.

The equations for fluid flow in turbulent regimes cannot be derived easily from first principles, and the engineering textbooks fall back on various empirical relationships. The starting point is the D'Arcy–Weisbach equation

$$h_f = \frac{fL\bar{v}^2}{4gr} \qquad (6.20)$$

The turbulent flow velocity varies with the square root of the hydraulic gradient. The quantity f is a dimensionless friction factor. In laminar flow regimes, the friction factor is inversely related to the Reynolds number

$$f = \frac{64}{N_R} \qquad (6.21)$$

For turbulent flow over a smooth surface, the friction factor is also determined only by the Reynolds number through the Prandl–von Karman equation:

$$\frac{1}{\sqrt{f}} = 2 \log (N_R \sqrt{f}) - 0.8 \qquad (6.22)$$

For turbulent flow over rough surfaces, the friction factor is determined only by the roughness

$$\frac{1}{\sqrt{f}} = 2 \log \frac{2r}{e} + 1.14 \qquad (6.23)$$

where e is the relief of the surface irregularities in the same units as the pipe radius. Values of e for carbonate rocks are likely to be similar to those for concrete, e(mm) = 0.3 and 3.0 (Daugherty and Ingersoll, 1954). The equations can be combined into a single expression that describes the transition from smooth to rough boundaries:

$$\frac{1}{\sqrt{f}} = 1.14 - 2 \log \left(\frac{e}{2r} + \frac{9.35}{N_R \sqrt{f}} \right) \qquad (6.24)$$

The velocity of flow is a maximum along the center line of the conduit and decreases to zero for the final layer of fluid in contact with the wall. In laminar flow, the velocity profile is parabolic. In turbulent flow in a smooth pipe there is a laminar boundary layer near the wall that results in a flattened velocity profile (Fig. 6.9). In turbulent flow over a rough surface, a thin laminar layer exists, across which there is a sharp discontinuity in the velocity.

The thickness of the laminar layer, δ, is given by the dimensionless equation

$$\frac{\delta}{2r} = \frac{32.8}{N_R \sqrt{f}} \qquad (6.25)$$

showing that the film thickness decreases with increasing Reynolds number. Comparison with experimental results shows that the tran-

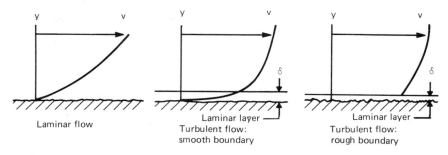

Figure 6.9 Velocity profiles for laminar and turbulent flow regimes.

sition from smooth pipe behavior to rough pipe behavior occurs at $e/\delta \simeq \frac{1}{4}$. When the surface roughness is less than ¼ of the thickness of the laminar layer, the conduit behaves as a smooth pipe. When $e/\delta > 6$, the laminar layer breaks down and the conduit behaves as a totally rough pipe. The friction factor is no longer a function of the Reynolds number.

The appropriate expression for the friction factor may be combined with the D'Arcy–Weisbach Eq. 6.20 to calculate the velocity for turbulent flow. Figure 6.10 shows the variation in mean flow velocity with conduit radius for a series of geologically reasonable hydraulic gradients. The roughness of typical limestone walls is taken to be $e = 1$ mm. This would be approximately correct if the roughness were due to sand grains or oolites etched out in relief. The onset of turbulence in cave conduits is not known. Certainly N_R is less than 2100; it may be as low as 10.

6.2.2 Open Channel Flow

Channels are usually classified as rigid or erodible. Rigid channels have fixed walls that are not modified by fluid flow. Flumes and spillways made of concrete or metal are examples. Erodible channels have walls of unconsolidated material that can be rearranged by the moving water. Most creeks and rivers flow on beds of alluvium, and thus have erodible channels. Most man-made channels have cross sections approximating simple geometric figures such as rectangles, circles, or trapezoids. The channel is characterized by a hydraulic radius, usually defined as the cross-sectional area divided by the wetted perimeter (Fig. 6.11), and a hydraulic depth equal to the actual water depth for rectangular channels.

Water flowing in open channels is subject to three forces: inertial forces resulting from the mass of water in motion, viscous forces resulting from shear within the water mass, and gravity forces. The latter arise because flow in an open channel is not confined and flow depth can vary with flow velocity. The ratio of inertial forces to vis-

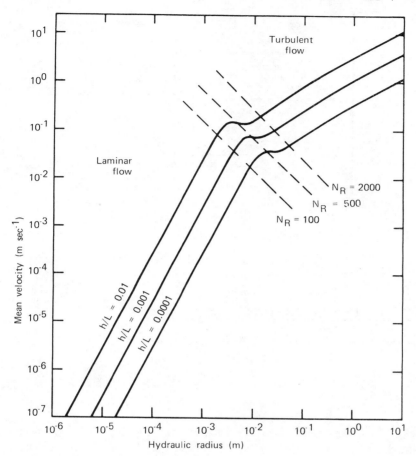

Figure 6.10 Mean flow velocity as a function of hydraulic radius for water at 10°C. Wall roughness factor, $e = 0.1$ cm.

cous forces is described by a Reynolds number in the same form as in conduit flow:

$$N_R = \frac{\bar{v} R \rho}{\eta} \qquad (6.26)$$

Because of the different definition of the hydraulic radius (see Fig. 6.11) the transition to turbulent flow takes place at $N_R = 500$ in smooth channels.

The balance between inertial forces and gravity forces is described by the Froude number, defined by

$$N_F = \frac{\bar{v}}{(gD)^{1/2}} \qquad (6.27)$$

166 / *Geomorphology and hydrology of karst terrains*

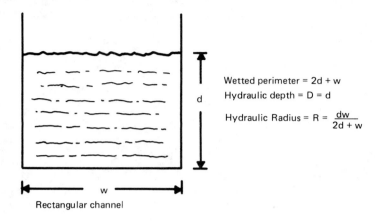

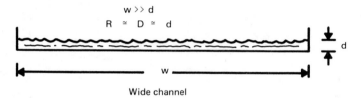

Figure 6.11 Defining properties of rectangular channels.

where D is the hydraulic depth. Flows for which $N_F < 1$ are termed subcritical or tranquil. Flows for which $N_F > 1$ are supercritical. Unlike the gradual transition between laminar and turbulent flow, the transition from subcritical to supercritical flow is abrupt, creating a displacement of the water surface known as a hydraulic jump. Fast-moving supercritical flows are deceptively smooth and shallow. They occur in both surface and subsurface stream channels, where steep bottom slopes (chutes) raise the velocity above the critical value. When supercritical flows are broken by a flattening of the channel or by an obstruction (such as a boulder in the stream bed), a large amount of energy is released at the hydraulic jump, which can be zones of intense erosive activity. The "white water" well known in steep surface streams results from hydraulic jumps.

For the special case of wide shallow channels, $w \gg d$, the hydraulic radius and the hydraulic depth become nearly equal and it is possible to plot a depth–velocity diagram (Fig. 6.12). The diagram shows the four flow regimes found in channels. Most surface and cave streams fall into the subcritical–turbulent regime. Some

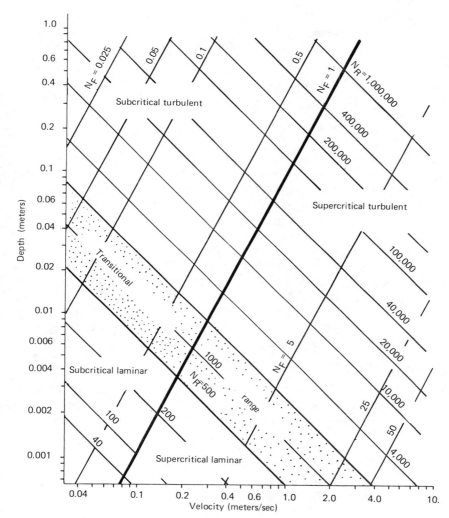

Figure 6.12 Fence diagram showing four regimes for open channel flow. The assumptions are that the channel is wide ($w \gg d$) and the temperature is 10°C. The grid contours give the values of the Reynolds number, N_R, and the Froude number, N_F.

slow-moving, ponded waters are subcritical–laminar, and a few high gradient channels have supercritical turbulent flows. An odd case is that of the water films streaming down the sides of vertical shafts (which can be thought of as wide channels standing on end). These fall into the supercritical–laminar regime. This particular flow regime is used later to help explain why vertical shafts are vertical.

Open channels cannot sustain a pressure head, and thus the flow velocity is determined by the slope of the channel and its roughness. These variables are most commonly related through the Manning equation:

$$\overline{v} = \frac{1}{n} R^{2/3} S^{1/2} \quad \text{(for units of meters)}$$

$$\overline{v} = \frac{1.49}{n} R^{2/3} S^{1/2} \quad \text{(for units of feet)}$$

(6.28)

The Manning equation has not been properly dimensionalized (see Chow, 1959, for discussion); the convention has been to let the roughness factor, known as Manning's n, have the same value in both English and metric units. R is the hydraulic radius and S is the channel slope, expressed in dimensionless units. The roughness factor is determined experimentally by measuring velocity, hydraulic radius, and slope for a variety of channels and then using these as reference channels for selecting n values. Table 6.3 lists values for some man-made channels that can be compared with cave channels as well as some values for surface stream channels. Barnes (1967) shows color photographs of a large selection of rivers with measured values of Manning's n, so that values for surface streams can be determined by visual comparison.

6.2.3 Porous Media Flow

Porous media aquifers are sandstones, sand and gravel-filled valleys, some basalts, and other rocks in which the primary storage and movement of water is through the pores between mineral grains.

Table 6.3
Manning's n for Various Channels

Channel	n
Cast iron conduit (flowing partly full)	0.014
Corregated metal storm drain	0.024
Concrete culvert	0.013
Gunite on excavated rock	0.020
Dressed stone in mortar	0.017
Channel with gravel bottom and sides of concrete	0.020
Channel with gravel bottom and sides of riprap	0.033
Brick in cement mortar channel	0.015
Small stream, straight channel, no riffles or pools	0.030
Small stream, some pools and shoals	0.040
Mountain stream with cobbles and boulders	0.050

Source: Selected values are from Chow (1959).

Because the scale of the pores is so small compared with the scale of the aquifer, the rock can be treated as a continuous medium of characteristic permeability and water movement as a continuous flow field.

For the limiting case of homogeneous, isotropic, porous media aquifers, the movement of water is guided by two principles: D'Arcy's law (Eq. 6.2) and the principle of continuity. The principle of continuity is a statement of the conservation of mass. Given a small volume of dimensions dx, dy, and dz, anywhere in the aquifer, water entering the cube is exactly balanced by water leaving the cube; no water is created or destroyed. The principle of continuity takes the form

$$\frac{\partial q_x}{\partial x} + \frac{\partial q_y}{\partial y} + \frac{\partial q_z}{\partial z} = 0 \tag{6.29}$$

or in vector notation

$$\text{div } \vec{q} = 0 \tag{6.30}$$

Because water flows in three-dimensional patterns, D'Arcy's law must also be rewritten in vector form, as

$$\vec{q} = -K \text{ grad } h \tag{6.31}$$

Equations 6.30 and 6.31 combine to yield the general equation guiding groundwater flow in porous media:

$$\frac{\partial^2 h}{\partial x^2} + \frac{\partial^2 h}{\partial y^2} + \frac{\partial^2 h}{\partial z^2} = 0 \tag{6.32}$$
$$\nabla^2 h = 0$$

which is the well-known Laplace equation. Like all differential equations, the Laplace equation has an infinite number of solutions. Application of the equation to a particular hydrologic situation is an exercise in selecting appropriate boundary conditions. Many of the implications of this equation were worked out in detail in Hubbert's (1940) classic paper.

When the nature of karst groundwater was debated in Europe and theories for the origin of caves were being argued in the United States there was much misunderstanding about groundwater flow paths. Many workers apparently believed that water flowed vertically through the vadose zone and then turned and flowed down the slope of the water table much like water running down a roof. Solutions of Laplace's equation for this problem show that water continues to follow vertical flow lines when it reaches the water table, and these only gradually curve and rise under a river bed or other zone of discharge (see Fig. 6.1). Early-published sketches (not calculations) of these flow lines led Davis to his deep phreatic theory of the origin of caves.

The application of porous media flow models depends on the validity of D'Arcy's law over all flow regimes. There is some evidence (Kraft and Yaakobi, 1966) of a threshold below which flow velocity is no longer proportional to hydraulic head, but this becomes important only in highly impermeable clays. The velocity range of D'Arcy's law has a very definite upper limit. At high velocities turbulent flow occurs, and the velocity is no longer proportional to the hydraulic gradient. Analysis of flow through granular materials, with the assumption that the interconnected pore spaces can be treated as bundles of small tubes, led to the conclusion that under geologically reasonable hydraulic gradients turbulence could be expected in fine gravels and coarser materials (Smith and Sayre, 1964).

Ward (1964) defined a Reynolds number for flow in granular materials by using the permeability as the hydraulic radius:

$$N_{R,K} = \frac{\rho(Nd^2)^{1/2}\bar{v}}{\eta} \tag{6.33}$$

By measuring the Reynolds number and porous media friction factor independently by laboratory experiments on a great variety of granular materials, Ward showed that the results deviate from the strict inverse relation required by D'Arcy's law at $N_{R,K} = 1$.

The nonlinear response of a coarse-grained aquifer is sometimes described by the Forchheimer equation, as

$$-\frac{dh}{dL} = aq + bq^2 \tag{6.34}$$

where h is the total piezometric head (pressure head plus gravity head). In effect, the Forchheimer equation simply adds laminar and turbulent flow terms. It can also be derived from first principles (Irmay, 1958). Ward's dimensional analysis shows that the Forchheimer equation can be written in dimensionally consistent form as

$$-\frac{dh}{dL} = \frac{\eta\bar{v}}{Nd^2} + \frac{0.550\rho\bar{v}^2}{(Nd^2)^{1/2}} \tag{6.35}$$

where 0.550 is a dimensionless constant determined by experiment, thus establishing both a and b parameters in Eq. 6.34.

Flow in fractures and solutionally modified aquifers deviates from strict D'Arcy behavior. However, it has not proved useful to construct a continuous flow field model, a nonlinear differential equation, with the Forchheimer equation.

6.2.4 The Problem of Fracture Flow

Laminar flow through parallel plane-wall fractures follows D'Arcy's law with the hydraulic conductivity (Witherspoon et al., 1979) given by

$$K = \frac{\rho g B^2}{12\eta} \qquad (6.36)$$

equivalent to Eq. 6.16. The fracture permeability is then $B^2/12$, where B is the full width or aperture of the fracture ($B = 2b$ in Witherspoon's notation). The discharge through a single fracture is given by the "cube law" (Witherspoon et al., 1980)

$$\frac{Q}{\Delta h} = \frac{C}{f} B^3 \qquad (6.37)$$

where f is a friction factor and C is a constant.

Although the flow properties of a single fracture can be accurately described, application to a full-scale fracture aquifer is difficult because of the following factors:

1. Irregularities and fracture aperture variations along the length of the flow path. The dependence of the flow on the cube of the aperture makes this a very sensitive parameter (Neuzil and Tracy, 1981; Smith and Schwartz, 1984).
2. The onset of turbulence in large fractures or at fracture intersections, which breaks D'Arcy's law (Castillo et al., 1972).
3. The strong anisotropy of most fracture aquifers, in that they tend to be dominated by single fracture sets.
4. The problem with modeling the entire assemblage of fractures that make up the secondary permeability of the aquifer when their individual geometries are not known (Wilson and Witherspoon, 1974; Endo et al., 1984).

6.3 GROUNDWATER IN CARBONATE AQUIFERS

6.3.1 The Hydrologic System

Karst aquifers are of diverse form and character, depending on their hydrogeologic setting. See, for example, Burdon and Papakis (1963), Stringfield and LeGrand (1969), Parizek et al. (1971), and LeGrand and Stringfield (1971, 1973). Two extremes can be recognized: conduit aquifers, in which the groundwater throughput is completely dominated by the conduit system, and diffuse flow aquifers, in which conduit systems are either absent or so poorly integrated that they have little influence on the groundwater circulation (Shuster and

White, 1971; Atkinson, 1977a). Many carbonate aquifers contain both elements. The flow system can be represented in a schematic form (Fig. 6.13) that permits water balance calculations and investigation of the chemical characteristics of the various types of water (see Chapter 7).

The top of the diagram is the land surface, which assumes three kinds of catchment areas: the karst surface itself, borderlands adjacent to the karst that may be the source for allogenic streams, and a caprock area standing above the karst, which may be the source of either sinking streams or direct vertical input into the karst aquifer below. Precipitation is partly lost by evapotranspiration; the remainder enters the aquifer as allogenic runoff, internal runoff, and diffuse infiltration.

Precipitation must first saturate the soil and fractured rock zone before moving deep into the vadose zone. Such water can be stored in what Williams (1983, 1985) calls the subcutaneous zone for a considerable period of time. It can be sampled and has a unique chemistry. Even et al. (1986) show that the holdup time in the subcutaneous zone can be decades.

Sinking streams that focus the water collected from a considerable area to a single swallow hole usually continue underground in a conduit, which may appear as an open channel or as a flooded pipe. Water draining from surface catchments or perched aquifers above the limestone, such as the karst of the Mammoth Cave area or of the Cumberland Mountains of Tennessee and northern Alabama, often takes a direct route through the vadose zone by means of vertical shafts. These may drain to an integrated conduit system or feed small springs directly.

The internal runoff feeder channels are generally thought to be the upstream tendrils of the conduit system, although there is no real evidence that the systems are always well integrated. Diffuse recharge moves uniformly downward through available joints, accounting for much of the infiltration from the karst surface. A few of the joints intersect conduits and discharge their water where it can be sampled. Most seepage waters miss the conduit system completely and provide an important source of recharge to the more dispersed, deeper water body labeled phreatic storage in Fig. 6.13. The balance between internal runoff and diffuse infiltration depends on soil permeability, depression density, and the efficiency of the closed depression drains.

What happens near the water table is fairly complicated and not very well worked out for real karst aquifers. Cave streams can certainly be observed and explored, and their sources and destinations can be determined by tracer techniques. In some caves one can follow the sinking stream underground and observe directly that it is a

Karst hydrology / 173

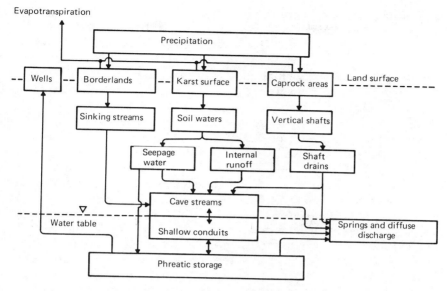

Figure 6.13 Internal flow system of a karst aquifer.

tributary to an open-channel conduit. Sometimes one can enter the mouth of a gravity spring and follow the feeder channel upstream, observing the trunk drainage system. In a few caves one can do both. Results from some good underwater exploration and surveying are beginning to reveal a picture of undulating conduits with some reaches above the water table and some reaches continuously flooded. Flooded conduits may exist in parallel with stream-carrying conduits. This situation is indicated in Fig. 6.13 by a double arrow between two boxes—a gross simplification of what, in real caves, is a highly complicated set of interconnections. The sketch may also be taken as the low flow condition of the system. During times of high flow, the water table rises and many of the passages become flooded.

Below and beside the conduit system lies a region of bedrock with water stored in small channels, fractures, and primary pores. Water from phreatic storage supplies cave streams during periods of low recharge. Phreatic storage is also the source of water for most wells drilled in carbonate aquifers. The conduits create a low gradient groundwater trough, so that water exchange between the cave stream/conduit system and the phreatic storage/diffuse flow system is lateral as well as vertical. In the headwater reaches of the system, large conduits are more sparse (unless there is a drainage conduit for a sinking stream catchment), and the diffuse flow system tends to dominate. Downstream, closer to the springs, the reverse is true.

The schematic diagram can be adapted to many real aquifers by

174 / Geomorphology and hydrology of karst terrains

adjusting the mix of water traveling along the various flow paths. To obtain a diffuse flow aquifer, such as some of the principal dolomite aquifers, simply shut off the conduit system altogether. Seepage waters and some internal runoff make their way directly to the phreatic storage, from which they may be discharged through diffuse flow springs. As will be seen, the chemical and physical response of the springs tells much about the systems that feed them.

A water balance calculation for the karst aquifer in the Mammoth Cave area (Hess and White, 1974) took advantage of the fact that the aquifer is underlain by impermeable shales that inhibit loss of water through deep flow. The aquifer is drained to Green River through some 100 springs of various sizes as well as by possible diffuse discharge in the river channel. The U.S. Geological Survey has gages on Green River at Munfordville, upstream from the Mammoth Cave area, and at Brownsville, downstream from the area. The only major intermediate tributary, the Nolin River, is also gaged. Mean flows were determined from the river records. The difference between the downstream and upstream gages, corrected for the contribution of the Nolin River, was taken as a measure of mean runoff from the karst (Table 6.4). Evapotranspiration losses were determined by balancing mean precipitation against runoff, but the results shown in Table 6.4 are in good agreement with calculations using the Thornthwaite potential (see, e.g., Gray, 1970). The evapotranspiration loss from the karst was some 19 percent less than for the remainder of the Green River basin, presumably because of rapid internal runoff into sinkholes.

Base flow measurements for the southcentral Kentucky karst were obtained by taking the monthly average runoff during September and October over 6 years of recordings. The low flow for the karst area was 1.58 m^3 sec^{-1}, or a low flow intensity of 0.0016 m^3 sec^{-1} km^{-2}. The discharges from the six largest springs that enter Green River from the karst area south of the river were measured, and from these data along with the estimated recharge areas, it was deduced that, to within experimental error, all discharge from the aquifer is through the springs. The diffuse component of the aquifer must drain laterally to the conduit system, which then carries the water along lines of concentrated flow to the base level surface stream.

6.3.2 Hydraulic Geometry of Conduit Systems

Although the conduit system in a karst aquifer is traditionally thought to contain groundwater, its hydraulic properties have more in common with the hydraulic properties of surface channels. Some

Table 6.4
Runoff Characteristics of the Central Kentucky Karst Aquifer

	Area (km^2)	Mean annual discharge (m^3 sec^{-1})	Runoff Intensity (m^3 sec^{-1} km^{-2})	Runoff Intensity (mm)	Evapo-transpiration (mm)	Years of record
Green River Basin at Brownsville	7154	115.15	0.0161	508	756	40
Green River Basin at Munfordville	4333	72.64	0.0168	529	735	45
Nolin River Basin	1831	23.22	0.0127	400	864	23
Central Kentucky Karst	989	19.29	0.0195	615	649	—

Source: Data from Hess and White (1974).

of the hydraulic properties that might be shared between underground conduits and surface streams are the following:

Channel width–channel depth relations
Sinuosity
Channel braiding
Ordered branching and stream length ratios
Distinct catchment area–discharge relationships

Open channels often adjust their widths to accommodate the available discharge, so that the ratio of width to depth remains constant. Cave channels also appear to correlate with discharge at least roughly. Small canyons have small discharges; wide rectangular channels have either larger discharges or evidence from scallop markings of past large discharges. Downcutting streams in canyon passages can act both vertically and horizontally, depending on gradient and discharge, allowing a balance between flow depth and passage width to be attained. Some sparse measurements on canyon passages in the Flint Ridge–Mammoth Cave System suggested the relation

$$\bar{v} = w^{-0.7} \tag{6.38}$$

between velocity and passage width (White and White, 1970). They also argue for a ratio of passage width to flow depth of 1:3.

Many tube and canyon conduits are sinuous. The meander pattern of surface rivers is known to be related to parameters such as channel width and river discharge (Leopold and Wolman, 1960). The bend spacing or meander wavelength is related to channel width by a simple power law:

$$L = Kw^n \tag{6.39}$$

where L = bend spacing and w = channel width in the same units. K and n are fitting parameters. Sinuous cave passages obey the same power law relation as surface rivers, with similar values for the constants (Table 6.5). Data for the sinuous caves of Missouri, a homogeneous data set, are plotted in Fig. 6.14.

Some meanders are incised in the sediment-covered floors of larger passages by the action of free-surface streams. As these streams downcut, the meandering pattern developed on the clastic sediment is superimposed on the bedrock, and a deep winding canyon results. Canyons, with incised features, can adjust their width/meander length ratios to suit the hydraulic requirements of discharge. Meanders migrate downstream as the channels deepen, and this effect is seen in the tortuous canyon passages of many alpine caves.

Table 6.5
Some Values for the Parameters of the Power Law[a] Describing Meander Bend Spacing, $L = KW^n$

Data set	K	n	Reference
World rivers	10.9	1.01	Leopold and Wolman (1960)
Indian rivers	6.6	0.99	Inglis (1949)
Missouri caves	6.8	1.05	Deike and White (1969)
Other caves	8.2	0.92	Deike and White (1969)
New York caves	7.4	1.15	Baker (1973)

[a]The parameters K and n are nearly dimensionless (strictly dimensionless if $n = 1$). The parameters given here were calculated for channel widths and meander bend spacings measured in feet.

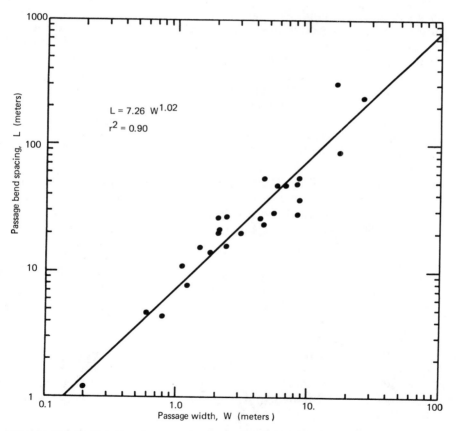

Figure 6.14 Relation between bend spacing and passage width for some caves in Missouri. Plotted from data of Deike and White (1969), which were drawn from published maps of the Missouri Cave Survey.

Meandering tubes are more of a problem. If the meandering amplitude is small, the meandering could have begun in a linear passage with enhanced solution on alternating regions of the wall, so that the passage widened into a meandering form. Tubular passages with large meander amplitudes are difficult to explain.

One strange discrepancy has been reported. Smart and Brown (1981) examined the meander wavelength–passage width relations for some measured meanders in Irish caves and found that the bend spacing varied inversely with the square root of the passage width. Clearly, more data are needed, especially accurately mapped passage widths that may be the largest source of error in data taken from published cave maps.

Maze caves are also analogous to surface channel forms. Network mazes are the conduit equivalent of swamps. Flow through all available mechanical openings is slow because it is throttled either by slow percolation through overlying clastic beds or by constraints on the discharge (Palmer, 1975). The strong structural control in most network mazes makes all of the islands nearly the same size. Anastomotic mazes and caves with loose maze patterns are more analogous to braided rivers (Howard et al., 1970; Howard, 1971).

Perhaps the most important evidence that conduit systems are really extensions of surface drainage systems comes from the specific discharge relationships. Because scallops serve as indicators of paleoflow, it is possible to learn something about the discharge, flow velocity, and required catchment areas for long-abandoned conduits. Many cave passages exhibit good scalloping. Passages with canyon cross sections frequently contain small scallops that extend uniformly up the walls. Some elliptical tubes are uniformly scalloped on floors, walls, and ceiling. Although uncommon, such tubes provide the best places for determining paleodischarge. Figure 6.15 shows three scalloped passages from the Flint Ridge–Mammoth Cave System (White and Deike, 1976). There is a place in Colossal Cave's Grand Avenue where the passage consists of an upper tube, uniformly scalloped, and a lower canyon—Bicycle Avenue—also uniformly scalloped. Figure 6.15 shows histograms of the scallop size distributions in comparison with the scallops in the main trunk of Great Salts Cave, which has the form of a large canyon. These can be used to calculate flow velocities using Curl's equations (Eqs. 3.3, 3.4, and 3.7), and from the known cross section of the passages, the total discharge can be calculated. Table 6.6 compares some values with other data compiled by Gale (1984).

The various literature on scallops carefully skirts the issue of flow variability. Curl's early (1966b) work was based in large part on observations in fluted cave passages. Flutes are thought to result from constant flow, and indeed, many fluted passages are situated so

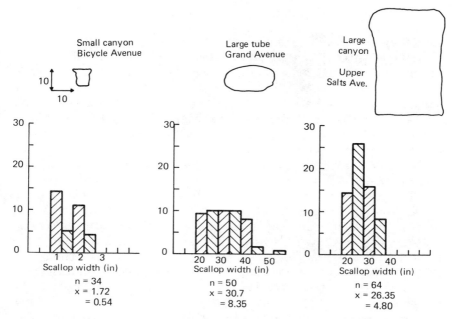

Figure 6.15 Scallop distributions in three passages in the Flint Ridge portion of the Mammoth Cave System.

Table 6.6
Calculated Paleodischarge for Some Cave Passages[a]

Passage	Scallop length (cm)	$\bar{v}$ (m sec^{-1})	A (m^2)	Q (m^3 sec^{-1})	
White Lady Cave, Little Neath Valley, UK[b]	4.1	1.21	7.6	9.14	Canyon
Poulnagollum Cave, Ireland[b]	7.5	0.43	1.1	0.48	Canyon
Roudsea Wood Cave, Cumbria[b]	24	0.10	2.4	0.23	Tube
North Entrance Streamway Friars Hole Cave, WV[b]	8.3	0.38	1.8	0.67	Tube
Grand Avenue, Flint–Mammoth Cave System, KY[c]	78	0.03	20.7	0.71	Tube
Robertson Avenue, Mammoth Cave, KY[d]	48	0.05	9.3	0.47	Tube

[a]Scallop length is given as the Sauter mean (see Eq. 3.3); $\bar{v}$ is the mean flow velocity; A is the cross-sectional area of passage; Q is paleodischarge.
[b]Data are from Gale (1984).
[c]Data are from White and Deike (1976).
[d]Data are from White and White (1970).

that they flood under constant head, thus maintaining a constant velocity. Most conduits are subject to variable head resulting from variations in seasonal runoff and, as described by the Bernoulli equation (Eq. 6.13), the velocities also fluctuate, although only as the square root of the head. If we take the simplest assumption—that the rate of removal of bedrock to form scallops is approximately proportional to the time of exposure to water of a particular chemistry and flow regime, then we conclude that the scallops record the most probable flow. If this is close to the mean flow, then one can calculate the catchment area from the runoff characteristics of the karst aquifer. The results for the Kentucky conduits are reasonable given the present-day hydrogeologic setting.

If flow velocities in canyon passages are calculated from scallop patterns and the flow velocities and channel cross sections are then used with the Manning equation to calculate the channel slope, one finds that shaft drain canyons usually give reasonable values for the slopes of free-surface streams. Calculated slopes of some of the larger canyons, however, are extremely low. It appears that some canyon passages actually formed as lakes, the water level controlled by flooded conduits at both ends (White and White, 1970).

6.3.3 *Water Tracing*

Information on the internal connections of unobservable portions of conduit systems can be obtained through tracing experiments. A tracer substance is added to streams at swallow holes or to cave streams, or is flushed down sinkholes by heavy rainstorms or by tank trucks of water. It is then picked up at accessible points in the underground system or at the discharge points. Many tracers have been used, including dyes, spores, salt, microorganisms, rare elements, and radioactive substances. Techniques and results are described by Maurin and Zötl (1967), Gospodarič and Habič (1976), and Aley and Fletcher (1976).

Dye tracing with fluorescein or rhodamine dyes is most straightforward. If the water is restrained to a system of conduits, and there is not too much ponding or absorption by clays or organic matter in the underground waterways, the colored water will be observed, thus establishing a connection between the injection point and the rising. It is not necessary to inject enough dye to visibly color the water, as was done in the early days of dye tracing. Both fluorescein and rhodamine dyes are strongly adsorbed on small packets of activated charcoal, which can be inserted in suspected resurgences as dye detectors. Later the dye can be elutriated from the charcoal by an ethanol solution of potassium hydroxide (KOH) and observed either by its color or by its strong fluorescence under ultraviolet light (the Dunn method; see Haas, 1959). Automatic samplers and spectrofluorometers allow one to determine tracer hydrographs, which yield more information about the conduit system than simple point-to-point connections (Crawford, 1979). Other dyes have been investigated, and the simultaneous use of several dyes, using colored filters to recognize them in the water, has been proposed (Smart and Laidlaw, 1977). Most organic dyes are toxic at high concentrations, and colored surface streams or public water supplies would not be welcomed by most citizens. Some states require permits to carry out legal dye tests. Smart and Laidlaw (1977) give a very useful comparison of the advantages and disadvantages of many tracer dyes.

The spore-tracing method developed by Zötl and his colleagues in Austria (Zötl, 1974) uses lycopodium spores, typically 25 to 30 μm in diameter, which have been dyed characteristic colors with biological stains. The detectors are plankton nets suspended in the suspected resurgences. Materials trapped on the plankton nets are examined under the microscope, where the observation of a single colored spore is deemed a positive test. The spore-tracing techniques have the advantage that several swallow points can be tested simultaneously by using different colored spores, and the tests are more reliable. There is no problem with adsorption on sediments, and no interference from natural organic materials, which sometimes tinge waters green much as fluorescein does. On the other hand, spores are more expensive, and the staining process is messy and time-consuming. Because the spores are tiny and easily dispersed, there is a very real problem of contamination.

Optical brighteners are good tracing agents (Glover, 1972; Quinlan and Rowe, 1977; Quinlan, 1981). They are inexpensive commercial products, which are detected in suspected resurgences by wads of unsized cotton clamped in holders to keep them in the main flow of water. They have the further advantage of working in muddy water. The cotton wad is simply washed free of mud and debris after it is collected. The presence of the brightener on the cotton is detected by an intense blue-white fluorescence under longwave ultraviolet light.

6.3.4 *The Karst Water Table*

Throughout the long history of karst research, the argument over the nature and even the existence of a water table in karst aquifers has persisted. Grund, Cvijič, Penck, and others of the Vienna School believed that sinking streams and cave streams drained down to a coherent body of *karstwasser* and denied the existence of integrated drainage systems. Katzer, Martel, and others regarded cave streams as underground rivers that were not necessarily connected, did not flow at common levels, and were not connected to any deeper groundwater body, and they denied that a true water table exists in karst.

Both views have persisted, with the former influencing the cave origin theories developed in the United States, and the latter underlying most French thinking on karst hydrology until fairly recently.

British writers have also tended to discount the notion that there could be a water table in highly karstified limestones. Sweeting (1972) gives a long and thoughtful account of the various positions. The Doolin Cave System, which contains air-filled passages running beneath the River Aille in County Clare, western Ireland, was often given as an example of no definable water table (Tratman, 1969).

The water table controversy can be resolved by examining the hydrology of the entire subsurface basin, rather than focusing only on the conduit system. Groundwater basins contain both diffuse and conduit components, and even those that discharge nearly all of their water through conduit systems have in-feeders through which water moves more slowly. Productive wells can be drilled in karstic aquifers if they intersect some of the fracture permeability. It is not necessary that the wells tap the major conduits. Water table maps can be prepared by contouring water levels in wells. A map of the Spring Creek Basin in central Pennsylvania (Giddings, 1974) shows a well-developed water table surface with a deep groundwater trough following the axis of a dry valley. The trough was interpreted as marking the route of the conduit system carrying subsurface flow from the valley uplands to a group of springs. More dramatically, Quinlan and Ray (1981) were able to contour the water table beneath the Sinkhole Plain south and east of Mammoth Cave National Park, certainly one of the most intensely karstic aquifers in the United States. Again, the route of major conduit systems was marked by well-defined groundwater troughs. Water table highs, in general, coincided with the boundaries of the groundwater basins that had been defined by extensive tracer experiments. The tracing experiments, combined with cave mapping, lay out the conduit portion of the basin; the water table surface tells something about the behavior of the diffuse portion of the aquifer. Both are present and the water table is well-defined.

The careful work of Zötl (1961) and his colleagues showed that in the alpine karst of the Austrian Alps there is also a deep circulating groundwater body of regional extent on which are superimposed a number of more open and more local drainage systems. The local systems tend to carry recharge from a small catchment through a conduit system to a single spring whereas the regional system releases water to many springs.

Much of the confusion over the existence or nonexistence of a karst water table seems to have arisen from misunderstanding of the water table concept itself. Water table surfaces are not static; they rise and fall in response to changes in recharge caused by storms and seasonal changes in precipitation. The water table surface in topologically or structurally complex areas may be quite irregular. The unique feature of karst aquifers is the rapid response of the conduit system compared with that of the diffuse system. A spring flood or a summer thunderstorm may fill the conduit system, causing water levels to rise by tens of meters in a few hours. In effect, the groundwater troughs fill up, bringing flood levels up to the level of the diffuse water table or mound the water above it. Because the conduit system has little hydraulic resistance, the flood waters also drain quickly. The diffuse system, with a much lower hydraulic conductivity, responds more slowly to transient events and often does not remain in phase with the conduit system. Perched streams also occur in the unsaturated zone. There is no requirement that every free-surface stream seen in a cave be related to a local water table.

6.4 RUNOFF CHARACTERISTICS OF KARST BASINS

6.4.1 *Hydrographs*

Subsurface basins with conduit drainage trunks respond to rapid changes in recharge much like surface basins. The hydrograph measured at the outlet of the conduit system is a useful probe of the aquifer flow system.

Figure 6.16 gives the essential features of the hydrograph. The most instructive hydrographs are those of abrupt, intense storms that inject sharp pulses of water into the karst system. There is a lag while the input pulse is transmitted through the system, and the water does not start to rise at the spring until some time after the storm. The lag time is not the time required for storm water to flow from the inlet point to the spring, but rather the time necessary to transmit the impulse. Rising water levels in the upstream portion of the subsurface basin increase the hydrostatic head, which drives water from deeper storage and causes spring flow to increase. The discharge rises abruptly above base flow and reaches a crest, equivalent to the flood crest of a surface stream, Q_{max}. After the discharge crests, the flow from the spring begins to decrease. However, the recession as the water drains from temporary storage is slower than the rise time. In an ideal system, the receding limb of the hydrograph is an exponential curve that eventually brings the discharge back to base flow, if later storms do not intercede.

A system dominated by allogenic recharge and with a well-developed conduit system has a flashy response to storms, whereas the response of an aquifer system with mostly diffuse flow, poorly developed conduits, and little allogenic recharge is much more subdued. Figure 6.17 compares the response of two basins of nearly equal area

184 / *Geomorphology and hydrology of karst terrains*

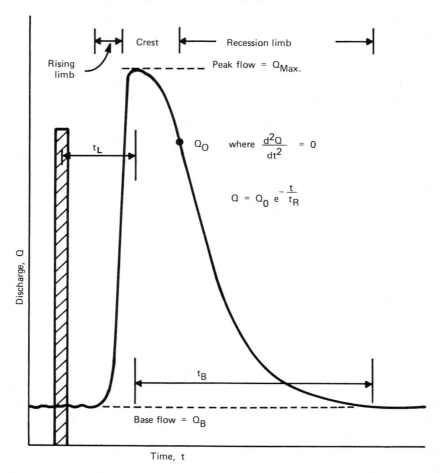

Figure 6.16 Single-storm hydrograph showing various features and numerical measures. The bar graph represents the storm pulse that triggers the hydrograph and starts the clock on the time axis.

to Hurricane Agnes, which swept over central Pennsylvania in 1972. The Rock Spring system is fed by runoff from mountain streams draining the ridges parallel to the carbonate valley. In spite of the huge flood discharge, Rock Spring returned to nearly base flow condition in 10 to 20 days. Thompson Spring obtains its recharge mostly from diffuse infiltration into a dolomite fracture aquifer. The flood crest was only about three times base flow, but the recovery required many months and discharge remained 50 to 75 percent above base flow throughout the late summer.

A measure of the "flashiness" of the response is given by the

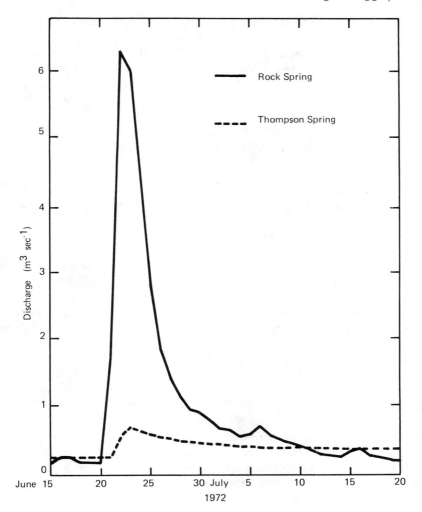

Figure 6.17 Hydrographs for Rock Spring (basin area 14.3 km^2) and Thompson Spring (basin area 11.2 km^2) of Centre County, Pennsylvania, in response to Hurricane Agnes of June 1972. During the storm, 250 mm of rain fell in Centre County over a 24-hour period. Discharge data are from Jacobson (1973).

ratio of peak discharge to base flow, Q_{max}/Q_B, for individual storms or for the annual peak discharge. Several time parameters can be extracted. The lag time, t_L, is the length of time between the storm pulse and the peak in the discharge hydrograph. It is difficult to measure accurately because of the relative rarity of intense, sharp impulse storms. The time for return to base flow, t_B, is a good mea-

186 / *Geomorphology and hydrology of karst terrains*

sure of aquifer response, but it is also difficult to measure because of the lack of precision in determining exactly when the discharge has dropped to base flow levels. Later precipitation events often distort the tail of the recession curve. The most useful and most easily obtained time parameter is the response time, t_R, which is determined by fitting the recession limb of the hydrograph to an exponential function.

The exponentially decaying recession limb has been described by several equivalent equations. Many surface water hydrologists use the equation

$$Q = Q_0 K^{-t} \tag{6.40}$$

(Lindsley et al., 1949), which has the odd property of making the characteristic recession constant the base of a system of logarithms. Burdon and Papakis (1963) use

$$Q = Q_0 e^{-\epsilon t} \tag{6.41}$$

where ϵ is called the exhaustion coefficient; $t_R = 1/\epsilon = \ln K$. Analysis of recession limbs often leads to two or more segments, a fast response due to conduit flow and a slow response due to diffuse flow.

Precipitation events are variable in intensity, spacing, and duration. If t_R is much less than the mean spacing of precipitation events, hydrographs will be flashy and the yearly record will consist of a sequence of sharp peaks. If t_R is on the same order as the mean spacing of precipitation events, individual storms will tend to overlap but seasonal changes in precipitation will appear. If t_R is much longer than the mean spacing of precipitation events, the hydrographs will be broad and relatively featureless. Large systems tend to have longer response times than small systems simply because of the longer times needed to transmit the storm impulse from input to output. Thus, short response times and high discharge ratios are indicators of conduit systems only for small basins. Table 6.7 gives some examples.

6.4.2 *Transmissivity and Storativity*

Although the hydraulic conductivity, K, describes both fluid and rock properties in response to hydraulic gradients, one often wishes to know the amount of water in storage and the ease with which the stored water can reach a well. The specific storage, S_s, of a saturated aquifer is the volume of water that a unit volume of the aquifer releases under a unit decline in hydraulic head. The pressure of water in the pore spaces stresses and expands the porous media aquifers, and the fluid itself is compressed to some degree. The specific storage, in units of m^{-1}, is a combination of these effects.

$$S_s = \rho g (\alpha + \Theta \beta) \tag{6.42}$$

Table 6.7
Discharge Parameters for Some Karst Springs

Spring	Q_{max}/Q_B	t_R (days)	Reference
Fast-response springs			
Rock Spring, PA	42	7.3	Jacobson (1973)
Penns Cave, PA	96	19	White and Stellmack (1968)
Davis Spring, WV	91	4.1	USGS
Intermediate-response springs			
Thompson Spring, PA	9.5	70	Jacobson (1973)
Tuscumbia Spring, AL	7.0	65	USGS
Aghia Eleousa, Greece	7.5	86	Aronis et al. (1961)
Slow-response springs			
San Marcos Springs, TX	1.6	—	Garza (1962)
Silver Spring, FL	1.5	—	Ferguson et al. (1947)
Ras-el-Ain, Syria	1.16	2070 (5.67 years)	Burdon and Safadi (1963)

where α is the compressibility of the aquifer, β is the compressibility of the fluid, and Θ is the porosity (defined in Eq. 6.1). These parameters are related to the aquifer by

$$T = Kb \qquad (6.43)$$

$$S = S_s b \qquad (6.44)$$

where T is transmissivity in square meters per second, S is storativity, dimensionless, and b is the thickness of confined aquifers, or the distance from the water table to the base of the aquifer for unconfined aquifers.

The transmissivity and storativity are usually evaluated in the field by pumping tests on wells (see Walton, 1962 or Freeze and Cherry, 1979). Aquifers with transmissivities greater than 0.015 m² sec⁻¹ are considered suitable for water supplies. Storativities are small in confined aquifers because the yield depends on release of compression of nearly incompressible rock and water. Much higher values, often called specific yields, are obtained for unconfined aquifers because removal of water from storage results in lowering of the water table, forming a cone of depression around the pumped well. The water removed from storage comes from actual dewatering of the pore spaces.

Application of these concepts to karst aquifers is fraught with difficulty. The well bore is a very small object on the scale of the heterogeneities of a karst aquifer, and one does not know what component of the subsurface drainage system is being probed. Values obtained from pump tests vary widely over short distances, depending on exactly where the wells are drilled. A well that taps a connection with the conduit system can produce very large quantities of water with negligible drawdown, leading to extremely large calculated transmissivities. A well drilled a few meters away in an unfractured block of limestone may have negligibly small yields. Pump tests on fracture aquifers are generally better, but still tend to range over several orders of magnitude. In general, pump test data are of marginal value in evaluating water resources in karstic aquifers, except for the diffuse flow part of the system.

Karst springs usually occur at the lowest elevation, the site of minimum head, in the karst system. As long as the natural discharge of the spring is maintained, water is being drained from what may be termed the dynamic storage of the aquifer. Measurements of transmissivity and storativity obtained from an analysis of spring discharge are characteristic of the aquifer as a whole and, to some extent, average over the heterogeneities of karst aquifers.

Burdon and Papakis (1963), Aronis et al. (1961), and Burdon and Safadi (1965) applied an approximate form of hydrograph recession analysis to the spring discharge in the arid karst of the countries around the Mediterranean Sea. Here the recharge to the aquifers is limited to a short period of the year. Most of the carbonate aquifers in these countries are of the large, slow-response type, so that only the seasonal trends in discharge are observed. The exhaustion coefficient, ϵ (days^{-1}), was obtained from

$$Q_i = \frac{Q_{max}}{(1 + \epsilon t)^2} \qquad (6.45)$$

where Q_i is the instantaneous discharge (in m^3 sec^{-1}), usually measured during minimum flow. The time, t, was taken as elapsed time (in days) from the time of maximum discharge or the cessation of surface runoff. The volume of water in dynamic storage is then given by

$$V_0 = \frac{86{,}400 Q_{max}}{\epsilon} \qquad (6.46)$$

where 86,400 is the number of seconds in a day.

Equation 6.46 is general and is obtained by noting that the instantaneous change of volume in dynamic storage is

$$-dV = Q_i dt \qquad (6.47)$$

Substituting Eq. 6.41 for Q_i and integrating to infinite time (which would result in zero discharge if the recession curve were not interrupted) leads immediately to Eq. 6.46.

If the aquifer has a sufficiently fast response that an exponential recession curve can be fitted, information on transmissivity and storativity can be obtained directly. Following Milanović (1981), assume one-dimensional circulation and laminar flow in an aquifer with constant horizontal cross section. D'Arcy's law and the continuity equation are in the form

$$Q = bT\frac{2z}{L} \tag{6.48}$$

$$Q = -SbL\frac{dz}{dt} \tag{6.49}$$

Differentiating Eq. 6.48 with respect to time and substituting into Eq. 6.49 gives

$$\frac{dQ}{Q} = -\frac{2T}{SL^2}dt \tag{6.50}$$

Integrating these gives

$$Q = Q_0 e^{-2T/SL^2} \tag{6.51}$$

from which the exhaustion coefficient is immediately identified as

$$\epsilon = \frac{2T}{SL^2} \tag{6.52}$$

where L is the horizontal distance from recharge point to discharge point.

Mijatović (1968) applied these procedures to the evaluation of the karst aquifers of the Dinaric karst and elsewhere. A considerably more elaborate analysis of discharge hydrographs from karst drainage including the effects of turbulent flow can be found in the works of Mangin (1970, 1971, 1973) and Bakalowicz and Mangin (1980).

6.4.3 Base Flow

Base flow in surface basins occurs during the dry seasons when there is no runoff and all flow in the stream channel is sustained by groundwater discharge. The same concept is relevant to subsurface basins. Spring base flow is sustained by the discharge of the diffuse flow system into the conduit system. Aquifers consisting of massive, low-porosity limestones with mainly conduit permeability have low base flows because little water is held in storage. Diffuse flow aquifers generally have base flows only slightly lower than the mean flows, because the response time for discharging water in storage is comparable to the time between wet and dry seasons.

Comparison of base flow intensity (base flow normalized to basin area) with area fraction of carbonate rocks for 57 Appalachian basins produced only a random scatter of points (E. L. White, 1977), although base flow intensities could be separated into two groups (Fig. 6.18). Basins with base flow intensity less than 0.002 m³ sec⁻¹

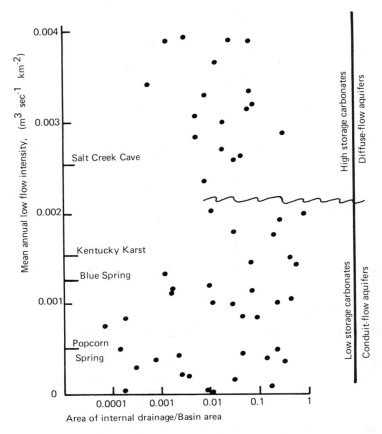

Figure 6.18 Comparison of base flow intensity with the fraction of the basin area that contributes concentrated input (allogenic recharge plus internal runoff). The choice of 0.002 m³ sec⁻¹ km⁻² as the division between high-storage and low-storage aquifers was based on statistical tests on the 57 basins. [Adapted from E. L. White (1977).]

km^{-2}, the low storage carbonate aquifers, were mostly those with well-developed conduit systems. Basins with base flow intensities greater than 0.002 m³ sec⁻¹ km⁻², the high-storage carbonate aquifers, were mainly diffuse flow systems.

6.4.4 Floods in Karstic Drainage Basins

The rapid response of the open conduit permeability serves to depress the crests of floods in associated surface basins. For a storm of a given intensity, the height of the flood crest in the surface channel depends on the efficiency of the surface channel system. If the stream channels are steep, and the arrangement of tributaries is such that their individual flood crests arrive at the main channel at the same time, the hydrograph of the main channel is marked by a very high peak flow with a rapid recession as the flood sweeps by. Low infiltration rates resulting from thin soils and impermeable bedrock enhance the flood peaks.

In karstic basins, many of the tributary streams terminate in swallow holes. The peak runoff from the tributaries flows into the swallow holes and is placed in temporary storage in the conduit sys-

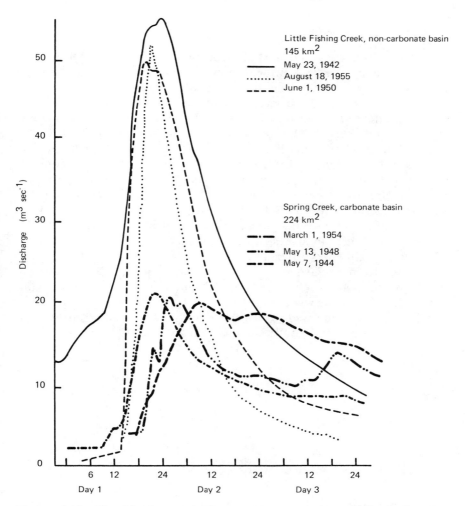

Figure 6.19 Flood hydrographs for two streams in central Pennsylvania. Fishing Creek has most of its catchment on clastic rock; Spring Creek derives most of its flow from a limestone and dolomite-floored valley. [Adapted from White and Reich (1970).]

tem. Although the conduit storage returns rather rapidly to the surface, creating the flood peak of the spring, the response times of the conduit systems are on the order of a few days, whereas the response times of the flood-prone surface basin may be only a matter of hours. Although the total runoff, the area under the hydrograph, is the same, the effect of temporary karst storage is to spread out the hydrograph. The peak flow is diminished and the recession tail is stretched out (Fig. 6.19) (White and Reich, 1970; E. L. White, 1976). In general karstic drainage basins should have less flashy floods than nonkarstic basins.

6.4.5 Flood Pulse Analysis

The flood discharge hydrograph from a karst spring or a karstic drainage basin contains information beyond what can be gleaned from an analysis of the recession limb. Lag time, return time, and overall shape of the hydrograph relate to both the physical layout of the drainage system and the input hydrograph of the allogenic recharge and storm recharge onto the karst surface. If both input and output hydrographs are known, it is possible, in principle, to extract information on the interior drainage system. The conduit system is taken as a transformation function that takes the input hydrograph into the output hydrograph. The principles have been set forth (Ashton, 1966; Wilcock, 1968), but a good deconvolution procedure for the output hydrograph has not been laid down. High-quality data for input storms and output floods on basins or aquifers with well-established physical flow paths seem to be rare.

Rather than precise mathematical mapping and deconvolution functions, one can also use cross-correlation and autocorrelation statistical analyses between input and output hydrographs. Brown (1972a) used this method to correlate discharge through the subsurface routes of the Maligne River Basin in Alberta, Canada.

7
Geochemistry of Karst Waters

7.1 CARBON DIOXIDE

Karst processes depend on carbon dioxide as the primary chemical driving force. The mass flux of CO_2 through the karst system is equally important as the water flux and the sediment flux. The ultimate sources of CO_2 are the atmosphere and various biological processes in the soil.

7.1.1 *Carbon Dioxide in the Atmosphere*

The concentration of CO_2 in the earth's atmosphere is close to 340 ppm by volume, corresponding to a CO_2 partial pressure of $10^{-3.47}$ atm. There has been a systematic increase in world CO_2 levels dating at least from the beginning of the Industrial Revolution (Fig. 7.1). Burning fossil fuels transforms carbon, long stored as coal and petroleum, to atmospheric CO_2. There is concern that increasing CO_2 levels will produce a greenhouse effect, which will raise global temperatures, melt arctic and antarctic ice, and raise the world's sea levels; however, there is no consensus on either the magnitude of the effects or the time scale on which they would occur (Clark, 1982). Superimposed on the trend line of Fig. 7.1 are annual oscillations from the seasonal drawdown of the atmospheric CO_2 reservoir by photosynthesis in the temperate climate zone.

7.1.2 *Carbon Dioxide in Soils*

Soil carbon dioxide is produced in the O and A horizons by respiration of the plant root system and by microbiologically mediated decomposition of organic debris. Carbon dioxide production varies with plant activity, soil moisture, soil temperature, and influx of

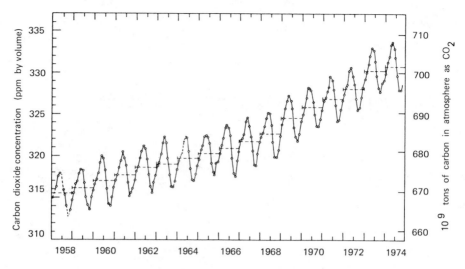

Figure 7.1 Carbon dioxide concentrations in the atmosphere about Mauna Loa Observatory in Hawaii, from 1958 to 1974. Adapted from Baes et al. (1976). For additional data, see Keeling et al. (1982).

organic material. The effective CO_2 concentration depends on the balance between production and loss upward to the atmosphere and loss downward as CO_2 dissolves in infiltration water. Carbon dioxide buildup is further controlled by the physical characteristics of the soils. Loose, homogeneous, sandy soils permit easy loss of CO_2 to the atmosphere, whereas impermeable clay-rich soils trap CO_2 and allow concentrations to build up. Measurements of carbon dioxide in soils vary from 0.03 volume percent, the atmospheric value, to 10 percent (Fig. 7.2). The variation of CO_2 concentration with climatic regime is masked by the variations of plant cover, soil moisture, and soil characteristics within any given sampling locality.

In temperate and arctic regions, where there is a well-defined growing season, carbon dioxide levels in the soil vary with the season by a factor of 10 to 100. Rightmire (1978), who followed seasonal changes in soil CO_2 near Washington, DC, found nearly constant low CO_2 levels from October through May. At the end of May, CO_2 concentrations rose rapidly, remained high throughout the summer, and decreased rapidly to background levels in October. Peak summer CO_2 pressures were above $10^{-2.0}$, whereas the winter levels were near $10^{-3.5}$. Rossi (1974) also found a distinct seasonal variation in karst soils in Madagascar. Here, soil moisture variations rather than temperature variations are probably responsible.

Quantitative prediction of soil CO_2 levels is difficult because of

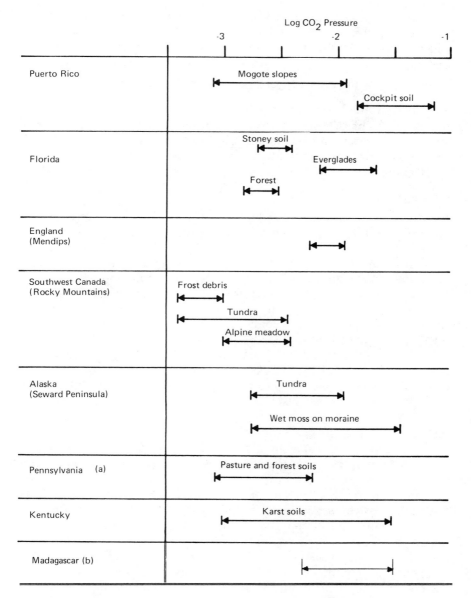

Figure 7.2 Ranges of CO_2 concentration in various soil atmospheres. All data from Miotke (1974, 1975), except (a) unpublished data of author and (b) Rossi (1974).

the large number of variables and their rapid variation in space and time. Recent research has been grappling with this problem. Crowther (1984) examined the tropical karst soils of Malaysia in some detail and found that the most important local variables were soil density and soil depth. These were described by the regression equation

$$\log V_{CO_2} = 1.146\rho_{soil} + 0.00698d_{soil} - 1.227 \tag{7.1}$$

which accounts for 86 percent of the variation. V_{CO_2} is the maximum volume percent CO_2, ρ_{soil} is the bulk density of the soil in grams per cubic centimeter, and d_{soil} is soil depth in centimeters. The highest CO_2 concentrations occur during long periods of wet weather, when high levels of soil moisture stimulate microbiological activity.

7.1.3 Carbon Dioxide in Cave Atmospheres

Relatively few data are available on the CO_2 levels in cave atmospheres. Miotke (1974) measured some locations in the Flint Ridge Cave System, mostly in the winter of 1971–72, and found values ranging from 0.04 to 0.15 volume percent. Ek et al. (1968, 1969) compiled systematic analyses for two caves in Belgium and five caves in Poland. There is a pronounced gradient even in these short caves, with CO_2 pressures increasing between the entrance and the back of the caves. Carbon dioxide concentrations are stratified with higher levels near the ceiling, especially in high alcoves and short side passages. Because CO_2 is denser than air, this inverted profile could be maintained only if CO_2 was continuously infiltrating through the roofs of the passages. CO_2 concentrations varied from 0.03 volume percent at the entrances to 0.3 volume percent in the high alcoves.

Carbon dioxide in cave atmospheres oscillates in an annual cycle just as the CO_2 in the soil atmosphere does. Figure 7.3 shows the CO_2 concentrations in a segment of cave where there is little exchange with the outside atmosphere. The seasonal variation is about an order of magnitude. Like the maximum in soil carbon dioxide, the maximum in the CO_2 concentration in the cave atmosphere coincides with the growing season but is slightly offset. The CO_2 level begins to rise in May and June, when plant growth and foliation are well underway. It reaches a maximum in September and October, when growth is slowing but introduction of organic matter, mainly leaf litter, is at a maximum. By December, with the full onset of winter and the freezing of the ground, CO_2 has dropped to background levels.

7.1.4 Geographical Variations in CO_2 Production

The most ambitious attempt to model CO_2 production is the study of Brook et al. (1983), who took available soil CO_2 data from 19 loca-

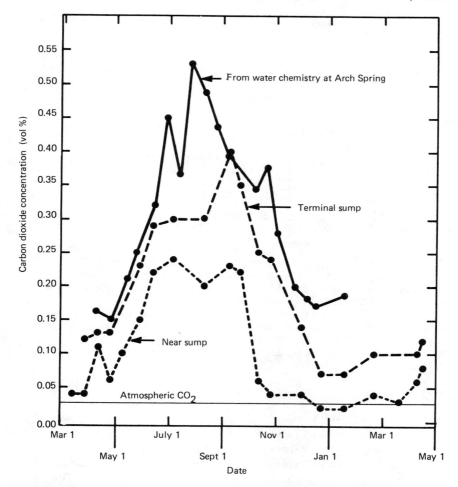

Figure 7.3 Carbon dioxide concentrations in Tytoona Cave in Pennsylvania. See Troester and White (1984) for further description of the sampling site.

tions throughout the world. Considering temperature, precipitation, potential evapotranspiration, and actual evapotranspiration as controlling variables, and using regression analysis, they decided that actual evapotranspiration best accounted for the observed variability. No attempt was made to include specific soil characteristics. Figure 7.4 shows the correlation, which was reduced to the regression equation

$$\log P_{CO_2} = -3.47 + 2.09(1 - e^{-0.00172 \text{ AET}}) \tag{7.2}$$

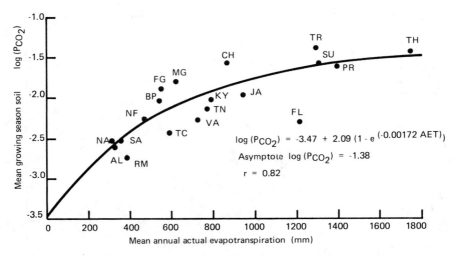

Figure 7.4 Relationship between P$_{CO_2}$ and the actual evapotranspiration for various areas in the world, according to the results of Brook et al. (1983). Location keys: NA = Nahanni karst, Canada; SA = Saskatchewan, Canada; RM = Rocky Mountains, Canada; NF = Newfoundland; BP = Bruce Peninsula, Canada; TC = Trout Creek, Ontario; AL = Alaska; VA = Reston, Virginia; TN = Sinking Cove, Tennessee; KY = Mammoth Cave, Kentucky; FL = south Florida; FG = Frankfurt-Main, West Germany; MG = Müllenbach, West Germany; JA = Jamaica; TR = Trinidad; PR = Puerto Rico; CH = Yunan, China; SU = Sulawesi; TH = Phangnga, Thailand.

where AET is actual evapotranspiration in millimeters per year. The CO_2 pressures are taken as those at the peak of the growing season.

Because the Brook-Folkoff-Box model uses widely available climatic data, the calculation can be modeled on a global scale (Fig. 7.5). The result is intriguing. The high CO_2 regions include the eastern and central United States, where many of the world's longest caves are located, the Caribbean limestone islands, with their extensive cone and tower karst, the karst regions of Mexico and Central America, the karst of southeast Asia, including south China, and the limestone islands of the South Pacific. Other regions of high CO_2 in central Africa and South America coincide with the igneous rocks of the continental shields. Referring to the "karst coordinates" discussed in Chapter 4 (Fig. 4.1), the global model certainly agrees in a general way with the requirements for intensive karst development.

7.1.5 *Other Sources of Carbon Dioxide*

It remains an open question whether the carbon dioxide cycle described in the previous sections is complete. Atkinson (1977b) compared soil CO_2 levels of the Mendip karst with CO_2 levels calcu-

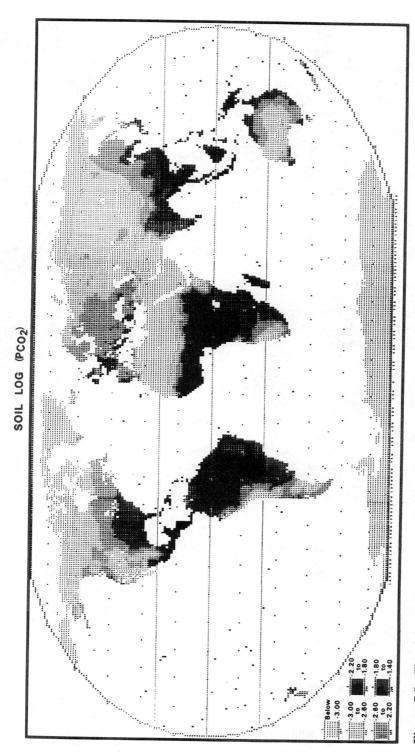

Figure 7.5 The global pattern of mean growing season carbon dioxide pressure predicted by Eq. 7.2. Intervals are indicated as log P_{CO_2} in atmospheres. From Brook, Folkoff, and Box (1983); courtesy of Dr. Brook.

lated from the chemistry of the water and found that usually P_{CO_2} in the water was greater than P_{CO_2} of the soils. He hypothesized that the CO_2 deficit was made up by the oxidation of organic matter swept into the aquifer. Much of the oxidation was thought to take place in the vadose zone, where free oxygen is available. This hypothesis is supported by the very detailed measurements on the changes in gas composition with depth in the Ogallala (noncarbonate) aquifer of the southern High Plains (Wood and Petraitis, 1984). Carbon dioxide pressures continued to increase to depths of 35 m at the same time the oxygen pressure decreased. Wood and Petraitis examined various alternative hypotheses and concluded that oxidation of infiltrated organic matter provided the best explanation for both chemical and isotopic data.

Bad-air caves are not common but they do occur. Some, such as the Zbrasov Caves in Czechoslovakia, contain high concentrations of CO_2 fed by underground hot springs. More commonly, cave passages that trap in-washed organic material and have poor air circulation build up high CO_2 concentrations. The exploration of Mystery Cave, Missouri, was hampered by high CO_2 levels from decaying logs and organic-rich muds that had built up behind a sump (Walsh, 1972). The Bungonia Caves in Australia (James, 1977) contain some of the highest CO_2 levels reported in cave atmospheres, reaching the 5 volume percent level in Grill Cave, Drum Cave, and Putrid Pit. James ascribes the high CO_2 levels to microbiologically enhanced decay of organic matter under moist conditions at the exceptionally high temperature of 18°C. High CO_2 levels can be a hazard to cave exploration. A few percentage points of CO_2 will bring on headaches and an itchy acid taste in the mouth and mucous membranes. Five percent CO_2 induces intense headache and very heavy breathing. Concentrations between 5 and 10 percent result in unconsciousness and eventually death (James et al., 1975).

7.2 HYDROCHEMISTRY OF KARST AQUIFERS

7.2.1 *Chemical Characteristics of Karst Waters*

The flow system sketched earlier (Fig. 6.13) offers many opportunities for the analysis of karst water and thus for following the evolution of aquifer chemistry. The parameters that have proved of most interest (defined in Chapter 5) are hardness, saturation indices of calcite and dolomite, and the theoretical carbon dioxide partial pressure. Table 7.1 gives some averaged chemical data for a collection of water analyses taken mostly from Pennsylvania.

The most obvious general feature in Table 7.1 is the variability in the chemistry of the different types of water. Drake and Harmon's (1973) discriminant analysis of a similar set of data showed that the

Table 7.1
Chemical Characteristics of Karst Waters from the North Central Appalachian Mountains

	Hardness (ppm $CaCO_3$)	SI_c	log P_{CO_2}	$\dfrac{P_{CO_2}}{P_{CO_2} \text{ (atm)}}$	Number of samples
Surface recharge	20	−2.66	−2.73	6.2	25
Soil water	35	−2.58	−0.97	357	11
Internal runoff	82	−0.76	−2.48	11	78
Drip water (Diffuse Infiltration)	145	+0.61	−2.37	14	12
Conduit springs	112	−0.89	−2.38	14	24
Diffuse flow springs	200	−0.24	−2.22	20	46
Wells	288	−0.13	−2.00	33	52

Data from Harmon et al. (1973) and Drake and Harmon (1973)

water types were statistically distinguishable using the saturation index and CO_2 partial pressures as variables.

Two kinds of recharge to the karst aquifer are represented. Both allogenic recharge (sinking streams) and soil waters are highly undersaturated, as they should be since the water has not yet contacted the carbonate rocks, but the soil waters contain much higher levels of carbon dioxide. To give a better sense of scale, the ratio of CO_2 pressure in the water to that of the surface atmosphere is also listed. This gives, in effect, a measure of the enhancement of CO_2 above atmospheric background. The distinction made previously between internal runoff and diffuse infiltration is strongly reflected in the chemistry. Internal runoff waters [Thrailkill's (1968a) "vertical flows"] are strongly undersaturated where sampled in the underlying cave passages. Drip waters are supersaturated, as they must be to deposit calcite. However, the numerical values in Table 7.1 may be exaggerated. Carbon dioxide can be lost during the time it takes to accumulate enough drip water for measurement, thus raising the calculated supersaturation. The CO_2 pressures in the two kinds of water are rather similar, as would be expected since they have ultimately the same source on the land surface. However, CO_2 pressure in both kinds of vertically moving water is much lower than that in the soil waters. Only a small portion of the CO_2 generated in the soil appears actually to be transported into the karst aquifer. Much must be lost by upward diffusion into the atmosphere.

Water is discharged from the aquifer through springs and wells.

All of these waters have a high hardness, but the hardness of the diffuse flow springs and the wells is about twice that of the conduit springs. All three types of water remain undersaturated with calcite. Conduit springs are highly undersaturated, as might be expected from the rapid throughput and open conduit system. Diffuse flow springs and wells are more nearly saturated, also as expected from the much longer residence times. Carbon dioxide pressure increases from conduit springs to diffuse flow springs to wells, but the contrast is only a factor of 2. The mix of diffuse infiltration water increases in the same order, again suggesting that CO_2 pressures measured on drip waters are too low because of degassing during water collection and measurement.

7.2.2 Relation of Water Hardness to CO_2 Availability

The actual concentration of dissolved carbonates relates closely to the availability of CO_2. The relation of hardness to CO_2 is illustrated with the anticlinal limestone valleys of central Pennsylvania (Fig. 7.6). The three distinct aquifers are the Cambrian Gatesburg dolomite that crops out on the anticlinal axis, a mixed sequence of dolomites and dolomitic limestones grouped with the dominant Beekmantown dolomite, which forms broad bands on both sides of the anticlinal axis, and finally 500 m of upper Ordovician low-magnesium limestones that form narrow bands near the base of the mountain ridges. The Gatesburg is sandy, porous, and vuggy and is overlain by thick sandy soils. The Beekmantown and associated rocks are massive, fractured carbonates, with loamy soils of intermediate thickness. The Champlainian limestones contain most of the caves in the valley and most of the surface karst, including the swallow holes of many small allogenic streams that head on the surrounding mountains. Springs and wells occur in all three aquifers.

The observed average hardness varies from a little above 100 ppm in the Gatesburg springs and wells and the limestone conduit springs to a little above 300 ppm in the Beekmantown dolomite wells and limestone wells. Allowing for the artifact introduced by expressing dolomite waters as $CaCO_3$ hardness, it can be seen that the dissolved carbonates in these aquifers increase with increasing CO_2 partial pressure and that most are near saturation. The limestone conduit springs lie off the trend because most of these waters are highly undersaturated with respect to both calcite and dolomite. The range in observed concentrations is higher for the wells than for the springs because the data shown are spatial averages of one sample each taken from many wells, whereas the data for the springs are averages of many samples taken at different times from a few springs.

The Gatesburg soils are barren, with little organic matter and

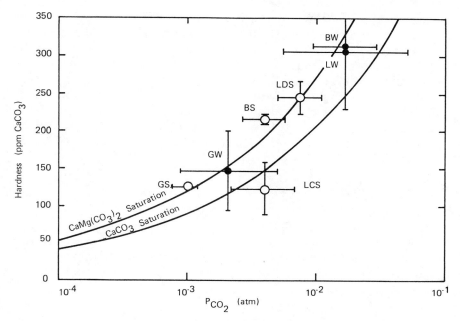

Figure 7.6 Calcium plus magnesium hardness expressed as $CaCO_3$ for various wells and springs in central Pennsylvania, as a function of carbon dioxide partial pressure. Points represent arithmetic means of hardness and geometric means of P_{CO_2}; range bars represent 1 standard deviation on each side of the mean. Code for localities: GS = Gatesburg Springs; GW = Gatesburg wells; LCS = limestone conduit springs; BS = Beekmantown springs; LDS = limestone diffuse flow springs; LW = limestone wells; BW = Beekmantown wells. Well data are from Jacobson (1973); spring data are from Shuster and White (1971).

sparse vegetative cover, which is strongly reflected in their groundwater chemistry. The Beekmantown dolomite soils provide some of the best farm land in central Pennsylvania, and the Champlainian limestones are similar. Higher CO_2 concentrations are found in the wells because these are fed by the more diffuse infiltration part of the groundwater flow system.

7.2.3 Chemical Evolution Through the Flow System

There has been much discussion about whether carbonate aquifers are open or closed systems. An open system is one in which the CO_2 in solution in the groundwater remains in equilibrium with the gas phase in the vadose zone. In open systems, the CO_2 consumed in the process of dissolving the limestone or dolomite wall rock is continuously replenished from the gas reservoir. The reservoir itself is replenished from CO_2 generated in the overlying soil. In contrast, a closed system is one in which water enters the aquifer with a certain quantity of dissolved CO_2, which is then gradually consumed as the water reacts with carbonate rock while moving along the flow path. A convenient model that distinguishes the two cases was devised by Langmuir (1971) (Fig. 7.7) using the directly measured parameters of pH and bicarbonate ion concentration. By using bicarbonate concentration rather than hardness, one avoids the problem of artifacts created by expressing dissolved Mg^{2+} in dolomite aquifers as $CaCO_3$. If the waters are exactly at saturation, the relations between bicarbonate concentration and pH are given by the equations

$$-\log m_{\text{HCO3}} = \tfrac{1}{2}[\text{pH} - \log\left(\frac{2K_c}{K_2}\right) + \log \gamma_{\text{Ca}}\gamma_{\text{HCO3}}] \tag{7.3}$$

$$-\log m_{\text{HCO3}} = \tfrac{1}{2}[\text{pH} - \log\left(\frac{4K_d^{1/2}}{K_2}\right) + \tfrac{1}{2} \log \gamma_{\text{Ca}}\gamma_{\text{Mg}}\gamma_{\text{HCO3}}^2] \tag{7.4}$$

where the symbols have the meanings defined in Chapter 5. If the system is open, the uptake of HCO_3^- as the system approaches saturation is described by

$$m_{\text{HCO3}} = \frac{K_{\text{CO}_2}K_1 P_{\text{CO}_2} 10^{\text{pH}}}{\gamma_{\text{HCO3}}} \tag{7.5}$$

Closed systems reach equilibrium at much lower HCO_3^- concentrations, and the waters follow the curved path given by

$$m_{\text{HCO3}} (2 \times 10^{-\text{pH}} + K_1) + \text{constant} \tag{7.6}$$

The constant can be determined numerically by the initial pH/HCO_3^- conditions that prevail where the water enters the closed portion of its flow path.

The definition of theoretical CO_2 pressure (Eq. 5.43) is a de facto assumption of open system conditions. It calculates the carbon dioxide partial pressure of the gas phase which would be in equilibrium with the water being analyzed. If the aquifer is a closed system, much of the original CO_2 would be consumed in reaction with the limestone, the pH/HCO_3^- conditions would drift to the right following the appropriate closed-system arrow in Fig. 7.7, and the calculated P_{CO_2} would be that of an open system with a much lower CO_2 pressure. It is not possible, from spring data alone, to distinguish between outlet points for closed-system aquifers with high CO_2 concentrations in the recharge area and open-system aquifers with low CO_2 concentrations in the recharge area.

Waters can be sampled along their flow path through the aquifer and their pH and HCO_3^- concentration plotted on Fig. 7.7. If the system is open, the data will follow a straight line appropriate to the CO_2 pressure of the gas reservoir. If the system is closed, the data should follow the closed system lines, which curve to the right from

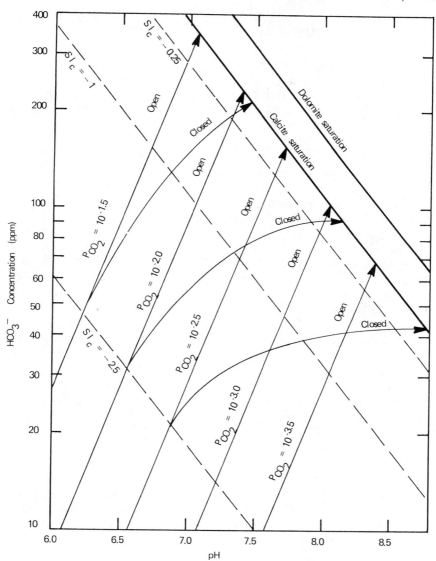

Figure 7.7 Pathways to equilibrium for undersaturated karst waters from various initial CO_2 pressures. The saturation lines are described by Eqs. 7.3 and 7.4; the open system by Eq. 7.5, and the closed system by Eq. 7.6.

the initial CO_2 partial pressure. In practice, the test is not easy to apply, first because of the difficulty in obtaining samples at intermediate points along the flow paths, and second because mixing waters from different sources with different initial CO_2 concentrations confuses the interpretation. Langmuir's (1971) original study of the carbonate groundwaters of central Pennsylvania concluded that these aquifers behaved as open systems. Harmon and Hess (1982) analyzed waters along the flow path through the Butler Cave–Sinking Creek System in westcentral Virginia and found strong evidence for an open system. On the other hand, Pitman (1978) showed evidence that the drainage from the tower karst of south Thailand flows through closed pathways from the recharge area to the springs.

7.2.4 Time Variations in Saturation and Hardness

Water moving through a diffuse flow system, in which the pathways are solutionally widened fractures and bedding plane partings, has time to come into chemical and thermal equilibrium with the bedrock. As a result the water discharging from diffuse flow springs varies little with the season or with discharge. Water moving through conduit systems, in contrast, has a residence time of hours or days, which is insufficient for either thermal or chemical equilibrium. As a result, waters discharging from conduit springs have highly variable chemistry (Fig. 7.8). Diffuse flow springs are nearly at saturation with calcite whereas conduit flow springs are usually undersaturated.

The variations in hardness show a pronounced seasonal effect, with higher values in the summer and lower values in the winter. Some of the variation may relate to the seasonal peak in available carbon dioxide, but much of it seems to result from dilution by a higher runoff during winter and spring. Bassett and Ruhe's (1973) study of the Lost River watershed in southern Indiana showed that hardness is directly related to discharge (Fig. 7.9).

$$Hd = 402.7 - 113.2 \log Q \qquad r^2 = 0.81 \qquad (7.7)$$

where Q is discharge in cubic feet per second and Hd is the hardness in ppm of $CaCO_3$. Bakalowicz (1976) found an even more precise match between hardness variations and the detailed fine structure of a discharge hydrograph. Simple dilution accounts for a substantial part of the hardness variations in conduit springs.

Shuster and White (1971) argued that the coefficient of variation of the hardness was a diagnostic criterion for diffuse or conduit flow. They claimed, based on examination of some 14 springs in central Pennsylvania, that a coefficient of variation greater than 10 percent indicated a conduit spring whereas diffuse flow springs generally had a coefficient of variation of less than 5 percent. Results of classification on the basis of chemistry were in general agreement with the known hydrogeology of the springs. Earlier results by Pitty (1968, 1971) related the overall carbonate content of springs in the English Pennines to the flowthrough time of the karst aquifer. Ter-

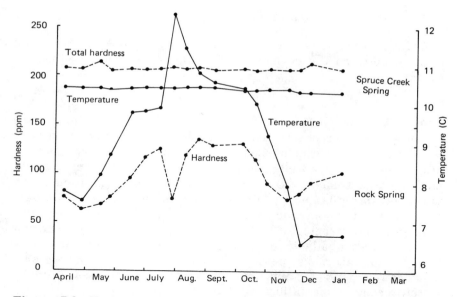

Figure 7.8 Variations in hardness and temperature for Spruce Creek Spring, a diffuse flow system, and Rock Spring, a conduit flow system. Data are from Shuster (1970).

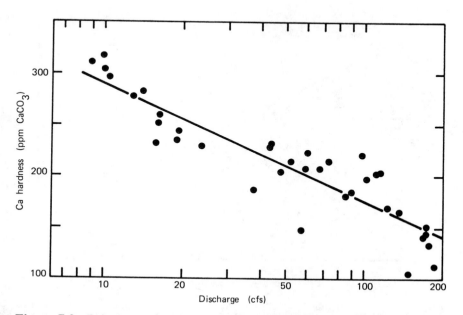

Figure 7.9 Relation of hardness to discharge for the Orangeville Rise in the Lost River drainage basin of southern Indiana. [Adapted from Bassett and Ruhe (1973).]

nan (1972) combined these ideas (Fig. 7.10) and showed a power law relation between the coefficient of variation and the flowthrough time:

$$CV = 35t^{-0.524} \qquad (7.8)$$

where CV is the coefficient of variation (standard deviation/mean) expressed as a percentage, and t is the flowthrough time, in days.

The characterization of the type of flow system by hardness variations seems to work well for small drainage basins in temperate climates. Large basins (>100 km^2) showed smaller variations even when the spring is known to be fed by conduits, because of longer travel times, the contribution from the diffuse flow part of the system, and the averaging of water chemistries from different parts of the basin. Data available for tropical karst suggest that these waters have lower variations, possibly because the higher temperatures enhance reaction rates and thus shorten the time necessary for the water to come into equilibrium with the bedrock.

7.2.5 Aquifer Response to Storms

Because hardness can be calculated from conductivity (Section 5.4.1) and conductivity can be measured continuously, it is possible to con-

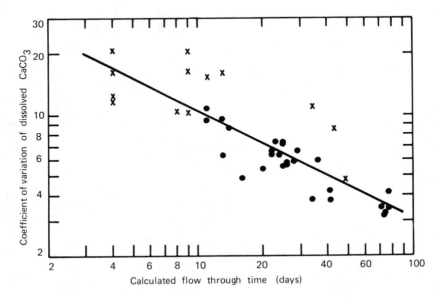

Figure 7.10 Relation of the coefficient of variation in hardness for various springs in the Pennine Hills, England, to calculated flowthrough time. (x) = Fountain's Fell springs; ($\bullet$) = High Mark springs; r^2 for the least-squares log–log fit = 0.84. [Adapted from Ternan (1972).]

Geochemistry of karst waters / 209

struct "chemical hydrographs" that carry additional information on aquifer behavior. The aquifer is treated as a black box, with a storm pulse as input and either discharge or some chemical parameter as output. The physical response makes up the subject of flood pulse analysis discussed in Chapter 6. Relatively little use has been made of the chemical response. Figure 7.11 shows two chemical hydrographs from the Turnhole Spring drainage basin in the Mammoth Cave area. The upper hydrograph corresponds to a period in which

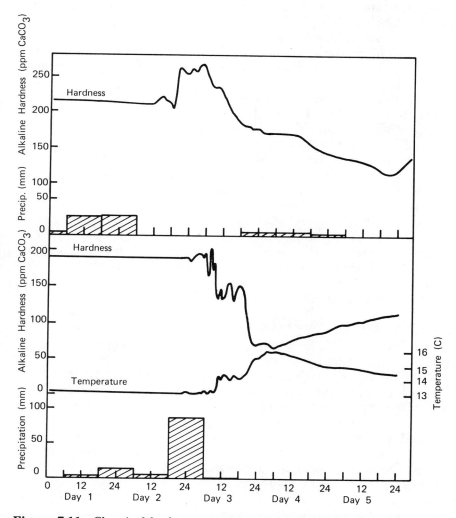

Figure 7.11 Chemical hydrographs for two storm pulses in the Turnhole Spring drainage basin in Mammoth Cave National Park, Kentucky. The bar graphs show precipitation. [Adapted from Hess and White (1974).]

the antecedent recharge had been very small. The ground was dry and flow from the springs was low. There came an extended period of rain, which provided a lot of water to the aquifer over a period of several days. The hardness of the water was initially high and remained high during low flow conditions. After a lag of only a few hours, the hardness observed at the Owl Cave sampling site increased by approximately 20 percent, reached a broad maximum with some ill-defined structure, and then began to decrease. The hardness decreased over a three-day period, after which it began to rise again and eventually recovered its prestorm value after two weeks. The leading pulse of increased hardness appears because water stored in the diffuse part of the aquifer is flushed out by increased hydrostatic head in the recharge area. The pressure pulse resulting from increased recharge moves through the aquifer much faster than the water itself.

The lower hydrograph was initiated by a very sharp and intense storm pulse in which 75 mm of rain fell within a period of 12 hours. There was no significant rise in hardness above the normal background. Approximately 12 hours after the input pulse, the hardness began to decrease and went through a complex series of minima and maxima over a period of about 24 hours. A total of seven peaks in the hardness curve, mimicked with reduced amplitude by the temperature curve, was observed before the hardness fell to the minimum value only 30 hours after the rainfall pulse. Again, a long gradual recovery to initial hardness levels followed over a period of two weeks. The fine structure in the chemical hydrograph was interpreted as the arrival of water from various tributaries into the main drainage conduit. Sharp, well-defined rainfall pulses are required to bring out the fine structure.

7.2.6 Seasonal Variations in Carbon Dioxide

The seasonal oscillation in CO_2 partial pressure observed in soils and cave atmospheres also appears in the chemistry of karst waters (Fig. 7.12). The use of karst springs as sampling points has the advantage of averaging the details that cause variations in P_{CO_2} of water following different flow paths through the aquifer. The CO_2 content of most karst waters is generally independent of discharge and is not closely related to dilution and flushing-out effects. Unlike the high variability in hardness in conduit-fed springs, the calculated P_{CO_2} tends to reflect mainly the seasonal variations. Indeed smoother P_{CO_2} time curves were obtained from conduit springs than from diffuse springs. Diffuse flow springs have residence times that are long with respect to the spacing between the seasonal maxima and minima; thus, P_{CO_2} calculated from the water chemistry may not keep up with the seasonal cycles.

Geochemistry of karst waters / 211

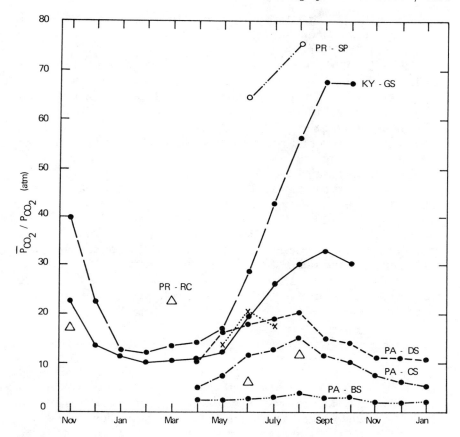

Figure 7.12 Carbon dioxide enrichment in karst springs. The data were averaged within each month when more than one observation was available. PR-SP = Puerto Rico, San Pedro Spring; PR-RC = Puerto Rico, Rio Camuy Resurgance (Troester and White, 1985). KY-GS = Kentucky, Graham Spring; KY-TS = Kentucky, Turnhole Spring (Hess, 1974). PA-DS = Pennsylvania, average data for three diffuse flow springs; PA-CS = average data for three conduit flow springs; PA-BS = Pennsylvania, Big Spring, discharging from Gatesburg dolomite aquifer (Shuster, 1970). B-CS = Belize, Boiling Hole Resurgence of Caves Branch System (Miller, 1981).

Figure 7.12 shows data from several karst regions. The springs in Pennsylvania are from the same set used in Fig. 7.6. The maximum in dissolved carbon dioxide occurs in August, about the middle of the growing season, for both diffuse flow springs and conduit-fed springs. The uncharacteristically low CO_2 content of Big Spring, like its unusually low hardness, is explained as a result of low CO_2 pressures in the source area, although a closed system along deep flow paths would also contribute. The Kentucky examples are large regional springs; Graham Spring, which shows the greatest seasonal oscillation, obtains much of its recharge from diffuse infiltration and internal runoff from the Sinkhole Plain in the southcentral Kentucky karst. The Turnhole Spring obtains recharge from sinking streams and from runoff of sandstone-capped ridges, which provide lower CO_2 concentrations. The tropical springs have CO_2 levels comparable to those of springs in temperate climates. The Boiling Hole resurgence in Belize and the Rio Camuy resurgence in Puerto Rico both drain large underground river systems. Open channel flow in large cave passages promotes CO_2 degassing and lowers the CO_2 pressure in the water. The San Pedro Spring in Puerto Rico, which emerges from a rise pool at the contact between the limestone and an alluvial valley fill, may be more typical of what is expected from diffuse infiltration and internal runoff in tropical karst.

The explanation for the seasonal oscillation is that the water is reflecting CO_2 generation in the soil, which in turn depends on the growing season and plant and microorganism activity. If this is correct, seasonal oscillations should not occur in the tropics, where the growing season is continuous. The sparse data available support the hypothesis.

7.3 GEOGRAPHICAL VARIATIONS IN KARST WATER CHEMISTRY

Karst geomorphologists have long sought explanations for the varied karst landforms in terms of the chemistry of carbonate rock dissolution. The task has proved much more difficult than was anticipated. A pavement karst in a northern climate and a cone and tower karst in a tropical climate are dramatically different landforms. The chemical differences are much less obvious. To a first order the chemistry of karst waters is controlled by the geological setting (independent of time). On the constant background are superimposed short-term stochastic fluctuations resulting from hydrologic factors, cyclical fluctuations imposed by annual temperature cycles, growing season, and seasonal patterns in precipitation. Whatever is left may be due to regional climatic effects.

7.3.1 *Regional Controls on Karst Water Chemistry*

Historically, hardness has been the parameter measured, and the literature contains many attempts to correlate hardness with climatic and other variables. Unfortunately for these rather naive correlations, the measured hardness depends on the interaction of the largest number of variables including the following:

1. The availability of CO_2 in the recharge area, as indicated by Fig. 7.6.

2. Whether the flow system acts as an open or a closed system. This, in turn, depends on many details of the local geology.
3. The state of saturation of the water when it emerges from the spring. This depends on the stage of development of the conduit system.
4. The proportions of allogenic recharge, diffuse infiltration, and internal runoff that mix within the aquifer.
5. Details of bedrock geology, including limestone/dolomite ratios, lithology, and the presence of gypsum or other interbedded soluble rock.

The Ca/Mg ratio depends almost entirely on the limestone/dolomite ratio in the aquifer. Springs issuing from dolomite have a Ca/Mg ratio of 1, whereas limestone springs have Ca/Mg ratios in the range of 3 to 10. The Ca/Mg ratio for mixed rock sequences ranges from 1.5 to 3. However, the travertines deposited in caves, including caves in dolomite, are generally Mg-poor calcites. Thus, the diffuse infiltration waters may be enriched in Mg^{2+} if they migrate through cave passages.

The calcite saturation index depends almost entirely on residence time. Open conduit systems permit rapid transmission of water, undersaturated waters appear at the springs, and the system is very sensitive to flushing by storm runoff. The laboratory studies of calcite dissolution kinetics show that times up to ten days are required for reactions to reach equilibrium. Aquifers with residence times longer than this are likely to discharge water with saturation indices above -0.3. Dolomite waters, in contrast, may take months or years to reach equilibrium, and waters undersaturated with dolomite are common.

7.3.2 Climatic Controls on Carbon Dioxide in Karst Water

The carbon dioxide in karst water is controlled by CO_2 production in the near-surface environment. Near-surface processes are strongly influenced by climatic variables (or one could say that CO_2 production *is* a climatic variable). Soil character, vegetative cover, growing season, winter snow pack, and the quantity and distribution of rainfall are all part of the melange of variables that might be considered part of the climatic setting.

The requirements for good P_{CO_2} calculations are high-quality analyses of bicarbonate ion, temperature, total dissolved solids, and precise, field-measured pH. Such data were rarely available before about 1970, although there was an abundance of hardness data. By 1975 a considerable data base had been assembled with water analyses from many locations in North America, from Canada to Mexico (Fig. 7.13). The calculated carbon dioxide partial pressures did,

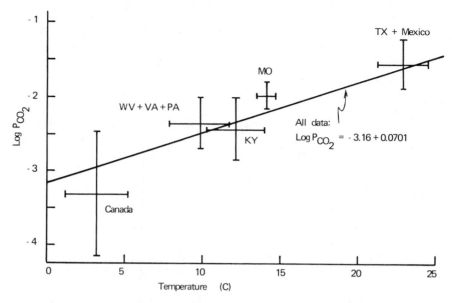

Figure 7.13 Variation of mean CO_2 pressure with temperature. [Adapted from Harmon et al. (1975b).]

indeed, increase from north to south and could be described by the regression equation (Harmon et al., 1975b)

$$\log P_{CO_2} = -3.16 + 0.070T \tag{7.9}$$

where T is the mean annual temperature, in degrees Celsius.

A lively debate ensued over the proper form of the P_{CO_2}-temperature relationship, particularly over the kinds of springs that were most suitable for analysis (Brook et al., 1977). Drake (Drake and Wigley, 1975; Drake, 1980, 1983) extended the calculation to model the P_{CO_2} of the source from the calculated P_{CO_2} of the water. This required making assumptions about the open or closed character of the aquifer and of other influences on soil P_{CO_2}. The current model proposes that the temperature dependence of CO_2 production can be described by

$$\log P^*_{CO_2} = -2.0 + 0.04T \tag{7.10}$$

with a correction factor

$$P_{CO_2} = \left[\frac{(0.21 - P_{CO_2})}{(0.21)}\right] P^*_{CO_2} \tag{7.11}$$

to account for the inhibition of CO_2 production by high CO_2 pressures.

All in all, it is not likely that the last word has been written on carbon dioxide modeling. Somehow the models for interpreting the calculated CO_2 pressures in carbonate groundwater must be connected with the models for CO_2 production in the soils and other sources, as discussed in Section 7.1. The model would need to contain at least the following terms:

$$P_{CO_2} = A_s(t)E_sE_a \qquad (7.12)$$

P_{CO_2} is here taken to be the calculated carbon dioxide pressure in groundwater. A_s is the source term, which must describe the CO_2 production in the soils and in the upper reaches of the aquifer. A_s must contain a description of soil characteristics, and it is a sinusoidal function of time in those regions with a distinct growing season. One model for A_s is Eq. 7.1. A_s can vary over a wide range, both locally and regionally, as shown by studies of bicarbonate ion concentration in North American groundwaters (Trainer and Heath, 1976) and in Brook and co-workers' detailed investigation of CO_2 in an arctic environment. E_s is the efficiency with which CO_2 is transferred from the source to infiltrating groundwater. It is a function of soil characteristics and precipitation patterns. A model for the product $A_s \cdot E_s$ is the Brook-Folkoff-Box model for global CO_2 production, using the actual evapotranspiration as the climatic variable (Eq. 7.2). E_a describes the losses (or transmission efficiency) within the aquifer between the recharge area and the water sampling point. Some CO_2 is lost through reaction with carbonate wall rock if the system is closed. CO_2 can also be lost to cave atmospheres if the water moves as open-channel flow in a well-ventilated cave passage. The objective of future research is to transform the general terms in Eq. 7.12 into a form that will permit predictive calculations.

7.3.3 Karst Denudation

The notion of an overall deflation or denudation rate for limestone surfaces arose from observations of glacial erratics perched on top of limestone pedestals (Fig. 7.14). It was deduced that the pedestal represented the lowering of the surrounding limestone surface since the melting of the glacial ice that had deposited the erratic. Denudation rates for England and Ireland where such pedestals occur were calculated to range from 10 to 40 mm/ka (Jennings, 1985).

The French geomorphologist, Jean Corbel, in the late 1950s introduced climate as a central consideration in karst geomorphology (Corbel, 1957; 1959a, b). Corbel calculated the average lowering of the limestone surface from measurements of hardness of springs and surface streams and advanced the controversial thesis that denudation rates were higher in cold climates because of the higher solubility of CO_2 in cold water.

216 / *Geomorphology and hydrology of karst terrains*

Figure 7.14 A residual limestone pedestal supports a glacial boulder in County Leitrim, Ireland. The height of the pedestal is indicative of the amount of regional denudation since the boulder was emplaced.

Karst denudation is usually measured by an overall mass balance between water input and dissolved carbonate output. Corbel's original equation was

$$D = \frac{4ET}{100} \quad (7.13)$$

where E is precipitation in decimeters and T is the mean hardness in milligrams per liter. This simple equation assumes constant limestone density of 2.5 g cm^{-3}, and it makes no allowance for runoff from nonkarstic rocks. Later workers seized on these points, and revised equations were proposed by Williams (1963) and Pulina (1972), among others.

The Corbel equation is oversimplistic, however, in assuming that some sort of grand mean can characterize entire regions without taking account of the tremendous variability of the measured hardness over the seasons and the variability with differences in topography, rock type, plant cover, and altitude. Marian Pulina in Poland, in particular, emphasized the necessity of detailed analysis of individual small drainage basins and of using runoff rather than precipitation as the measure of water throughput. Pulina's denudation equation is

$$D = 12.6 \frac{(T - T_a) Q}{A} \quad (7.14)$$

where T is carbonate hardness in milligrams per liter, T_a is hardness carried into the basin from nonkarstic rocks, Q is the mean discharge, and A is basin area.

An alternative method for measuring denudation rate is the microerosion meter devised by High and Hanna (1970). The device consists of three hardened steel pins permanently emplaced in the limestone bedrock. These support the legs of a micrometer frame, which is then used to measure the retreat of the exposed limestone surface. Precisions on the order of 10 µm are possible. Measurements over a period of a few years, therefore, give the erosion rates directly. However, direct measurements can only be made on exposed limestone surfaces and much of the important denudation takes place beneath the soil cover.

Many of the published denudation rates through 1975 were critically reviewed and compared by Smith and Atkinson (1976). They found that they could easily distinguish erosion in arctic climates, but the distinction between denudation rates in temperate and tropical climates was not nearly as pronounced as expected from the landforms. Figure 7.15 compares some denudation data with Smith and Atkinson's regression lines. For a similar plot based on somewhat different data, see Jennings (1983). The most important variable is the water throughput, measured either by runoff or by ($P - E$), the precipitation minus the evapotranspiration loss. The effects of other climatic variables are concealed in the scatter of the data points.

The annual runoff from a drainage basin can be calculated from the hydrograph by

$$R_f = \frac{1}{A} \frac{K}{t_R} \int_0^t Q(t)dt \qquad (7.15)$$

where R_f is runoff expressed in millimeters per year, A is basin area in square kilometers, K is a conversion factor $= 10^{-3}$ for the units given, t_R is the period of record in years, $Q(t)$ is the instantaneous discharge in cubic meters per second, and the time unit for integration is seconds. An exact expression for denudation is then given by the convolution of the runoff hydrograph and the hardness curve.

$$D = \frac{1}{N_L A} \frac{K'}{\rho} \frac{1}{t_R} \int_0^t Q(t)H(t)dt \qquad (7.16)$$

where D is denudation rate in millimeters per thousand years, N_L is the fraction of the basin underlain by carbonate rocks, $K' = 10^{-12}$ for the units given, ρ = density of carbonate rock in grams per cubic centimeter, and $H(t)$ is the instantaneous (Ca + Mg) hardness in grams per cubic centimeter of $CaCO_3$.

The difficulty in applying Eq. 7.16 is that continuous records for

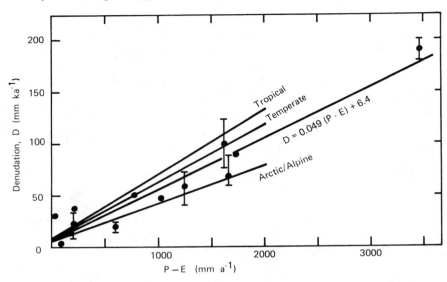

Figure 7.15 Some karst denudation rates from various climatic regimens. Data taken from Smith (1962), Canadian Arctic; Glazek and Markowicz-Lohinowicz (1973), Poland; Ogden (1982), West Virginia; Williams (1963), Ireland; Markowicz et al. (1972), Bulgaria; Williams and Dowling (1979) and Gunn (1981), New Zealand; Miller (1981), Belize; Sweeting (1979), Gunong Mulu, Malaysia (the extreme point on the right side of the diagram). The solid line with regression equation is linear least-squares fit to data given. The other lines are the climate-separated regressions on the data tabulated by Smith and Atkinson (1976).

carbonate hardness are rarely available, although the $H(t)$ curve could be determined in many regions by continuously monitoring specific conductance. More often available are spot analyses taken at intervals of one to two weeks. An approximate denudation rate can be calculated by averaging over the individual measurements (Drake and Ford, 1973; Ogden, 1982).

$$D = \frac{1}{N_L A} \frac{K'}{\rho} \frac{\Delta t}{t} \frac{1}{n} \sum_{i=1}^{n} Q_i H_i \qquad (7.17)$$

The exact measurement requires a runoff hydrograph, but over a one-year period or longer, the changes in storage in a karstic aquifer should average to zero; thus, either runoff or precipitation minus evapotranspiration could be used.

It is possible to calculate a theoretical expression for the denudation rate if it is assumed that the infiltrating waters reach equilibrium with the limestone (White, 1984). The assumption is most nearly valid in karst regions dominated by diffuse infiltration. The theoretical expression for denudation in terms of the chemical variables is

$$D_{max} = \frac{100}{\rho \sqrt[3]{4}} \left(\frac{K_c K_1 K_{CO_2}}{K_2}\right)^{1/3} P_{CO_2}^{1/3} (P - E) \tag{7.18}$$

where D_{max} is the denudation rate in millimeters per thousand years for the system at equilibrium. P_{CO_2} is given in atmospheres and P and E are expressed in millimeters per year.

Equation 7.18 has no adjustable parameters. It combines in a single statement the three climatic variables of temperature, CO_2 pressure, and precipitation. Even with the questionable assumption of equilibrium, the equation accounts for many of the observations on karst denudation rates.

1. Denudation is predicted to vary linearly with precipitation (or runoff) as implied by Fig. 7.15.
2. Denudation rate varies with the cube root of the CO_2 partial pressure. CO_2 pressures vary by about a factor of 100 from $10^{-3.5}$ on bare bedrock to $10^{-1.5}$ in organic-rich soils. Because of the cube root dependence, this amounts to about a factor of 5 in the denudation rate. This may account for the lack of distinction between temperate and tropical denudation rates where the contrast in P_{CO_2} is only about a factor of 10, thus a factor of 2 in the denudation rate. The distinctly lower dissolution rates in arctic and alpine climates may then be a result of both lower precipitation rates and the predominance of exposed bedrock.
3. The temperature dependence of the denudation rate is concealed in the temperature dependence of the equilibrium constants. Corbel was certainly correct in asserting that the solubility of CO_2 is higher in cold climates, but it is the combined temperature dependence of four equilibrium constants that describes the temperature dependence of the denudation rate. As the temperature rises from 5 to 25°C, the denudation rate decreases by only about 30 percent. Temperature is thus the least important of the three climatic variables with regard to equilibrium processes.

8
Soils, Sediments, and Depositional Features

Caves are favorable environments for the deposition of sediments of many kinds. Table 8.1 classifies and compares analogous surface sediments. Autochthonous sediments are those derived from local processes within the cave; allochthonous sediments are those transported from elsewhere. There are two main paths for transportation of material into caves: vertically (gravitational fills) from the land surface above and laterally by the action of streams or, more rarely, ice or wind. Because the cave environment provides constant temperature and high relative humidity, and maintains these conditions over long periods of time, it is ideal for the chemical deposition of minerals, which often occur in spectacular crystal forms. Much of the charm of caves derives from the secondary chemical crystallizations. Collectively these are called speleothems, a term coined by Moore (1952) to replace the older term, *cave formations* (Table 8.2).

Sinkholes and other surface depressions act as sediment traps. Thus, one sometimes finds fossilized sinkhole fillings buried under later sedimentary rocks (paleokarst) (Quinlan, 1978; Ford, 1984). Likewise, caves that have formed in old karst landscapes are filled with sediments, sometimes recrystallized and sometimes invaded by ore-forming fluids. Karst cavities, in fact, sometimes house important ore deposits (Bradbury, 1959; Zuffardi, 1976).

8.1 SOILS AND SEDIMENTS ON KARST SURFACES

Soil formation on any rock surface in any climate is a process by which the minerals of the bedrock, which are often not in chemical equilibrium with the local conditions of temperature and water content, are broken down into phases that are stable under ambient con-

Table 8.1
Classification of Cave Sediments

	Cave Sediment	Surface Equivalent
Clastic sediments		
Autochthonous sediments	Weathering detritus	Eluvium—soils
	Breakdown	Scree
	Organic debris	
Allochthonous sediments	Entrance talus	Colluvium
Gravitational	Infiltrates	
Transported	Fluvial deposits	Alluvium
	Glacial deposits	
	Aeolian deposits	
Chemical sediments	Travertines	Freshwater limestones
	Evaporites	Evaporites
	Phosphate and nitrate minerals	
	Resistates	Resistates
	Ice	Ice

ditions. Some components are preferentially leached and removed; others are left behind in modified form. Carbonate soils have two features that distinguish them from soils formed on other rocks: (1) A much larger proportion of bedrock is removed in solution. The soils are really a residue rather than a breakdown product. (2) Because of the interior drainage of karst areas, soils are often well-drained. Leaching can take place with greater efficiency than in many soil environments.

8.1.1 *Chemical and Mineralogical Composition of Carbonate Rocks*

Carbonate rocks are composed primarily of calcite and dolomite, with varying amounts of other minerals. The calcite generally contains from 2 to 4 percent by weight $MgCO_3$ (Goldsmith et al., 1955) in Mesozoic and Paleozoic carbonate rocks. Higher magnesium concentrations usually imply a mixture of calcite and dolomite, although high-magnesium calcites may occur in young sediments. The insoluble fraction is composed primarily of SiO_2, Al_2O_3, and Fe_2O_3. The silica may be present as chert, silicified fossil fragments, detrital or authigenic sand grains, or as a component of other silicate minerals. Alumina and iron may be present in clay minerals or as hydrated oxides. Other minor noncarbonate minerals such as phosphates (glaucophane), sulfides (primarily pyrite), feldspars, and a certain amount of organic matter may also be present. However, silica in its various forms and the clay minerals make up the bulk of the insoluble fraction of most carbonate rocks.

Table 8.2
Forms of Speleothems

Form	Size Range	Comments
Dripstone and flowstone forms (subaerial, hydraulic control)		
Stalactite	cm to meters	Pendulant. Flow through central canal. Many minerals; calcite common.
Stalagmite	cm to ten meters	Columnar or mound-like. Layered structure. No canal. Many minerals including calcite, gypsum, iron oxides, and ice.
Drapery	cm to meters	Furled sheets. No canal. Formed by dripping and trickling water. Mostly calcite.
Flowstone	meters to tens of meters	Layered deposit formed by flowing sheets of water. Many minerals. Calcite common. Also aragonite, ice, silica, and other minerals.
Crusts	meters to tens of meters	Wall coatings formed by flowing and seeping water. Usually of gypsum.
Erratic forms (subaerial, crystal growth control)		
Shield	10 cm to 5 meters	Disk-like objects. Medial crack acts as canal. Water flow under pressure. Mainly calcite.
Helictite	1 to 20 cm	Curved stalactite-like forms. Canal. Calcite, rarely aragonite.
Botryoidal forms	few cm	Bead-, knob-, or coral-like objects project from walls. Layered.
Anthodite	1 to 20 cm	Radiating clusters of acicular crystals. Sometimes dendritic. Aragonite.
Oulopholite	1 to 50 cm	Curved, radiating bundles of bladed or fibrous crystals, flower-shaped. Gypsum, epsomite, or mirabilite.
Moonmilk	—	Loose powder or wet pasty mass. Calcite, aragonite, or hydrated magnesium carbonate minerals.
Subaqueous forms		
Rimstone dams	meters to tens of meters	Natural dams of travertine in cave streams and pools. Mainly calcite.

Table 8.2
(cont.)

Form	Size Range	Comments
Subaqueous forms (cont.)		
Concretions	mm to cm	Concentrically layered, unattached structures. Both smooth (cave pearls) and rough surfaces occur. Mainly calcite.
Crystal linings	cm to meters	Crystal coatings in pools and entire caves. Scalenohedral calcite most common.

Free silica in carbonate rocks occurs as crystalline quartz or as chalcedony. Sand and silt are composed of small quartz crystals. Chert (or flint) nodules and layers are composed of chalcedony, a low-density, extremely fine-grained form of quartz in which the individual crystallites have a fibrous habit (Frondel, 1962).

Iron occurs mainly as goethite (FeOOH) or as the poorly crystallized mixture of hydrated iron oxides known as limonite. In extremely dehydrating environments, limonite and goethite decompose to form hematite (Fe_2O_3). The iron minerals occur as sparse opaque grains in most carbonate rocks.

Of the layer silicate minerals commonly found in carbonate rocks (Table 8.3) kaolinite and illite are most common. An analysis of the acid-insoluble fraction of some 27 limestones and dolomites from Missouri (Robbins and Keller, 1952) revealed only quartz, kaolinite, illite, and some montmorillonite. Limestones of Mississippian age from the Cumberland Plateau of Tennessee were found to contain

Table 8.3
Layer Silicates Found in Carbonate Rocks

Mineral	Composition	Basal Spacing (Å)
Kaolinite	$Al_4(Si_4O_{10})(OH)_8$	7.15
Illite	$K_{1+x}Al_4(Si_{7-x}Al_{1+x}O_{20})(OH)_4$ (essentially weathered muscovite mica)	9.95
Montmorillonite	$(Ca,Na)(Al,Mg,Fe)_4(Si,Al)_8O_{20}(OH)_4$	Variable
Chlorite	$(Mg,Al,Fe)_{12}(Si,Al)_8O_{20}(OH)_{16}$	14.3
Vermiculite	$(Mg,Ca)_{0.7}(Mg,Fe^{3+},Al)_6(Si,Al)_8O_{20}(OH)_4 \cdot 8H_2O$	28.9

illite, chlorite, and vermiculite with some montmorillonite as the principal layer silicates. Kaolinite was not common in these samples (Peterson, 1962).

8.1.2 Karst Soils

The soils that mantle carbonate rock terrains have various sources:

1. Residual soils composed of the insoluble fraction of the carbonate rock.
2. Colluvial soils drifted over the carbonate terrain from noncarbonate areas on higher slopes. This source of soil is especially important in the Appalachian karst valleys, which are bordered by steep ridges of sandstone and shale.
3. Alluvial soils carried out onto the karst by preexisting or contemporaneous surface streams. Upland soils in karst regions are frequently quite different from the soils on the floodplains of incised streams.
4. Glacial till or morainic material, either residual from Pleistocene glaciation or transported and modified by later processes.

The primary modification of the insoluble fraction of carbonate rock is the preferential leaching of silica. Silica is surprisingly soluble, particularly if it is in an amorphous form. Silica is taken into solution as silicic acid complexes, but the solubility is not very sensitive to pH (Morey et al., 1964). Experiments on leaching various clay minerals (Carroll and Starkey, 1959) show that only silica is leached when carbonated water is percolated through a clay bed. The leachate contains no aluminum. See Drever (1982) for a detailed discussion of aluminosilicate solution chemistry.

The amount of leaching of limestone soils varies considerably with climate and also with the drainage of the soils. Well-drained upland soils in areas of doline or cutter development are significantly more leached than more poorly drained soils. In northern climates, with lower soil temperatures and shorter growing seasons, the silica and clays from the limestones are preserved. Iron minerals are completely hydrated and brown soils with much quartz and clay are common. The Hagerstown soils formed on the Ordovician limestones of central Pennsylvania are typical (Braker, 1981). In more southern (but not necessarily tropical) climates leaching of silica and more dehydration of the iron minerals occur. The products are deep red soils known as terra rosa. The Cumberland soils of the Mammoth Cave area are typical (Latham, 1969). There is a similar contrast between the soils of England and those of southern France and Spain (Khan, 1960). In the tropics the leaching of silica may be nearly complete. Quartz and chalcedony are removed, the clay minerals are broken down, and iron occurs in the form of finely dispersed hematite. Aluminum is immobile and remains in the soil in the form of a mixture of gibbsite (Al(OH)$_3$), boehmite (AlOOH), and diaspore (AlOOH), known as bauxite.

The colors of soils are described by the Munsell system (Munsell Color, 1976) or by measuring the intensity of light reflected from them as a function of the wavelength of the light. Reflectance spectra are useful because they are more quantitative and may also include the infrared region. They can be used to classify soil types, identify certain constituents—particularly iron and organics—and compare laboratory soil samples with field-scale measurements obtained by airborne multispectral remote sensing (Mathews et al., 1973a,b).

The reflectance characteristics of surface soils and cave sediments from the Mammoth Cave region (White, 1977a) can be displayed by plotting the ratios of reflected light intensity, blue/red against green/red (Fig. 8.1). This procedure allows soils to be grouped into common spectral types and the cave sediments to be compared with the surface soils.

8.1.3 *Sinkhole Fillings*

Closed depressions act as sediment traps. The shallow depths and broad flat floors on many sinkholes is almost certainly indicative of a thick sequence of sediments nearly filling the bedrock basin.

Ruhe (1975, 1977) and his students core-drilled a sinkhole in the Lost River drainage basin of southern Indiana. The maximum relief of the closed depression was 8 m, but the sediment depth at the center of the depression was 13 m of sand, silt, and clay. The sediments were depleted in carbonate minerals to a depth of 8 m. The bottom 5 m contained carbonate material. The soil waters from the upper part of the sedimentary column had pH values from 5.8 (likely because of atmospheric CO_2) to 4.2 to 4.6, values that would be expected if the soil CO_2 was ten times the atmospheric level. The pH rises sharply when carbonate material appears in the soils, and the reported values are close to what would be expected from saturated waters. A process of carbonate dissolution occurs in the clastic sedimentary mass as well as at the soil/bedrock contact.

Aubert (1966) dissected a doline in the Swiss Jura by digging several trenches across it (Fig. 8.2). In this glaciated alpine environment, much more broken rock was found within the section. Unlike many dolines, this one contained a 1-m solution opening at the bottom, blocked with broken rock. Several episodes of in-filling and slumping were evident, and the time sequence was not worked out, although Aubert argues for a pre-Wisconsin age.

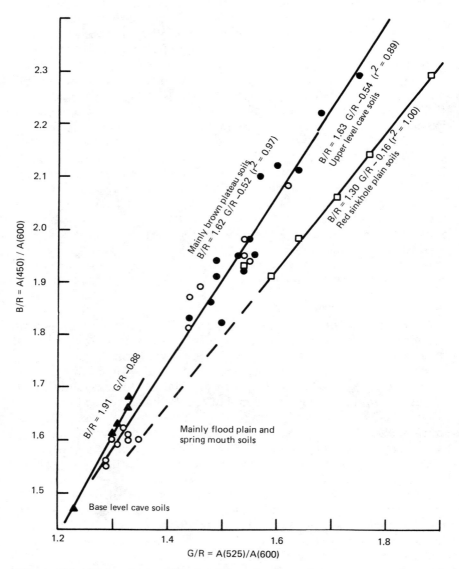

Figure 8.1 Reflectance ratios for Kentucky soils and cave sediments. B/R is the ratio of blue reflected light to red reflected light; G/R is the ratio of green reflected light to red reflected light. A(450) is -log (intensity of reflected light from sample)/(intensity of reflected light from standard) measured at a wavelength of 450 nm.

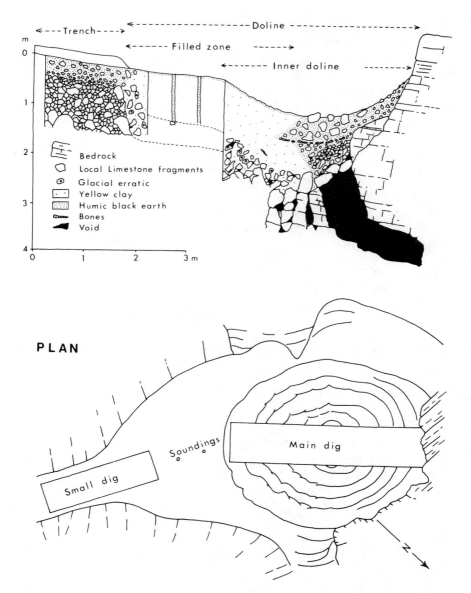

Figure 8.2 Profile and plan of a doline in the Swiss Jura. [Adapted from Aubert (1965) by J. Gunn; courtesy of Dr. Gunn.]

8.2 AUTOCHTHONOUS SEDIMENTS IN CAVES

8.2.1 *Weathering Detritus*

Weathering detritus is the insoluble residue left by the dissolution of the limestone bedrock. The mineralogy and composition of the weathering detritus depends on the bedrock's composition. The very coarse fraction is composed mainly of the chert nodules that occur profusely in many limestones. The intermediate-size fraction is usually quartz sand, although dolomite sand occurs in caves in the Black Hills of South Dakota and in some Missouri caves. The finer silt and clay-sized weathering detritus is extremely varied. Fine-grained quartz, sericite, and clay minerals are probably the most common. Analysis of red clays from three Missouri caves (Deike, 1960b) showed that they were composed of quartz, illite, and kaolinite, with traces of hematite—almost exactly the same composition found for the insoluble fraction of the limestone.

Some clay minerals may be authigenic, having formed directly in the cavern environment. Large amounts of endellite that occur in the New Mexico Room of Carlsbad Caverns are claimed to be authigenic (Davies and Moore, 1957).

The accumulation of weathering detritus in caves is analogous to the formation of residual soils on the land surface. In climates where extensive silica leaching and alteration do not take place, the two types of material derived from the same limestone should be indistinguishable. Residual soils are also transported into caves as infiltrates and as fluvial sediments. They are mixed with the weathering detritus formed in the subsurface to create the "cave mud" found in most caves.

8.2.2 *Organic Debris*

Caves form convenient dwelling places for bats and certain species of birds. In many caves, particularly in the tropics and in the American Southwest, the droppings from roosting bats and birds have built up into an organic sediment many meters thick. This material, known as guano, is often mined for use as fertilizer because of its high phosphorus and nitrogen content. Decomposition of the guano releases nitrate, ammonia, and urea. Phosphorus-rich leachates from the guano can react with the limestone wallrock to produce a suite of phosphate minerals in the contact zone between the guano and the wall. Other minerals form by direct reaction with bat urine. Guano dropped over preexisting calcite speleothems causes a pronounced corrosion of the calcite, with formation of such phosphate minerals as hydroxyapatite. Badly corroded speleothems are frequently seen in tropical bat caves. Table 8.4 lists some of the cave minerals that have been described from guano deposits.

Table 8.4
Cave Minerals Derived from Guano Deposits

Mineral	Formula	Cave	Reference
Urea	$(NH_2)CO$	Wilgie Mia Cave, Western Australia	Bridge (1973a, 1975)
Uricite		Dingo Donga Cave, Western Australia	Bridge (1974)
Guanine		Murra-el-elevyn Cave, Western Australia	Bridge (1974)
Phosphammite	$(NH_4)_2HPO_4$	Unknown cave, Western Australia	Bridge (1973a)
Biphosphammite	$(NH_4)H_2PO_4$	Murra-el-elevyn Cave, Western Australia	Bridge (1973b)
Struvite	$NH_4MgPO_4 \cdot 6H_2O$	Skipton Cave, Victoria, Australia	Bridge (1971)
Hannayite	$(NH_4)_2Mg_3H_4(PO_4)_4 \cdot 8H_2O$	Murra-el-elevyn Cave, Western Australia	Bridge (1973b)
Niahite	$NH_4(Mn,Mg,Ca)PO_4 \cdot H_2O$	Niah Great Cave, Sarawak, Malaysia	Bridge and Robinson (1983)
Newberyite	$MgHPO_4 \cdot 3H_2O$	Skipton Cave, Victoria, Australia	Bridge (1971)
Mundrabillaite	$(NH_4)_2Ca(HPO_4)_2 \cdot H_2O$	Petrogale Cave, Western Australia	Bridge and Clark (1983)
Brushite	$CaHPO_4 \cdot 2H_2O$	Pig Hole Cave, Virginia	Murray and Dietrich (1956)
Monetite	$CaHPO_4$	Caves of Mona Island, Puerto Rico	Kaye (1959)
Whitlockite	$\beta\text{-}Ca_3(PO_4)_2$	Caves of Mona Island, Puerto Rico	Kaye (1959)
Ardealite	$CaHPO_4 \cdot CaSO_4 \cdot 4H_2O$	La Guangola Cave, Italy	Balenzano et al. (1984)
Hydroxyapatite	$Ca_5(PO_4)_3(OH)$	Caves of Mona Island, Puerto Rico	Kaye (1959)
Dahllite	$Ca_5(PO_4,CO_3)_3(OH,F)$	et-Tabun Cave, Israel	Goldberg and Nathan (1975)
Crandallite	$CaAl_3(PO_4)_2(OH)_5 \cdot H_2O$	Caves of Mona Island, Puerto Rico	Kaye (1959)
Taranakite	$H_6K_3Al_5(PO_4)_8 \cdot 18H_2O$	Pig Hole Cave, Virginia	Murray and Dietrich (1956)

8.2.3 Breakdown

Mechanical fracturing of the limestone beds that make up the cave roof and walls produces a rough jumble of angular rock fragments known as *breakdown*. Breakdown piles are unsorted and highly permeable. Layering is indistinct to nonexistent.

The "particles" that make up the breakdown sediment can be classified, following Davies' categories (1949), according to their relationships to the parent beds:

> *Block breakdown* consists of masses of rock in which more than one bed has remained as a coherent unit.
> *Slab breakdown* consists of fragments of single beds.

Chip breakdown consists of bits of bedrock derived from the fragmentation of individual beds.

Block breakdown can be massive, relatively equant blocks measuring up to tens of meters on a side. The blocks are usually bounded by bedding planes along the bedding and by joint planes across the bedding. Slab breakdown has a plate shape, with the plate thickness being determined by the thickness of the beds. Widths of individual slabs vary from tens of centimeters to many meters. Chip breakdown typically ranges from centimeters to tens of centimeters. The shape of chip breakdown is variable and dependent on the process that created the breakdown. Roughly equant, very angular chunks result from shattering of bedrock by frost pry or crystal wedging and from rock with many closely spaced joints. A flatter, irregular form results from fragmentation of slab breakdown because of a somewhat greater tendency for the slabs to break along bedding planes. Very thin shards of breakdown, arising from spalling of bedrock under pressure and from mineral replacement and crystal pressure, also occur.

The weight of rock above a solution cavity causes the beds to sag slightly into the opening. If the beds are flat-lying, there is a concentration of shear stress at the passage walls. Each of the overlying beds is supported slightly by those below it, and the zone of shear stress curves inward and closes to form a dome-shaped surface in the bedrock (Fig. 8.3). If the beds do not have sufficient strength, they shear and collapse into the cavity, where some of the material may be removed by solution. The cavity stabilizes when the rock has collapsed to the surface of maximum shear stress.

A simple beam model was used by Davies (1951) to analyze cave breakdown. If the cave is in flat-bedded rock, elastic sag separates the individual beds of the roof. These then act as beams with a span equal to the passage width and a thickness equal to the bedding thickness. Since the width of the beams does not enter the final formulas, the calculation is independent of the extent of continuous bed along the passage. If the roof span is intact, it is treated as a fixed beam; if it is broken or strongly jointed, it may be treated as a cantilever beam. The main assumption is that the strength of the rock within the bed is much greater than that across bedding plane partings. Ceiling beams are clamped by the weight of the bedrock behind the passage walls. The beams are not free to rotate, so simple jointing in a thick bed does not necessarily transform a fixed beam into a cantilever. The joint faces are compressed and compressive strength is much higher than shear strength.

The resistance to fracture at the point where the beam is attached to the wall is determined by the maximum flexural or bend-

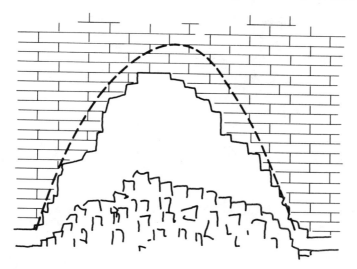

Figure 8.3 The "tension dome" or zone of maximum shear stress induced by the presence of a cavity *(dotted line)*. The stress is relieved by the collapse of the mass of rock within the stress zone. If the collapse material is removed by solution, a breakout dome results.

ing stress that will be tolerated by the limestone bed. The critical beam thickness that will just support its own weight is

$$t_{crit} = \frac{\rho l^2}{2S} \tag{8.1}$$

where S = flexural stress in the extreme fiber, t = beam thickness, ρ = density of beam material, and l = beam length, or roof span. The analogous formula for a cantilever beam is

$$t_{crit} = \frac{3\rho l^2}{2S} \tag{8.2}$$

The critical ceiling bed thickness (Fig. 8.4) divides the diagram into regions of stable ceilings and unstable ceilings. A 30-m passage width, rather large as cave passages go, requires a 70-cm-thick ceiling bed to be stable. Thus, simple breakdown mechanics provide an upper limit for the size of cave passages. Any process that transforms a fixed beam into a cantilever greatly increases the critical thickness and thus induces breakdown.

Caves in folded limestones typically exhibit different patterns of breakdown from those of caves in flat-lying limestones. Many cave passages are oriented along the strike. Beds are tipped at various angles. Strike joints tend to be perpendicular to the bedding and are

232 / Geomorphology and hydrology of karst terrains

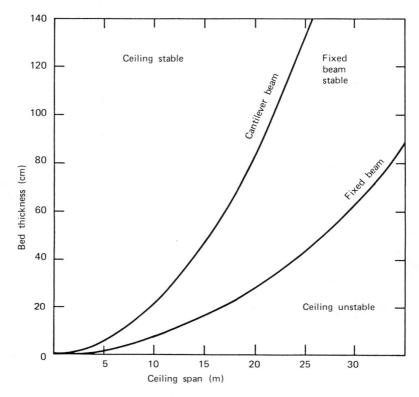

Figure 8.4 Critical breaking strength for ceiling beams in cave passages. Calculations based on data for Ste. Genevieve limestone of southern Indiana, shear strength = 18.2 MPa and density = 2.65 g cm^{-3}, based on means of four samples measured by the Indiana Geological Survey.

thus inclined in the opposite direction. These conditions are ideal for dropping blocks. The ceiling blocks are oriented in such a way that both bedding plane and joint face are in tension. Curiously, breakdown is not common in strike-oriented passages. Breakdown is so rapid that most of the unstable ceiling falls during the initial excavating stages of cavern development and is removed in solution or buried in the fluvial sediments. All that remains is a characteristic tent-shaped passage cross section.

Breakdown in later stages occurs when the passages cut across bedding planes. Beams across the portion of the passage running perpendicular to the strike are more stable, which perhaps also accounts for their smaller cross sections. Later development of a zone of weakness would result in the perpendicular passage ceiling beams falling where they would remain because water would no longer be in the passage to dissolve them.

8.2.4 Geologic Processes that Trigger Breakdown

Breakdown is a sequential process. Weakened cave ceilings and walls remain stable until later processes exaggerate the weaknesses to critical levels, causing collapse (White and White, 1969).

1. *Removal of buoyant support.* Many cave conduits are excavated under completely water-filled conditions. Major breakdown occurs when the water levels are lowered below the passage horizon. Each unit volume of limestone in a submerged cavern roof experiences an upward force equal to the weight of water it displaces. Assuming an average density for limestone of 2.5 g cm^{-3}, the buoyancy of the rock in water contributes 40 percent of the ceiling support. When the passage is drained, ceiling beams that were within this percentage of their ultimate strength fall. Cave passages exist whose ceilings show the blocky outlines of fracture surfaces, yet no breakdown remains on the floor of the passage. Unless it is buried in fluvial sediments, such a passage can be assumed to have collapsed when streams still occupied the floor, so that the fallen blocks were removed in solution.

2. *Base level backflooding.* Once a cave passage has emerged from the phreatic zone, it soon stabilizes. If the cave lies close to major surface streams, however, the annual fluctuations in stream level cause water to backflood into the cave. The flood water is an active agent that can open joints and cause further roof collapse. Examples occur in the Mammoth Cave System, where breakdown blocks are buried in alternating layers of sand and clay. The ceilings above the blocks are rounded, indicating additional solution after the blocks fell.

3. *Undercutting by free-surface streams.* As the base level is lowered still further, it becomes possible for the passage to be used as a route for free-surface streams. Since free-surface streams often have higher gradients than subwater table streams, they may carry a coarser sediment load and can also remove the fills deposited by earlier processes. Free-surface streams can activate breakdown in two ways: (1) as the fill is removed, partially supported ceiling beams are exposed to additional load and may fall; (2) the walls may be undercut, effectively increasing the length of beams and causing additional collapse. The process is particularly effective if the streams meander, giving them a cutting edge against the walls (Fig. 8.5).

4. *Breakdown activated by vertical shafts.* As protective caprocks are breached or valley sides retreat, vertical shafts develop that cut down through the limestone and intersect underlying cave

Figure 8.5 Slab breakdown in Pohl Avenue, Flint Ridge Cave System of Kentucky. A ceiling channel cut by a free-surface stream, probably an evolving shaft drain, cut from wall to wall across the passage, permitting a series of thin ceiling beds to collapse into the passage.

passages. If a shaft passes near a cave passage, beams may be weakened or converted from fixed beams to cantilevers. Shaft formation also removes preexisting breakdown, thus removing support from the ceiling.

    5. *Breakdown activation through attack by fresh surface waters.* Breakdown plays a leading role in the final stages of passage degradation. When the land surface is eroded down to the level where surface waters can easily penetrate the cave system, solution attack on walls and ceilings is renewed. Massive breakdown piles frequently mark the location of tributary valleys or gullies on the land surface above. Small surface valleys are frequently located along fracture zones, which both provide a more highly permeable route for surface water to reach the cave and weaken the bedrock of the cave passage.

    6. *Frost wedging.* When a fixed volume of water freezes into ice, the expansion produces a force of 1780 atm at 0°C. This force is sufficient to fracture bedrock when water freezes in small cracks and voids. Frost wedging is also a very late-stage process and is more effective near cave entrances and in shallow passages that lie above the frostline. Frost wedging tends to produce a very angular chip breakdown that can be recognized in cave sedimentary sections.

7. *Breakdown activation by mineral deposition and mineral wedging.* Deposition of minerals such as gypsum in joints and bedding planes can also produce enough force to fracture the rock. Breakdown resulting from simple crystal prying is similar to that produced by frost wedging: equant, very angular fragments (Fig. 8.6).

A rather different form of mineral attack generates large quantities of breakdown in the Flint Ridge System and, to a lesser extent, in other caves. The gypsum and other sulfate minerals in the Mammoth Cave area originate from the oxidation of pyrite in the upper part of the overlying Big Clifty sandstone (Pohl and White, 1965). Sulfate-bearing solutions percolate downward into the cave passages, where they react in the limestone and form gypsum by direct in situ replacement of calcite. The molar volume of gypsum is higher than that of the calcite it replaces, thus producing an expansion pressure in addition to the chemical attack. Replacement within the bedrock causes thin, platy fragments of rock to spall off, resulting in a characteristic chip breakdown. Expansion within the beds and slippage along the softer gypsum stringers also generate a peculiar curved slab breakdown (Fig. 8.7).

Figure 8.6 Wall deterioration and incipient breakdown created by emplacement of gypsum in joints and bedding plane partings. Turner Avenue, Flint Ridge Cave System in Kentucky.

236 / *Geomorphology and hydrology of karst terrains*

Figure 8.7 Curved breakdown slabs caused by in situ replacement of calcite by gypsum within limestone beds. Turner Avenue, Flint Ridge Cave System in Kentucky.

8.2.5 *Breakdown Features*

Breakdown talus piles are a ubiquitous feature of many limestone caves. These range in extent from scattered slabs of breakdown on the cave floor to breakdown "mountains" tens of meters in height. Breakdown talus piles give cave floors very irregular profiles.

Terminal breakdowns are breakdown piles that completely block cave passages. Terminal breakdown often contains sandstone and other rocks as well as limestone fragments, if the collapse extends upward to an overlying caprock. Terminal breakdown is the principal truncating mechanism for cave passages (Brucker, 1966).

Among the remarkable cavern features are the huge rooms that form as a result of major ceiling collapse. Some of these have floor dimensions of 100 m or more and ceiling heights of 50 to 100 m. Davies (1951) ascribed the origin of these major breakdown rooms to the collapse of tension domes and described the mechanics of their formation. The beehive shape of the rooms is the shape that best equalizes the stress distribution in the remaining bedrock (see Fig. 8.3).

Breakout domes occur over a continuum of sizes, from very large breakdown rooms to small roughly circular or elliptical breakdown areas sometimes only a few meters in diameter. There is usually a talus pile in the center of the breakout dome. Considering the low bulk density of loose piles of angular rock fragments, many breakout domes must have formed at the time when circulating water was available to remove some of the fallen blocks. The dome could then enlarge by a mechanism of solution with concurrent stoping of the sides. In the United States very large domes are found in Chief City in Mammoth Cave, Rothrock's Cathedral in Wyandotte Cave, the entrance room of Hellhole Cave of West Virginia, and the entrance room of Marvel Cave in Missouri.

8.3 GRAVITATIONAL SEDIMENTS IN CAVES

8.3.1 *Entrance Talus*

Transport of material mainly under the influence of gravity, assisted by slopewash, stoping, frost pry, solifluction, and other processes, occurs where cave passages intersect the land surface. The entrance area is characterized by talus fans of colluvial material from hill slopes, breakdown activated by the more active weathering near the surface, and other in-wash material including soils and rock debris from higher on the surface slopes. So different is the entrance talus from other cave sediments that Kukla and Ložek (1958) divide all cave sediments into an entrance facies and an interior facies. The entrance talus migrates inward as the hill slope retreats, removing old deposits and generating new ones with concurrent shortening of the cave. Both animals and early humans used the entrance areas of caves as shelters; thus, the unsorted, unconsolidated, and roughly stratified entrance talus is the prime site for excavation of their artifacts. Most cave "digs" are done in the entrance talus, not in the other sediments of the cave interior.

8.3.2 *Infiltrates*

Infiltrates are clastic sediments transported more or less vertically into the cave by the action of gravity. They differ from the entrance talus in that infiltration can take place through small crevices and openings, so that some sorting takes place. Major sources of infiltrates are surface soils washed, piped, or slumped into open crevices or down sinkholes. They are eventually discharged into the cavern system without much chemical modification. There is a continuum between the infiltrate sediments and the entrance talus type of material. When pits are open to the surface, quite large masses of material can fall into them. Such pits also serve as animal traps, and vertebrate fossils are frequently found in them.

8.3.3 *The Problem of the Red Unctuous Clays*

In addition to the infiltration of clastic sediments into air-filled crevices and pits, very fine-grained components of surface soils may migrate below the water table through small fractures and other openings. Certain of the deep phreatic theories of cave origin, particulary that of Bretz (1942) attached great significance to the "red unctuous clays" that were supposed to fill cave passages completely between the time that the caves were dissolved below the water table and the time that the sediments were removed by high-gradient, free-surface streams. Bretz pointed to the high-lying pockets of red clay in Missouri caves as evidence for his hypothesis. This class of what may be called "phreatic infiltrates" is not commonly found, or if it occurs it is well-mixed with other clastic material. A careful study of Missouri cave sediments (Reams, 1968) failed to turn up the supposed red unctuous clays. Instead, cave sediments derived from the local regolith, apparently deposited at shallow depths below present-day base level streams, were found.

8.4 FLUVIAL AND OTHER TRANSPORTED SEDIMENTS

8.4.1 *Fluvial, Glacial, and Aeolian Deposits*

Fluvial deposits, or stream-transported sediments, are the underground equivalent of alluvium. Fluvial deposits, which may be transported through water-filled conduits as well as in open channels, make up the greatest part of cavern sediment sequences. They are derived from many sources. Some are reworked autochthonous cave deposits, some are derived from rocks high in the stratigraphic section, and some are transported from nonkarstic borderlands by sinking streams. The material usually consists of stream-rounded rock fragments, sand, silt, and clay. The smaller grain-size material usually cannot be distinguished from weathering detritus or infiltrate sediment without extensive analysis.

Glacial deposits are uncommon in caves, except in entrance areas. Glacial tills and morainic materials are important sources of other cave sediments in glaciated regions. Ice sheets that overrode karst areas filled sinkholes with glacial tills that were later carried down into the cave systems, for example, in the Helderberg Plateau of New York (Mylroie, 1977). Although some caves are blocked with glacial ice, ice does not seem to move through caves.

Aeolian deposits are materials carried directly into caves by wind action. Such sediments occur primarily in the immediate vicinity of present or past cave entrances, although, as with the glacial deposits, wind-blown sediments on the surface can be carried into caves as infiltrates with little sorting or chemical alteration occurring. Loess occurs frequently in the alpine caves of central Europe (Schmid, 1958). It occurs mainly near cave entrances and was apparently blown into the caves during the Pleistocene glacial maxima. Reworked loess is a component of fluvial sediments in some caves of the Midwest.

8.4.2 Composition of Fluvial Sediments

Fluvial sediments in caves are generally unconsolidated, although they may be somewhat indurated by a minor amount of calcite cement. They consist, broadly, of a light fraction and a heavy fraction. The light fraction, which makes up nearly all of the bulk, can be classified roughly as coarse fraction, medium fraction, and fine fraction. Grain-size distributions have been measured for sediments from Mammoth Cave (Davies and Chao, 1959) (Fig. 8.8), several Texas caves (Frank, 1965), and selected Missouri caves (Reams, 1968).

The coarse fraction consists of materials in the cobble-to-boulder-size range. Mainly, these materials are chert nodules weathered from the limestone, chert fragments from the breakup of bedded chert, and sandstone cobbles or boulders that have been transported from overlying caprock or borderland clastic sediments. Occurrence of other rock types depends on the geologic setting of the cave. Sandstone boulders, 30 to 50 cm in diameter, are sometimes found in horizontal passages, hundreds or thousands of meters from any possible source—a testimony to the power of flowing water in conduit systems.

The medium-size fraction consists of gravel, sand, and silt. Gravels found in caves usually have a characteristic lithology that relates to the source material. Chert gravels occur in Missouri caves. Rounded quartz pebbles occur in many parts of the Flint Mammoth

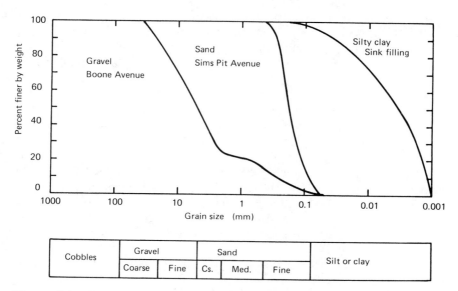

Figure 8.8 Grain-size distribution for some clastic sediments in Mammoth Cave, Kentucky. [Data are from Davies and Chao (1959).]

Cave System, derived from a basal conglomerate in the overlying sandstone. The sand and silt are usually quartz, although dolomite sands occur in the Black Hills Caves of South Dakota and in Missouri.

The fine fraction consists of very fine-grained quartz and clay. The clay minerals in the fluvial sediments are mostly the same as those that occur in the source rock, which may be the limestone, overlying soils, or nonkarstic rocks in the borderlands. Clay minerals are surprisingly sparse in most cave sediments. Some of the muddiest cave fills from southcentral Kentucky and from the Appalachians have turned out to be mainly quartz, with only small amounts of clay.

Heavy minerals usually constitute from 0.1 to 1 percent of the clastic sediments. The heavy mineral fractions from Texas caves (Frank, 1965) consisted of hornblende, magnetite, kyanite, epidote, zircon, garnet, tourmaline, staurolite, rutile, biotite, leucoxene, and limonite, with a distribution essentially matching that of the heavy minerals fraction of the sediments on nearby stream terraces. The heavy minerals from central Kentucky cave sediments (Davies and Chao, 1959) are interesting in that in addition to rounded, and therefore presumably transported, grains of goethite, zircon, tourmaline, rutile, garnet, and chomite, there were euthedral grains of authigenic brookite, anatase, celestite, and barite.

8.4.3 Depositional Sequences

Fluvial sediments in caves exhibit a complex stratigraphy of interbedded sands, gravels, cobbles, and clays, sometimes in a delicate layering. Sedimentary sequences have been measured in Kentucky (Davies and Chao, 1959), Missouri (Reams, 1968; Helwig, 1964), West Virginia (Wolfe, 1973), and Texas (Frank, 1965), as well as in Europe (Schmid, 1958), Australia, and New Zealand (Frank, 1975).

All such sediments appear to be channel deposits (Fig. 8.9). Layering and lithology vary greatly over distances of a few meters. Such rapid facies changes are best interpreted as evidence for deposition by streams subject to large annual fluctuations, accompanied by much reworking by later processes. Carwile and Hawkinson (1968) attempted to follow a single elliptical conduit, Columbian Avenue in Flint Ridge Cave System, by trenching the sediments to bedrock at intervals of 50 to 100 m along the passage. Adjacent columns did not correlate. It may be possible to assign regional significance to broader features of the sedimentary sequence, but the fine details are very chaotic.

It is apparent that cave streams alternate from aggrading to degrading streams. One finds old channel deposits, relic stream channels, and other features buried in later sediments and sometimes exposed in cross section. The time sequence of these events is

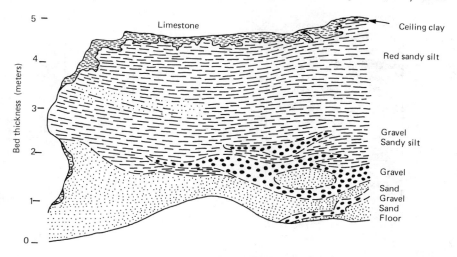

Figure 8.9 The clastic sediment plug occluding Dismal Valley, an upper-level trunk passage in Great Salts Cave, Kentucky. [Adapted from Davies and Chao (1959).]

obscure. Layers of travertine may be interbedded with clastic sediments (Fig. 8.10). In alpine caves Schmid (1958) has found layers of angular breakdown she attributes to deep frost action during ice advances.

8.4.4 Bedforms

Although cave passages are cut in bedrock, the free-surface streams frequently flow on pads of fluvial sediment that are exactly analogous to the alluvial pads that underlie surface streams. Like surface streams, underground streams have movable beds, and all bedforms except those caused by vegetation should be found underground.

Winnowing of fine-grained material near the surface of the bed leaves very coarse-grained material behind, and many underground streams flow on a layer of cobble armoring. The bed tends to be very flat, and braided flow patterns occur. Jones (1971) has referred to these fluvial sediment surfaces as the *underground floodplain*. Small streams tend to meander and form point bars. Alternating reaches of pools and riffles are common. These features tend to develop when the sediment load of the underground stream is in equilibrium with transporting capacity.

8.4.5 Mechanisms for Fluvial Sediment Transport

Although the clastic material deposited in caves comes from many sources, fluvial transport is the only significant mechanism for taking it out again. Caves act as sediment traps, and without an exca-

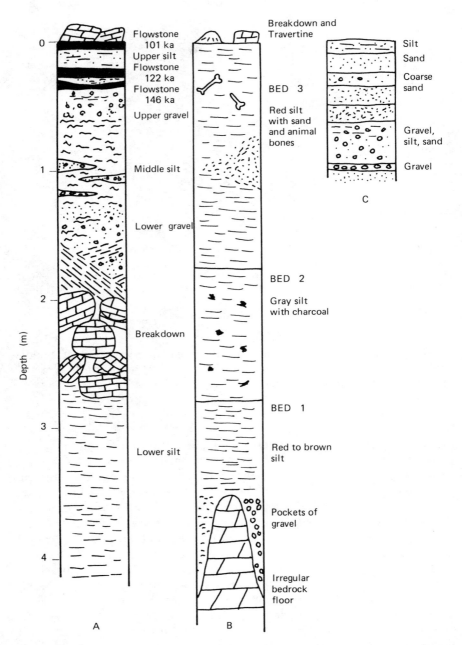

Figure 8.10 Stratigraphic sections for some stream-transported cave sediments: (A) Enigma Pit in Mystery Cave, Minnesota (Milske et al., 1983); (B) Carroll River Passage in Carroll Cave, Missouri (Helwig, 1964); (C) Milroy Cave, Pennsylvania (Ogden and Ebaugh, 1972).

Soils, sediments, and depositional features / 243

vating mechanism cave passages would simply choke up with sediments (White and White, 1968). There is a large body of literature on sediment transport, some of which is summarized in books such as those of Leopold, Wolman, and Miller (1964) and Graf (1971).

The transport of granular sediments by flowing water takes place by two mechanisms: bed load and suspended load transport. Figure 8.11 shows the situation for intermediate levels of discharge. At the bottom of the sediment pile, where the clastic material is in contact with the bedrock, the bed is stable and no material moves. At the top of the pile is the bed load. In the movable part of the bed, individual particles move by a saltation or hopping mechanism but are supported most of the time. The thickness of the mobile layer varies with discharge and boundary shear. At low flows the bed load may be only a few grains at the surface. Extreme flood events may set the entire sediment pile in motion, scouring the channel to bedrock. The suspended load extends upward from the bed surface. Since particles are kept in suspension by the turbulent motion of the water, there is a balance between the lifting action of the water and the fall velocity of the particles. The concentration of suspended sediment falls off rapidly with distance above the bed. The line of demarcation between the suspended sediments near the bottom of the water column and the bed load at the top of the sediment pile is vague. There

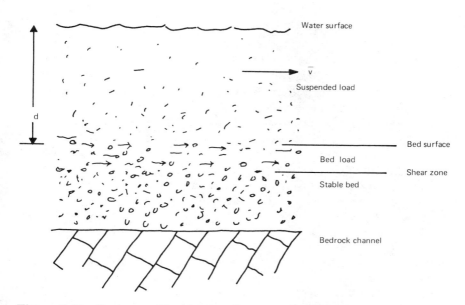

Figure 8.11 Features of bed load and suspended load in a clastic sediment transportation system.

is a transition zone several grain diameters thick where the type of transport is not well defined.

Vanoni (1966) expresses the relationship of the critical boundary shear to the mean sediment diameter as

$$\frac{\tau_c}{(\rho_s - \rho)\,\overline{d}\,/\rho} = f\left(\frac{U_* \overline{d} \rho}{\eta}\right) \qquad (8.3)$$

where τ_c = critical boundary shear, ρ_s = density of sediment, ρ = density of fluid, $\overline{d}$ = mean sediment grain diameter; $U_* = \tau_c/e$ = friction velocity, where e measures bed roughness, and η = viscosity of fluid. The equation is known as the Shields relation. It can be expressed in any consistent set of units because both sides are dimensionless. The left side can be thought of as a particle Froude number and the right side as a function of the boundary Reynolds number.

Although the threshold for bed movement is rather well described by the critical boundary shear, it is not easy to convert this into a critical threshold velocity because of the ambiguity in calculating bottom velocities from average channel velocities. The bottom velocity depends on both the bed roughness and the hydraulic depth of the channel. On a hydraulically rough bed, Einstein (1964) gives the velocity in turbulent flow as

$$\frac{v_y}{U_*} = 5.75 \log \frac{30.2 y}{d_{65}} \qquad (8.4)$$

where v_y is the water flow velocity at a distance, y, above the boundary with the bed, where the friction velocity is U_*; d_{65} is the grain size at which 65 percent by weight is finer. Representing the complex distribution of sediment particle sizes by a single number is accomplished by a variety of statistical measures of which $\overline{d}$, the mean particle size, d_{65}, d_{50}, and d_{90} commonly appear.

Figure 8.12 shows the relationships in terms of the much-used Hjulstrom curve (Vanoni, 1966; Novak, 1973). The broad minimum in the Hjulstrom curve shows that fine sand is the easiest material to move. The curve rises at large particle sizes because the weight of cobbles and boulders holds them in their sockets. The curve rises at small grain sizes because silt and clay minerals are cohesive and tend to stick together. At velocities above the Hjulstrom curve, the bed material is destabilized and transported. In the area labeled "transportation," material already in motion is transported but the shear is not sufficient to break up fine-grained cohesive sediments. When velocities fall into the area labeled "sedimentation," the boundary

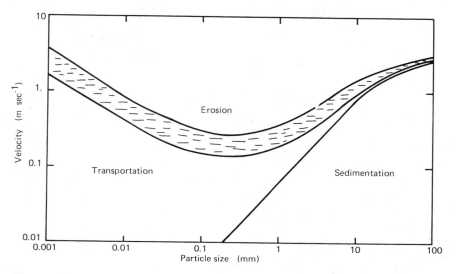

Figure 8.12 Critical threshold velocity for bed load movement. The hatched band is Hjulstrom's curve, which is determined experimentally. The solid line is the Shields Curve, which describes conditions necessary to keep material moving once the more cohesive small particles have been torn loose from their bed. [Adapted from Vanoni (1966) and Novak (1973).]

shear on the bed is below the critical threshold, the bed stabilizes, and sediment motion ceases.

The calculation of bed-load discharge from natural channels is difficult because of the distribution of grain sizes and the irregular form of the bed. Most of the bed load formulas were developed by empirical analysis of data obtained with laboratory flumes using uniform sand sizes. The Meyer-Peter formula is typical (Graf, 1971; Colby, 1964), although other equations have been used (Einstein, 1964).

$$\rho g \frac{R}{d} \frac{Q}{Q_b} \left(\frac{K_B}{K_G}\right)^{3/2} \frac{dz}{dl} - 0.047(\rho_s - \rho) = \frac{\rho^{1/3}}{4d} \left(\frac{\rho_s - \rho}{\rho_s}\right)^{2/3} \left(\frac{S_b}{B}\right)^{2/3} \quad (8.5)$$

where R = hydraulic radius = D for wide channels, d = median grain diameter, Q = discharge (in cubic meters per second), Q_B = discharge responsible for bed transport (in cubic meters per second), K_B = channel roughness = $1/n$, where n = Mannings n (Eq. 6.28), K_G = grain roughness = $26d_{90}^{1/6}$, dz/dl = channel slope, S_b = total bed-load discharge as dry weight (in metric tons per second), B = channel width (in meters). The Meyer-Peter formula contains the critical threshold below which no bed transport occurs.

Particles in suspension are continuously falling and depend on the forces of turbulent flow to sweep them into suspension again. The fall velocity is limited by the drag between the particles (usually assumed to be spherical) and the fluid. The velocities are described by Stokes law (Einstein, 1964; Vanoni, 1963):

$$v_s = \frac{\rho_s - \rho}{\rho} \frac{g\rho}{18\eta} d^2 \qquad \text{for laminar flow} \qquad (8.6)$$

$$v_s = \left(\frac{4}{3} \frac{\rho_s - \rho}{\rho} \frac{gd}{C_d}\right)^{1/2} \qquad \text{for turbulent flow} \qquad (8.7)$$

where v_s = fall velocity (in centimeters per second), d = grain diameter (in centimeters), g = gravitational acceleration (in centimeters per second square), η = viscosity of fluid (in grams per centimeter—second), and C_d = drag coefficient, which depends on Reynolds number but has a numerical value of about 0.5 over a wide range of N_R.

Computing total suspended load is difficult because the suspended sediments are unevenly distributed through the water column. Einstein (1964) gives the distribution as

$$\frac{C_y}{C_a} = \left[\frac{a(D-y)}{y(D-a)}\right]^z ; \qquad z = \frac{v_s}{0.4U_*} \qquad (8.8)$$

where C_y = concentration of sediment at distance, y, above level, a, and D = flow depth. The total sediment load is then obtained by integrating over the depth of the water column from the reference level, a, taken as the top of the boundary layer at the top of the bed load and multiplying by the width of the channel:

$$S_s = B \int v_y C_y \, dy \qquad (8.9)$$

where v_y is given by Eq. 8.4 and C_y is obtained from Eq. 8.8. B is the width of the channel. The integration can be done graphically (see Einstein, 1964, or Graf, 1971) or by computer.

The sediment transport theory described earlier is that usually given in the civil engineering literature and is based on various empirical equations that have been validated by both laboratory flume experiments and engineering practice. An alternative and more fundamental approach relating the transport of sediment to the stream power is derived from considerations of energy and momentum conservation (Bagnold, 1966, 1968, 1977).

Transport of sediments through caves is highly episodic. During periods of low flow, when most caves are visited by observers, cave streams and springs are clear—there is little evidence for suspended load—and the sediment beds appear to be stable. Floods drive large pulses of bed load and also carry slugs of suspended load. Because caves are not readily observed when in flood, little is known of the actual transport rates or how the loads are distributed among flood pulses of various magnitudes.

8.5 CARBONATE SEDIMENT DEPOSITS

8.5.1 *Mineralogy*

The secondary cave carbonate minerals are listed in Table 8.5. Of these, calcite is common, aragonite is found frequently, and the others are rare. Massive speleothems are usually composed of calcite, rarely aragonite. Magnesium carbonate is found as residual deposits, often as the loose pasty masses known as moonmilk. Speleothems in dolomite caves are usually composed of calcite; dolomite is rarely found in speleothems. Hill (1976), W. B. White (1976b), and Hill and Forti (1986) provide many detailed descriptions of speleothems.

8.5.2 *Chemistry of Deposition*

The basic mechanism of calcite deposition is illustrated in Fig. 8.13 and 8.14. The principal source of carbon dioxide is the soil. The source of the calcite is the limestone dissolved at the soil/bedrock contact by the CO_2-rich waters. When the diffuse infiltration waters emerge from the ceilings of caves, the solutions are far out of equilibrium with the CO_2 pressure of the cave atmosphere (typically in the range of $10^{-2.5}$ atm). Excess CO_2 is degassed with concurrent deposition of $CaCO_3$. Because the reaction is driven by CO_2 degassing, travertine can deposit in caves with 100 percent relative humidity

Table 8.5
Carbonate Minerals Found in Caves

Mineral	Composition	Form	Reference
Calcite	$CaCO_3$	Dripstone, flowstone, eccentric forms	a,b
Aragonite	$CaCO_3$	Dripstone, anthodites	a,b
Dolomite	$CaMg(CO_3)_2$	Alteration products, moonmilk	c,d
Huntite	$CaMg_3(CO_3)_4$	Moonmilk	d
Magnesite	$MgCO_3$	Moonmilk	d
Hydromagnesite	$4MgCO_3 \cdot Mg(OH)_2 \cdot 4H_2O$	Moonmilk	a,b,d,e
Nesquehonite	$MgCO_3 \cdot 3H_2O$	Moonmilk	d,e
Monohydrocalcite	$CaCO_3 \cdot H_2O$	Knobular crusts	e

Sources: (a) W. B. White (1976b); (b) Hill (1976); (c) Thrailkill (1968b); (d) Baron et al. (1959); (e) Fischbeck and Müller (1971).

248 / *Geomorphology and hydrology of karst terrains*

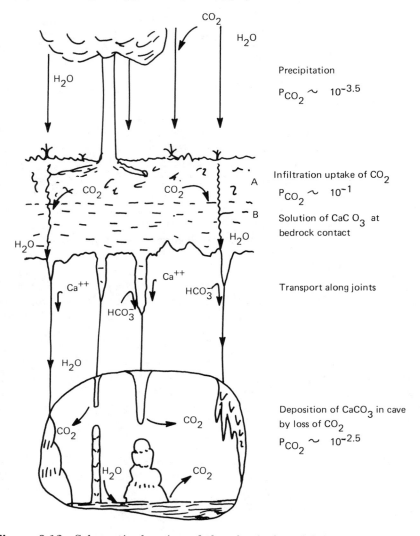

Figure 8.13 Schematic drawing of the physical model for calcite speleothem deposition from concepts of Holland et al. (1964).

such as those with no entrances or those in which air circulation is blocked by sumps. Under these conditions the $CaCO_3$ concentration decreases until the CO_2 pressure in the water at the growing crystal becomes equal to that of the cave atmosphere. If the relative humidity is less than 100 percent some evaporation will also take place and further deposition will result. In general, crystal perfection is much greater when growth occurs without evaporation.

The solubility curve for aragonite is also shown in Fig. 8.14. Ara-

Soils, sediments, and depositional features / 249

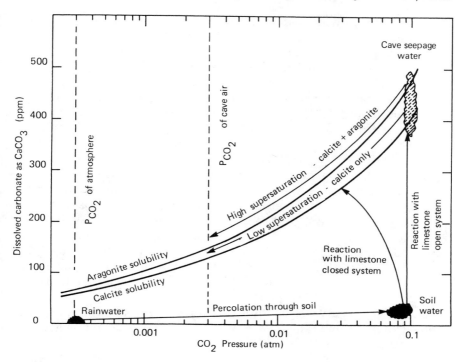

Figure 8.14 Solubility curves for calcite and aragonite, showing the geochemical pathway of seepage water that picks up $CaCO_3$ from the base of the soil and redeposits it as speleothems in caves.

gonite is more soluble than calcite over the entire range of temperatures and CO_2 pressures, and never appears as a thermodynamically stable phase. However, aragonite is a common mineral in caves. The explanation appears to be that the nucleation and growth of calcite is easily poisoned by impurities such as Mg^{2+}. If calcite growth is inhibited, a small additional CO_2 loss will bring the seepage water into supersaturation with aragonite, which precipitates easily at low supersaturations (Curl, 1962; Berner, 1975).

Climate appears to have a profound influence on the quantity of speleothems in caves. Tropical caves often contain massive calcite deposits, and aragonite is rare. Temperate caves have less secondary calcite, and aragonite is more common. The rate of secondary calcite deposition must be controlled by the water and CO_2 supply. Factors that could lead to more prolific growth of calcite in tropical caves include the following:

1. Higher CO_2 contents in the soils, particularly in crevices that are the inlet points for the diffuse infiltration.

2. Longer growing seasons. Year round plant growth and soil bacteria activity enhances the overall CO_2 supply.
3. High precipitation rates in many tropical regions.
4. Higher temperatures. Although the solubility of CO_2 is lower at higher temperatures for any given partial pressure, high temperatures increase kinetics, thus bringing the soil closer to equilibrium. Higher temperatures also promote more rapid decay of organic material in the soil.

8.5.3 Carbonate Speleothems

Table 8.2 lists the general speleothem forms. Most of these can be formed from carbonate minerals, but dripstone and flowstone forms are the most common. Dripstone is deposited from seepage water entering the cave passage through joints or bedding plane partings. The emerging water accumulates into a drop, which degasses CO_2 and deposits a minute ring of calcite before the drop breaks free and falls. Each succeeding drop adds another increment to the ring, which elongates into a hollow tube known as a straw stalactite (Fig. 8.15).

Curl (1972) analyzed the forces acting on the pendant droplet and showed that straw stalactites have a minimum diameter, which is described by a dimensionless Bond number

$$B_0 = \frac{d^2 \rho g}{\sigma} \tag{8.10}$$

where d = minimum diameter of stalactite, ρ = density of fluid, g = acceleration due to gravity, and σ = surface tension. The experimentally determined Bond number for minimum-diameter stalactites is 3.50, which, for a nominal cave temperature of 10°C and normal earth-surface gravity, gives a minimum diameter of 5.1 mm—close to the observed values.

The fast growth direction in calcite is along the c-axis of the trigonal crystal. Of all possible nuclei formed in the initial drop, those that accidentally have their c-axis aligned vertically act as seeds for succeeding rings of crystal growth; as a result, most straw stalactites are composed of only one or a few grains of calcite, all oriented with their c-axes along the straw axis (Moore, 1962). Once a straw stalactite is established, some flow usually occurs down the outside and some leaks through the walls from the central canal to produce new calcite growth on the outside (Fig. 8.16). If growth is very slow, the entire stalactite may be monocrystalline (Andrieux, 1962, 1965a). At slightly faster growth rates, new calcite grains are nucleated on the surface, oriented with the fast-growth direction perpendicular to the straw. The result is a stalactite with wedge-shaped grains arranged radially around the primary straw with its central canal. At still

Figure 8.15 Straw stalactites growing from solution percolating down a curved ceiling joint. Hostermans Pit in Pennsylvania. [Photo by T. P. O'Holleran.]

faster growth rates, the orientations of individual grains are lost. Polycrystalline stalactites may be of any grain size, and some have enclosed layers of clay or sand that mark events of flooding or other breaks in the depositional sequence. Stalactite growth from both tip and sides results in speleothems with concentric shells; in cross section these are seen as concentric growth rings.

When drops fall to the floor they deposit mound-shaped structures known as stalagmites. Stalagmites grow vertically from the thin moisture films formed by the falling drops. They have no central canal; in longitudinal section they appear as a series of superimposed caps that may exhibit growth banding (Fig. 8.17). Except for a rotational symmetry about the vertical axis, stalagmites are not subject to other constraints and are found in many complex shapes.

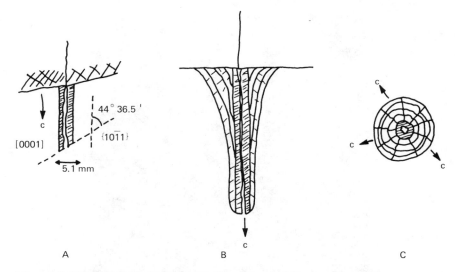

Figure 8.16 Crystallographic relationships in (A) straw and (B) overgrown stalactites. The tip of a straw is often marked by the [10$\bar{1}$1] cleavage plane of calcite. (C) Shows the cross-section of (B). Most stalactites have random orientations of calcite crystals.

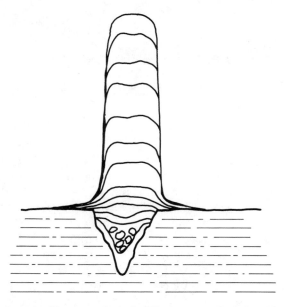

Figure 8.17 Longitudinal section of cylindrical stalagmite showing sequential growth caps and a "root" beginning as a drip pocket in sediment with a nest of cave pearls.

Smooth cylindrical columns form when both water feed and calcite growth rates are constant. Mound-shaped forms and terraced stalagmites are also common, suggesting more complex variations in growth rate (Franke, 1965).

Mass balance considerations require that a cylindrical stalagmite's growth rate and equilibrium diameter be related (Curl, 1973).

$$\frac{dz}{dt} = \frac{4C_0 q}{\pi \rho_c d^2} \tag{8.11}$$

where d = stalagmite diameter (in centimeters); C_0 = concentration of calcite in solution available for deposition (in grams per cubic centimeter) (less than the total concentration if deposition can take place only to the limit imposed by the CO_2 pressure of the cave atmosphere); ρ_c = density of calcite (2.71 g cm^{-3}); q = flow rate of water feeding the stalagmite (in cubic centimeters per second); and dz/dt is the vertical rate of growth of the stalagmite (in centimeters per second). In principle d, C_0, and q can be measured and the equation integrated to give the age of the stalagmite as a function of its height.

Curl (1973) was able to show that there are two limiting flow regimes for stalagmites. At the limit of rapid flow, the equilibrium diameter varies with the square root of the flow rate, as indicated by Eq. 8.11. At the slow flow limit, the equilibrium diameter is independent of the flow rate, and the minimum diameter for stalagmites is given by

$$d = \left(2\frac{V}{\pi \delta}\right)^{1/2} \tag{8.12}$$

where V = volume of the water drops and δ is the thickness of the moisture film left on the surface of the stalagmite by the falling drops. The minimum diameter is on the order of 3 cm, which is in agreement with observations in caves.

Flowstone, the most massive of the secondary cave deposits, forms from water moving as sheet flow down walls and over floors (Fig. 8.18). The deposits are layered. Many are terraced in complex patterns. Water that emerges along a line of joints or drips over the edges of ledges deposits forms of great complexity. Exact shapes are determined by the rate of water flow and the path they follow. Draperies are an intermediate form between dripstone and flowstone.

Most dripstone and flowstone deposits are colored shades of yellow, orange, tan, brown, occasionally deep blood-red, and even more rarely green or blue. There appear to be three sources for the color of calcite speleothems (Gascoyne, 1977; White, 1981).

1. Most of the common oranges and browns are stains from organic compounds brought in from the surface. These are

Figure 8.18 Flowstone cascade in Thorn Mountain Cave, West Virginia.

probably fulvic and humic acids, although no exact identification has been made.
2. Deep browns and blacks are often caused by oxides and hydroxides of iron and manganese. Some dripstone is mud-brown because iron-pigmented clay is deposited on the calcite surface by occasional flooding. Iron pigmentation, however, is much less common than was previously supposed.
3. The rare yellows, greens, and blues are caused by the iron-group ions such as Cu^{2+} and Ni^{2+} substituting for Ca^{2+} in the calcite structure.

Shields are unusual massive speleothems that, in a sense, are the two-dimensional analogs of stalactites. They are plate-shaped structures that protrude from cave walls. Shields vary from 10 cm to several meters in diameter, with thicknesses of a few tens of centimeters. They consist of two matching concentric plates with a medial crack analogous to the central canal of stalactites. The perimeter of the shield is often fringed by stalactites, apparently fed by the medial crack. Shields occur in only a few caves, of which Grand Caverns in Virginia and Lehman Caves in Nevada are the best known. Kundert (1952) showed that the orientation of the plates in Lehman Caves follows the regional joint system through which water moves under pressure. Moore (1958) suggested that earth tides keep the medial crack open.

Erratic speleothems are formed when the forces of crystal growth dominate the effects of flowing or dripping water. The shapes of erratic speleothems are determined by both mineralogy and growth rate. Those composed mainly of carbonate minerals are helictites, anthodites, and botryoidal speleothems.

Helictites are similar to stalactites in that they have a central canal and grow from the tip by water fed from some source in the bedrock (Fig. 8.19). Some helictites have radii much less than the minimum radius for stalactites and thus must grow without droplet formation. Some are thicker because of calcite growth over the primary crystallization (Moore, 1954; Andrieux, 1965b; Kempe and Spaeth, 1977). The French call the long slender helictites with diameters less than the stalactite minimum, filiform, and the more massive varieties, vermiform (Géze, 1957).

Anthodites are clusters of radiating crystals (Fig. 8.20). The shapes are dictated by the growth habit of aragonite, which seems to form either the entire anthodite or at least the cores of the radiating masses. Some anthodites are composed of calcite, which has formed from the transformation of the original aragonite.

Botryoidal forms include a diverse collection of knobular, corallike, and beadlike deposits. Some, composed entirely of calcite grow-

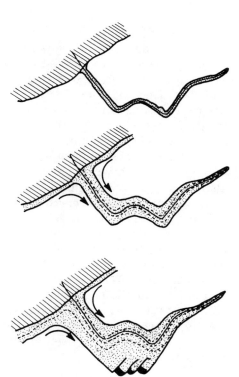

Figure 8.19 Cross sections of filiform and vermiform helictites with their central canals. [Adapted from Andrieux (1965b).]

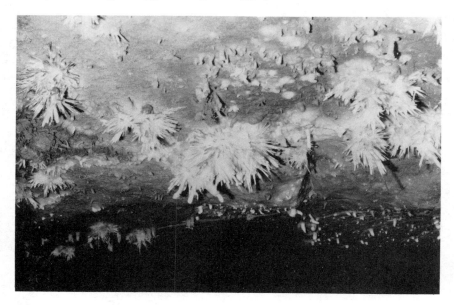

Figure 8.20 Anthodites from Skyline Caverns in Virginia, the type locality for the original description of this speleothem (Henderson, 1949).

ing in successive layers from small projections, appear to be related to fast-moving sheets of supersaturated water. Large gargoylelike forms are sometimes formed on walls of pits by splashing water. These are also layered, but tend to be spongy and include much sand and clay. They are sometimes called spattermites. A third variety is the cave coral, which occurs as irregular, often roughly knobular forms, sometimes with tufts of aragonite or blobs of moonmilk on them (Fig. 8.21). The mineralogy is often mixed calcite and aragonite, and sometimes other minerals.

Moonmilk is a white unconsolidated deposit of carbonate minerals. When wet, it resembles cottage cheese; when dry, it is a loose white powder with an earthy texture. Most of the magnesium carbonate minerals listed in Table 8.5 are found as moonmilk. Calcite is a common constituent of moonmilk in caves at high altitudes. There is some evidence that calcite forms in cold cave environments whereas hydromagnesite and, to a lesser extent, the other moonmilk minerals form in more temperate environments. The presence of blobs of moonmilk in association with tufts of aragonite crystals suggests that it is the residual magnesium left behind by the evaporating solutions, because the Mg^{2+} ion has a very low solubility in the aragonite structure. It has been argued that some moonmilk has a bacteriological origin.

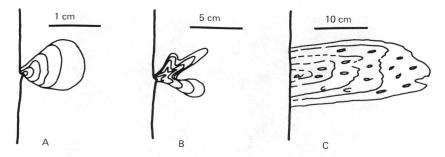

Figure 8.21 Several styles of botryoidal deposits: (A) globulites, which are spherical, beadlike, or popcornlike forms; (B) branched or stalklike deposit (cave coral), which often contains aragonite or moonmilk; (C) spattermite type of spongy, layered deposit with much sand, clay, and void spaces included.

Calcite has a tendency to build up around pools of water to form rimstone dams, which sometimes reach heights of several meters. Continued degassing of CO_2 from the pool surface produces supersaturation, and the interiors of many cave pools are lined with well-formed scalenohedral calcite. Water dripping into shallow cave pools induces layered concretions to build up around sand grains, bat bones, and other nuclei to form the loose spherical objects known as cave pearls (Baker and Frostick, 1947, 1951). Some caves flooded with calcite-supersaturated water thus have crystal linings. The caves of the Black Hills in South Dakota (Tullis and Gries, 1938; Deal, 1962; White and Deike, 1962; Palmer, 1981b) are the best known examples. Conditions under which crystal-lined caves can develop are very restricted. The water must become highly supersaturated to develop large masses of crystals and yet must be sufficiently free of impurities to allow individual calcite crystals to develop without interruption.

8.5.4 Travertine and Tufa Deposits

Karst water emerging from springs contains considerable dissolved calcite at high carbon dioxide partial pressures, although many of these waters are only slightly supersaturated or even slightly undersaturated. A combination of increased temperature as the water warms to surface ambients, and aeration as the surface stream flows over a rough bed, causes loss of CO_2, supersaturation, and sometimes a deposit of calcite in the form of calcareous tufa or travertine. When these deposits are on steep slopes, they often form spectacular travertine dams analogous to rimstone pools (Fig. 8.22).

Very large supersaturations—$SI_c = +1$ or greater—seem to be

Figure 8.22 Travertine falls on the Krka River in Yugoslavia.

necessary to produce travertine in surface waters. Many karst springs feed surface streams that have saturation indices in the range of 0 to 1, but large travertine deposits seldom occur. Suarez (1983) found that the Colorado River was supersaturated throughout both diurnal and seasonal cycles, but no travertine was observed. Massive surface travertines occur mainly in semitropical to tropical environments, where the differential in both temperature and P_{CO_2} between the spring and the atmosphere is larger.

8.6 EVAPORITE DEPOSITS IN CAVES

8.6.1 *Mineralogy*

In most karst groundwaters the order of abundance of anions is $HCO_3^- > SO_4^{-2} > Cl^- > NO_3^-$. In wet caves and caves with near 100 percent relative humidity, only carbonate minerals are deposited. The more soluble sulfates, chlorides, and nitrates remain in solution and are carried out of the system. When the relative humidity drops to the range of 75 to 90 percent evaporation can take place. Dry, even dusty, caves often have relative humidities in this range, and water-soluble minerals are deposited. Evaporite deposits in caves are much like evaporite sediments in desert areas, dry lakes, and similar environments, except that their mineralogy is less complex because of the small number of ions in solution.

Most of the evaporite minerals are sulfates (Table 8.6), although halite (NaCl) occurs in a few caves. Gypsum is by far the most common. Because of its relatively low solubility, gypsum is subject to less stringent environmental constraints than the magnesium and sodium salts, which are highly soluble in water.

Table 8.6
Evaporite Minerals Found in Caves

Mineral	Composition	Form	Reference
Gypsum	$CaSO_4 \cdot 2H_2O$	Dripstone, wall crusts, "flowers," bladed crystals, mixed in clastic sediments	a,b
Epsomite	$MgSO_4 \cdot 7H_2O$	Dripstone, "flowers," in clastic sediments	a,b
Hexahydrite	$MgSO_4 \cdot 6H_2O$	Loose crusts on floor	c
Mirabilite	$Na_2SO_4 \cdot 10H_2O$	Dripstone, "flowers," loose tufts of crystals	a,b
Blödite	$MgSO_4 \cdot Na_2SO_4 \cdot 4H_2O$	Loose crusts on floor	c
Celestite	$SrSO_4$	Wall crusts	b
Halite	$NaCl$	"Flowers" and crusts	d

Source: (a) White (1976b); (b) Hill (1976); (c) Freeman et al. (1973); (d) Dunkley and Wigley (1967).

8.6.2 Chemistry of Deposition

Sources of sulfate ion, and of gypsum and other sulfate minerals, include at least the following:

1. *Evaporation of sulfate-bearing groundwaters.* Dissolution of gypsum beds elsewhere in the stratigraphic section provides sulfate in solution that can be carried into caves and deposited by simple evaporation of the solutions. The sulfate minerals in many caves of the American Southwest seem to originate in this way. George (1974) has argued that much of the gypsum found in caves of the Interior Lowlands of Kentucky and Tennessee also originates from lateral transport of solutions derived from evaporite beds in the Mississippian limestones.

2. *Oxidation of pyrite.* Pyrite (FeS_2) is a common accessory mineral in many limestones and in shale beds in the limestone. When pyrite is exposed to the oxygen-rich, moist atmosphere of caves it reacts:

$$FeS_2 + \tfrac{7}{2} O_2 + H_2O \rightleftharpoons Fe^{2+} + 2 SO_4^{2-} + 2 H^+ \qquad (8.13)$$

$$Fe^{2+} + H^+ + \tfrac{1}{4} O_2 \rightleftharpoons Fe^{3+} + \tfrac{1}{2} H_2O \qquad (8.14)$$

$$Fe^{3+} + 3 H_2O \rightleftharpoons Fe(OH)_3 + 3 H^+ \qquad (8.15)$$

$$SO_4^{2-} + 2 H^+ + H_2O + CaCO_3 \rightleftharpoons CaSO_4 \cdot 2 H_2O + CO_2 \qquad (8.16)$$

Minor gypsum speleothems in many Appalachian caves seem to come from this source. The sulfate ion produced by pyrite oxidation is mobile, but the ferric iron immediately precipitates as $Fe(OH)_3$. Pohl and White (1965) claimed that the sulfate minerals in the Flint

Ridge–Mammoth Cave System originated through a two-step process. Pyrite oxidation occurs in the overlying Big Clifty sandstone, and the iron remains there. Acid sulfate solutions migrate downward into the limestone, but do not immediately react because of the backpressure of carbon dioxide built up in joints in the bedrock. Open cave passages act as sinks for carbon dioxide, allowing the reaction to proceed. Thus, gypsum is formed in situ by direct replacement of calcite in the wall rock, generating the expansion pressure that results in mineral-activated breakdown.

3. *Hydration of anhydrite.* Some limestones contain thin lenses or nodules of anhydrite ($CaSO_4$). When the rock is exposed to weathering, anhydrite reacts with water to form gypsum, which is then redistributed through soils and over the walls of cave passages. Some anhydrite nodules have been found in the Greenbrier limestone of the West Virginia karst.

4. *Oxidation of H_2S.* There is some evidence that gypsum in Kane Caves of the Bighorn Basin, Wyoming (Egemeier, 1981), and Carlsbad Caverns and other caves of the Guadalupe Mountains (Davis, 1980) is formed by the oxidation of H_2S transported into the caves from some deep source, perhaps a petroleum reservoir. When H_2S reaches the oxygen-rich environment of the upper phreatic and vadose zones, it oxidizes, sometimes with elemental sulfur as an intermediate step. The sulfuric acid reacts with calcite to form gypsum:

$$2\ H_2S + O_2 \rightleftarrows 2\ S^0 + 2\ H_2O \tag{8.17}$$

$$H_2S + 2\ O_2 \rightleftarrows 2\ H^+ + SO_4^{2-} \tag{8.18}$$

$$H_2O + 2\ H^+ + SO_4^{2-} + CaCO_3 \rightleftarrows CaSO_4 \cdot 2\ H_2O + CO_2 \tag{8.19}$$

Elemental sulfur has been found associated with gypsum in Cottonwood Cave in the Guadalupe Mountains (Davis, 1973). Either this is evidence for an H_2S oxidation mechanism, or the depositional environment of the Cottonwood Cave sulfate minerals must have been much more reducing than they generally are in caves.

8.6.3 Evaporite Speleothems

Dripstone and flowstone made of sulfate minerals are uncommon, although spectacular ice-clear stalactites of epsomite and mirabilite are sometimes found.

Gypsum often appears as a granular crust (Fig. 8.23) formed by water seeping uniformly from pores in the wall rock. This is an ongoing process; old crusts become detached, fall off, and new crusts form behind them.

Gypsum flowers (oulopholites) grow in the form of curved radiating masses of very long, narrow gypsum crystals (Fig. 8.24). The

Soils, sediments, and depositional features / 261

Figure 8.23 Gypsum crust in Cumberland Caverns, Tennessee.

Figure 8.24 Gypsum flowers, or oulopholites in the Cumberland Caverns of Tennessee.

flowers consist of masses of individual crystals rather loosely bound together. Growth originates at the base, where the nutrient solution emerges from the wall, and faster growth in the center of the bundle causes the radiating petal-like growth pattern. Epsomite and mirabilite grow in the form of flowers.

Gypsum is found buried in soils and sometimes spread over the surface of soil as long, narrow "gypsum needles." These are usually massive, water-clear crystals. The crystals are multiply twinned, and a reentrant angle twinning may be responsible for the growth of the needles.

Sulfate minerals occur in many other complex forms and styles, such as "angel hair," and "gypsum rope," as well as in starburst patterns and other forms brought about by details of crystal growth. See Hill and Forti (1986) for more detailed descriptions.

8.7 OTHER CAVE MINERALS

8.7.1 *Phosphates and Nitrates*

Most phosphate minerals from caves are associated with bat or bird guano (Table 8.4), although some may have an inorganic origin. Alkaline earth and aluminum phosphates are highly insoluble; phosphate from surface sources is usually precipitated in the soils and is only rarely carried into caves. Small veins of a mixture of crandallite, taranakite, and other phosphate minerals were found in the sediments of Butler Cave in Virginia (White, 1982).

Nitrate minerals are widely dispersed in the sediments of dry caves and were the source of nitrate for gunpowder in the early nineteenth century. Saltpetre caves are widely distributed throughout the United States (Hill et al., 1981). Nitrate is thought to occur in cave sediments as nitrocalcite [$Ca(NO_3)_2 4H_2O$] (Hill, 1981c), but this salt deliquesces readily, and the nitrates are more likely to disperse in solution in the adsorbed water of the cave sediments. In the original saltpetre mining operations, cave sediments were leached to obtain nitrate solutions, which were reacted with wood ashes as a source of K^+, and then boiled down to fractionally crystallize KNO_3 (Eller, 1981). The nitrate minerals are usually associated with water-soluble sulfate minerals in sediments, and one double salt, darapskite [$Na_3(NO_3)(SO_4)H_2O$], has been reported (Hill and Ewing, 1977).

Hypothesized sources for nitrate deposits include bat guano, action of nitrogen-fixing bacteria in cave sediments, and leachate from surface soils. Hill (1981a,b) argues that nitrates are leached from surface soils, particularly in the eastern U.S. hardwood forests, carried into caves by percolating seepage waters, and concentrated in cave sediments by evaporation.

8.7.2 Resistate Minerals

Iron is immobile in near-neutral, oxidizing environments. Concentrations of Fe^{3+} in groundwater are generally less than 1 ppm. Ferric iron can be formed in situ in caves by means of the oxidation of pyrite where it is immediately precipitated as $Fe(OH)_3$, usually as the amorphous form known as limonite. Limonite speleothems take the form of thin stains and sheets on cave walls and sometimes stalactites and stalagmites (White et al., 1985). Limonite stalactites can form from highly acid drip waters where cavernous limestones are in contact with overlying shales and sandstones. Limonite is also carried into caves in suspension as a slurry of finely divided particles or colloids.

Stream pebbles in many caves are coated with thin, black layers often called manganese oxide. More rarely, massive deposits of black manganese oxides are found, such as those in Jewel Cave in South Dakota (Deal, 1962). Moore (1981) found the most common manganese mineral to be birnessite ($CaMn_6O_{13}3H_2O$), although other manganese minerals occur. The manganese minerals are usually amorphous to x-rays and are best identified by infrared spectroscopy.

8.7.3 Ice

Ice forms in the entrance areas of many caves in temperate and alpine climates, creating spectacular, if temporary, speleothems. Some caves act as cold air traps, retaining ice far into the summer. Merriam (1950) and Halliday (1954) have compiled lists of the glaciers or freezing caves in the United States. Some caves in alpine areas retain perennial ice, for example, Fossil Mountain Ice Cave in Wyoming. This ice is correctly classified as a cave sediment because some of it may well have persisted since the Pleistocene period. The great ice caves of Austria—the Eisriesenwelt and the Dachstein Ice Cave—expose massive cliffs of delicately layered ice where air currents have ablated part of the deposit. Ice formation in caves is intimately related to air flow patterns, temperature, and rate of heat transfer into and out of the cave (Wigley and Brown, 1976). Ice deposition in caves was of great interest to early cave investigators, and lists of ice caves (e.g., Balch, 1900) and cave meteorology occupied much of the early textbooks (e.g., Kyrle, 1923).

9

Theories, Models, and Mechanisms for the Origin of Caves

9.1 CONCEPTS

Caves have excited curiosity since time immemorial, so it is not surprising that efforts have been made to explain them scientifically since the dawn of the geological sciences. Of all subjects in the geomorphology and hydrology of karst, cave origins carry the heaviest baggage of historical precedent and ancient theorizing and speculation. The baggage is not made lighter by the fact that few of the earlier theories have been completely superseded or discredited. Watson and White (1985) have analyzed much of earlier cave origin theory in some detail and only brief summaries are given here.

This chapter deals only with solution caves in carbonate rocks. Excluded momentarily are solution caves in gypsum and salt which will be discussed in Chapter 11 and various nonsolution caves such as lava tubes, glacier caves, and tectonic caves which will be discussed in Chapter 12.

It is generally agreed that caves are the voids that remain after portions of the carbonate rock have been removed by dissolution in circulating groundwater. The dissolution reactions are usually those with carbonic acid, although other acid sources, such as sulfuric acid from oxidation of pyrite or hydrogen sulfide gas, have been shown to be important in certain special cases. It is also generally appreciated that caves vary greatly in length, passage morphology, groundplan, and vertical extent. If some common mechanism of development underlies the solutionally widened fractures in the dolomite karst of Ohio, the vast labyrinth of tangled tubes in the Mammoth Cave area, the gigantic chambers in Gunong Mulu, and the kilometer-deep shafts and streamways in the Huautla Plateau of Mexico, the theory must have considerable flexibility.

Flexibility is provided by having a large number of variables that can be adjusted to the individual hydrogeologic environment. These can be drawn from Fig. 4.1, which laid out the overall framework for karst landscapes in terms of a seven-dimensional coordinate system. It is the diversity of possible "dial settings" of these seven variables that accounts for the observed diversity in cave patterns. It is the purpose of cave origin theory to connect the dial settings to observed caves as they have evolved through time.

9.2 HISTORICAL PERSPECTIVE

Efforts to explain or understand the existence of caves began in Europe well back in the last century. In the United States most of the systematic theorizing dates from the work of Davis, Piper, Swinnerton, and Gardner in the 1930s. Although most of the ideas propounded in the papers of the 1930s had been anticipated 30 years earlier in Europe, a direct comparison is not easy. The European geomorphologists were concerned with active drainage systems and saw caves only as the conduits through which the water moved. The American geomorphologists were talking about dry caves, abandoned conduit fragments remaining after the water had disappeared. They were also preoccupied with the water table and whether caves were formed above or below it. This emphasis was misplaced and led to a long sterile debate that diverted attention from more interesting questions concerning caves (Watson and White, 1985). The theories also suffered from inadequate understanding of the laws of fluid flow and almost total innocence of the chemistry of limestone solution. The chemistry was known in the 1930s (e.g., Pia, 1933), but like fluid mechanics, was not much studied by geomorphologists. A. C. Swinnerton was a notable exception.

9.2.1 *The Groundwater Debate*

Grund (1903) and Katzer (1909) each examined the behavior of groundwater in karst aquifers and came to diametrically opposite conclusions. Grund divided the karst aquifer into two zones separated by a water table. The saturated zone was thought to be largely stagnant water and its upper surface a water table of regional extent that sloped down to the sea. The upper zone was one of active water circulation through which water moved in open caves and conduits. Water that drained from poljes and dolines flowed through open caves (thereby forming the caves), down to the static water level.

In Katzer's view there was no integrated groundwater body in the karst. Streams that sank in ponors flowed through open channels or flooded conduits to springs at lower elevation. There was no concept of a water table; the channels were thought to be independent of each other and the bedrock between them to be essentially impervious. Von Knebel (1906) and Martel (1921) strongly supported this view. They made no distinction between vadose and phreatic zones.

The observation of both open and flooded conduits in the alpine karst of Austria, France, and other parts of Europe led other investigators, particularly Lehmann (1932), to the idea of pressure tube flow. Lehmann proposed that all that was needed for karst drainage was a system of open fractures (Lehmann thought they should be at least 1–2 mm wide). Recharge water from the high plateaus would force its way through these fractures, enlarging them first by solution and later by both solution and abrasion. Some pressure tubes descend deep within the karst mass and so are always flooded. Others are drained during seasons of low flow and can be explored at that time. Horn (1947) found the pressure tube concept useful for explaining caves in Norway, where extreme variation in flow rates derives from seasonal variations in melting of snow and ice. The concept of caves as pressure tubes determined by details of fracture systems, faults, and placement of noncarbonate rocks, remains a central theme in much European literature, especially that of France. See, for example, Cavaillé (1965), Géze (1965), and Vandenberghe (1964).

9.2.2 Vadose Theories

The central thesis of vadose theories of cave origin can be stated very simply: caves are formed by the action of underground streams flowing either at or above the water table (Fig. 9.1A). It is claimed that infiltrating surface water and water descending through sinkholes is very aggressive and that the aggressiveness is quickly lost as the water passes into the saturated zone. Solution, therefore, takes place most actively at the top of the saturated zone. The infiltrating waters move vertically downward, dissolving sinkhole drains, vertical shafts, and solution chimneys until they reach the top of the saturated zone. At this level, they turn and begin to flow horizontally along fractures and bedding plane partings, gradually enlarging them into channels carrying free-surface streams.

To Martel (1921) the existence of underground rivers was self-evident, as was their tremendous erosive power as they thundered down deep pits from the high plateaus to springs in the valleys below. The water simply picked the easiest path through the joints, fractures, and bedding plane partings. No special mechanisms, stages, cycles, or other details in the historical development of the cave were needed. Indeed, the development of cave passages was considered such a simple phenomenon that Trombe (1952), in his classic textbook, discussed the development of cave systems only in terms of shafts and steeply sloping stream passages.

Theories, models, and mechanisms for the origin of caves / 267

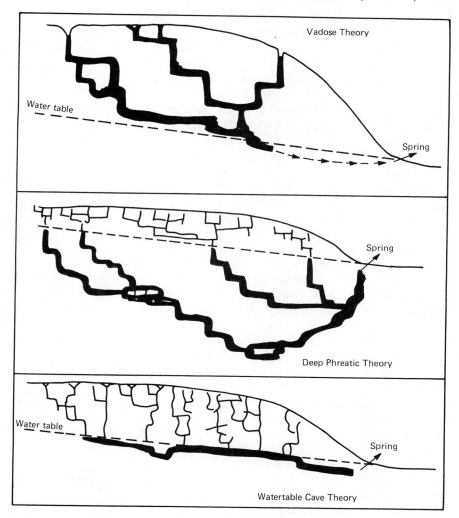

Figure 9.1 Zones of maximum cavern development, according to the vadose, deep phreatic, and shallow phreatic theories. [From D. C. Ford.]

Many of the early accounts of groundwater in carbonate rocks devoted a few pages to the origin of caves (Greene, 1908; Matson, 1909; Johnston, 1933). The ideas of Piper (1932) are perhaps the best articulated. Piper draws an analogy between the erosion cycle on the surface and the underground erosion cycle. Depending on the permeability of the limestone, the underground cycle either precedes or lags behind the surface cycle. In rocks with densely spaced, open joints, drainage may be rapidly diverted to the subsurface and a mature underground system may develop before the surface drainage pattern has time to develop. Conversely, if joints are sparse and tight, surface erosion may dissect the system before an integrated underground drainage can develop.

Gardner (1935) placed great emphasis on geologic structure, assuming that most large caves were formed on the up-dip side of surface valleys. He then claimed that caverns are formed by tapping zones of static water in certain favorable beds by downcutting surface streams. Vadose water was thought to flow down-dip along the bedding planes. As downcutting continues, more and more levels develop and piracy takes place from level to level.

Clyde A. Malott was a pioneer student of karst phenomena in the United States with special interest in southern Indiana. Most of the support for his "invasion" theory of cave origin is interweaved in a series of descriptive papers (Malott, 1929, 1932, 1949, 1951, 1952). The formal statement of his theory, appearing only as an abstract (Malott, 1937), captures the essence of all vadose theories.

The last exponent of simple vadose cave development was H. P. Woodward (1961). Woodward tied mature cave development to contemporary surface drainage and piracy of the flow underground in the manner of Malott. He further required an abrupt change in base level activated by surface stream piracies, climatic change, or sudden lowering of base level to activate the process.

9.2.3 Deep Phreatic Theories

The highly developed deep phreatic theory (Davis, 1930, 1931) follows directly from the ideas of Cvijič (1893) and Grund (1903), but with an emphasis on caves rather than hydrology. The theory has no precedents in earlier American writings.

Davis proposed that solution takes place in the phreatic zone during the old-age stage of the geomorphic cycle. Upon reaching the water table, groundwater continues toward its outlet along flow lines that curve deeply into the phreatic zone (Fig. 9.1B). Solution takes place uniformly along all possible paths. The force driving the flow is the hydrostatic head between base level and the highest part of the water table in the interstream areas. When rejuvenation takes place, as a result of regional uplift, the surface streams cut down into the old peneplain surface. The water table is lowered, and gradually the upper levels of the cavern system become air-filled. At this stage dripstone deposits begin to form. Thus, the theory provides a mechanism for both solution and deposition—a two-cycle theory.

Davis was impressed with the slow rate of solution and argued that only in the peneplain stage of the geomorphic cycle would there be enough time for caverns to grow. It is not clear from any of his writing how deep the deep phreatic flow was supposed to be. Davis

Theories, models, and mechanisms for the origin of caves

was following the emerging theory of groundwater flow, and implies that circulation could go to depths of thousands of feet below the peneplain surface. Caves, according to this theory, are relict features, formed at random depths beneath an old peneplain surface, and not related to any of the contemporary drainage.

The deep phreatic theory found its strongest defender in J Harlan Bretz (1942), who was the first to collect systematic field evidence to support the theory. From an examination of some 130 caves, Bretz cataloged a series of solutional features (described in Chapter 3), which he used to assign caves to a "vadose" or "phreatic" origin category.

The deep phreatic theory emerged from Bretz's analysis with an epoch of clay filling inserted between the two cycles proposed by Davis:

Stage I. Deep-seated solution by groundwater circulation beneath an old-age land surface.

Stage II. Fill epoch. Caverns lying beneath a peneplain are saturated with stagnant water. Fine clay and silt, filtering down from the land surface above, fills the caverns completely (or nearly so) with unctuous clays.

Stage III. Cavern system is drained of water. Invading vadose streams begin to remove clay fills. Sometimes this continues until the vadose streams flow on bedrock floors. In other cases most of the cave remains clogged with clay. Vadose features are superimposed on the phreatic record.

Bretz (1949, 1955) later examined other caves, coming each time to the same conclusion: most caves formed below the water table and were modified and cleared of fill only by the streams that now use them. However, a careful reading of the 1942 paper is instructive. There is much discussion of subwater table streams but no implication of flow at great depths below the water table. Although he was a staunch supporter of the Davis hypothesis, one gets the distinct impression that Bretz's deep phreatic was nowhere near as deep as Davis' deep phreatic.

9.2.4 Shallow Phreatic Theories

Like the vadose theories, the theories of cave origin from groundwater moving near the water table have American and European variants. The American variant stems from examination of caves in low-relief karst in nearly flat-lying limestones, the European variant from the alpine systems that produced the concept of pressure-tube flow.

Swinnerton (1932) was impressed by the observation that most cave passages are nearly horizontal and some seem to be stacked in levels (tiered caves). Tiered caves occur in folded limestones where the horizontal passages cannot be regarded as simply a result of low dip—an observation that bothered Davis and was somewhat brushed aside by Bretz. To account for passages that truncate the geologic structure to maintain a horizontal level, Swinnerton hypothesized that they form in the interval between the high and low stands of the water table as it fluctuates between wet and dry seasons (Fig. 9.1C).

It quickly became apparent that the same arguments would apply if solution took place immediately below the water table. Davies (1960) proposed a revised shallow phreatic hypothesis that consisted of four stages:

Stage I. Random solution at depth. Primitive tubes, pockets, and other small solution openings form at random depths in the aquifer.

Stage II. Integration and mature development of solution openings. This occurs at the top of the water table during a long period of uniform elevation and flow in the water table. A coincidence in cave passage elevations with nearby river terraces suggests a causal relationship between the position of the water table and the position of local base level. This relationship also ties the evolving cave pattern to the evolving history of the drainage basin in which the cave is located.

Stage III. Deposition of fill. Clastic materials are deposited after the integration and mature development of the cave passages.

Stage IV. Uplift and erosion. With the end of stable conditions the cavern passages are raised above the level of the water table and become air-filled. In this stage the passages are modified by deposition of speleothems and erosion of fill material. The ultimate development in this stage leads to total destruction of the cavern by collapse and erosion.

As more data accumulate, it appears that many caves have developed by some sort of shallow phreatic mechanism. The three-dimensional networks of tubes and shafts required by the most extreme interpretation of the deep phreatic theory are seldom found. Caves in the folded Appalachians were shown to retain a horizontal aspect and to terminate in up-dip and down-dip directions, as would be expected if they were controlled by some aspect of base level but not if there was extensive three-dimensional circulation (White, 1960). Davies (1960) and White and White (1974) found that caves could be related to river terraces in the Potomac River Valley. The same kind of correlation was found in Yorkshire, England (Sweeting,

1950), and in the caves of the Demanova Valley in Czechoslovakia (Droppa, 1957).

9.2.5 Introduction of Hydrologic and Kinetic Models

The interpretations of cave origin, in both Anglo-American and European writings, were cast in terms of the flow paths of groundwater and surface streams, of local base levels, and of the evolution of surface landscape. It was apparent to other investigators that account must be taken of the hydraulics of fluid flow and the chemistry of carbonate dissolution.

Rhoades and Sinacori (1941) noted that the flow lines in a homogeneous isotropic aquifer converge near the point of discharge. Because the total flux between adjacent flow lines must be constant, as the flow lines converge, the velocity of flow must increase. This concentration of flow near the zone of discharge, they argued, causes caves to form first in this zone. Once a cave has enlarged, it acts as a drain or short circuit to the flow net, which is then modified to converge toward the upstream end of the advancing conduit. The cave passage thus migrates headward by tapping more and more flow lines as it goes.

Clifford Kaye (1957) was one of the first to recognize that cave development is a kinetic phenomenon. Kaye dissolved specimens of limestone and marble in dilute hydrochloric acid and showed that the rate of solution increased with the velocity of fluid flow. He then postulated a principle of self-acceleration of cave conduits. Of the various pathways through an aquifer, some have somewhat less resistance to flow than others and thus flow velocities are higher along these paths. If the rate of solution increases with flow velocity, these favored routes will dissolve faster. As they become larger, flow velocities increase further, thus increasing the rate of solution still more. In this runaway process the favored paths enlarge at the expense of others and ultimately become the observed conduits. White and Longyear (1962) extended this concept by arguing that the flow would become turbulent at some critical velocity. Turbulent mixing would also increase the rate of solution. They calculated the critical size for the onset of turbulence to be 5 to 10 mm.

Thrailkill (1968a) analyzed the flow of fluid in a network of pipes using the Hagen-Poiseuille equation (Eq. 6.17) for laminar flow and the D'Arcy-Weisbach equation (Eq. 6.20) for turbulent flow. The pipes in the network were of fixed and equal size. Water was inserted into one corner of the network and allowed to discharge from the opposite corner. Under these rather restricted circumstances, Thrailkill found that the flow rates were essentially the same in all pipe segments and that the same distribution was obtained regardless of whether the flow was laminar or turbulent. The pipe-network

272 / Geomorphology and hydrology of karst terrains

calculations suggested that the flow paths should not be dramatically different from those observed in porous media aquifers. Injection of water uniformly over the karst surface (Fig. 9.2) would produce deeply curving flow paths, since the flow lines are required to be perpendicular to the water table surface. Recharge into sinkholes or swallow holes that locally penetrate an otherwise capped aquifer builds up groundwater mounds from which emerge flow lines, some of which are parallel to the water table beneath the caprock. This arrangement was offered to explain the nearly horizontal tiers of passages in Mammoth Cave and its associated cave systems. In the case of capped aquifers with multiple inputs, the additional hydrostatic head from downstream inputs depresses the flow lines from inputs farther upstream.

9.3 ANALYSIS OF CAVE DEVELOPMENT IN TERMS OF GEOLOGIC VARIABLES

The earlier theorizing was focused on the caves themselves, their patterns, solutional sculpturing, and sediment in-fillings. The position of the cave in the drainage basin, its recharge sources, and the hydrogeologic setting were often introduced as afterthoughts or not at all. An important conceptual breakthrough turned this perspective inside-out, taking the drainage basin and hydrogeologic setting as given conditions and interpreting the resulting cave patterns as their necessary consequences. Using the geological variables sketched in Fig. 4.1, one can go quite far in explaining cave patterns without considering the details of chemical reactions along the flow path.

9.3.1 *Structural Setting and Hydraulic Pathways: The Ford–Ewers Model*

D. C. Ford and R. O. Ewers, in a series of papers (Ford, 1965b, 1968, 1971; Ford and Ewers, 1978; Ewers, 1982), systematically con-

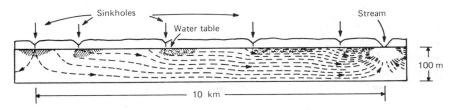

Figure 9.2 Expected flow paths in an open aquifer, a capped aquifer with one input, and a capped aquifer with several inputs, according to the cave origin model of Thrailkill (1968a).

structed one of the most comprehensive models to relate observed cave patterns and profiles to stratigraphy and structure in a manner consistent with fluid mechanics and solution chemistry although these latter variables do not enter the theory specifically. The key variables are the relief, the concentration and pattern of joints, fractures and bedding plane partings, and the regional dip and pattern of folding. The emphasis is on the selection of hydraulic paths and their enlargement to form caves.

Instead of a single geomorphic type of cave, Ford and Ewers offer several varieties of vadose cave, four "states" of phreatic cave, and two special cases—the isolated pocket cave and the artesian cave.

The basic concept in the model is that an optimum hydraulic path is selected from all possible fractures, faults, and bedding plane partings. This path need not be, and indeed usually is not, a curvilinear flow line. Figure 9.3 shows the phreatic states. State 1, the bathyphreatic cave, occurs when fracture frequency is very small, and as a result features a deep loop beneath the piezometric surface. Their example is the cave system of the Sierra de El Abra, Mexico, where both descending and ascending tubes have been identified (Fish, 1977).

State 2 is cave development through an aquifer of low fracture frequency, so that the optimum flow path follows a sequence of loops. With higher fissure frequency, the flow path comes up to points of zero pressure at the tops of the loops, the locus of which defines the piezometric surface. The Ford–Ewers model assigns no special role to the position of the water table in cave genesis. Instead, the water table is regarded as a result of a particular flow path unless it is fixed by other barriers within the hydrogeologic setting. The amplitude of the State 2 loops may range from a few tens of meters to a few hundred meters. The type example is the Hölloch Cave System in Switzerland, where the loop relief is several hundred meters.

State 3 is an intermediate case that appears with low to moderate fissure frequency. It consists of long reaches of water table cave with occasional phreatic loops.

State 4 caves occur at moderate to large fissure frequencies and are the true water table caves. A continuous grade between input and discharge point defines a drain of negligible flow resistance, and thus a low gradient water table. At high fissure frequency, all parts of the aquifer are in good hydraulic communication. Deep flow paths either do not form or are quickly closed by sediment.

The phreatic loops of states 2 and 3 are modified as the water table is lowered (Fig. 9.4). Entrenchment by free-surface streams downcuts across the tops of the old phreatic loops. These appear in caves as canyon passages, which give way to tubes at both ends. It

274 / *Geomorphology and hydrology of karst terrains*

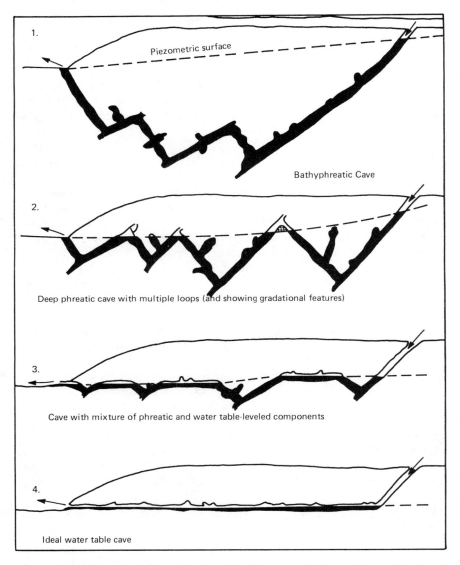

Figure 9.3 States of phreatic cave development. [Adapted from Ford and Ewers (1978).]

should be kept in mind that the drawings shown here to illustrate the flow paths have considerable vertical exaggeration. Particularly in limestones of low dip, the loops and their modifications may be very subtle, with a relief of only a few meters over hundreds or thousands of meters of passageway. The lower phreatic loops act as sediment traps. If these fill as downcutting proceeds, and the cave is still

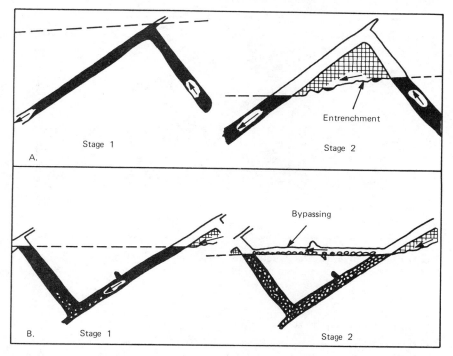

Figure 9.4 Modification of phreatic loops (A) by entrenchment and (B) by bypassing. [Adapted from Ford and Ewers (1978).]

subject to annual fluctuations in water level, bypass tubes may develop across the top of the loops. Once the bypass is available for the main water flow, the older loop may be filled with sediment. One is suspicious of places where a spacious phreatic conduit big enough for easy walking suddenly gives way to a considerably smaller bedding plane crawl.

Phreatic loops are also subject to a process described by both Ford and Ewers (1978) and by Renault (1967–1968) as *paragenesis*. Figure 9.5 illustrates this process. A cave passage at the bottom of a phreatic loop carries a substantial sediment load. As the cross section of the phreatic loop becomes larger, flow velocities decrease because the hydrostatic head across the loop has not changed and the sediment-carrying capacity of the conduit decreases. Sediment, accumulated on the floor of the phreatic passage, insulates the floor from further solution, but solution attack on the ceiling continues. The passage is thus dissolved upward, remaining always at the top of a growing pile of clastic sediment. A balance is developed between the cross section of the open portion at the top of the paragenetic passage and the sediment load. If the passage gets too big, more sed-

276 / *Geomorphology and hydrology of karst terrains*

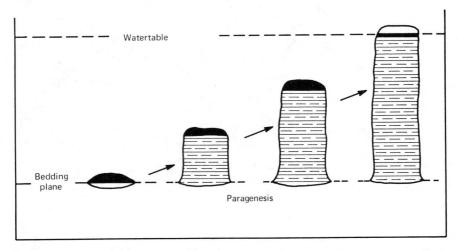

Figure 9.5 Formation of a sediment-filled canyon by upward solution of a subwater-table tube, the process of paragenesis.

iment is dropped until the opening is throttled down to where the velocity is sufficient to move the remaining sediment load. Upward solution continues until the passages reach the piezometric surface.

Figure 9.6 illustrates two types of vadose caves. The first is the drawdown vadose cave. These develop where there is a catchment on an adjacent nonkarstic border land so that water collected in this catchment is emptied into the limestone. Initially the water table must be close to the land surface, but as the cave system develops, the water table quickly drops until it is either a piezometric surface across the top of a series of phreatic loops or oriented along a water table cave. The rapid discharge through the cave system produces drawdown in the same sense that drawdown is created by a pumped well. However, the injection of water continues at the swallow hole because there is no way for the input to readjust itself across the impermeable barrier. The cave may be enlarged considerably between the swallow hole and the phreatic tube, most of which was formed by free-surface streams entirely within the vadose zone. As with the phreatic caves, the relief and gradients of the drawdown vadose caves can be quite variable.

Once an open underground system has been established, further inputs of surface water can create predominantly vertical caves through the vadose zone. Ford and Ewers call these invasion vadose caves. Vertical shafts in the plateau karst of the eastern United States, and the deep vertical caves of many alpine regions are forms of vadose invasion caves.

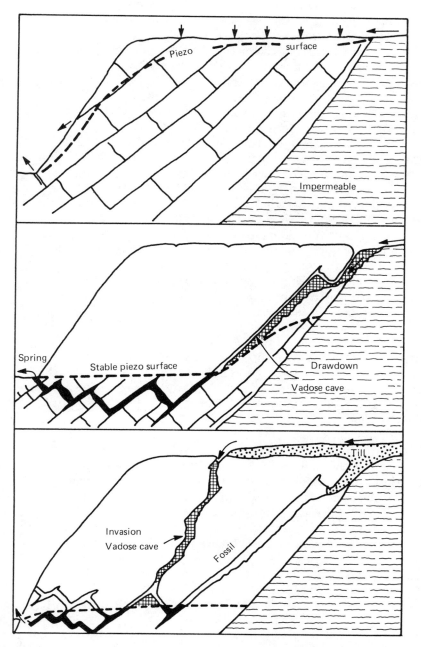

Figure 9.6 Two types of vadose cave that can develop from same initial condition (top drawing) according to Ford and Ewers (1978).

Nonintegrated caves are isolated cavities of various sizes found within the phreatic zone. They are not part of an integrated drainage system, and the flow of water through them is limited by the flow of water in the bedrock that surrounds them. Cavities encountered by deep-drilling operations are thought to be of this type (when they cannot be ascribed to paleokarst). One must invoke special chemical mechanisms to account for solution at a depth where the only circulation is through diffuse flow.

True artesian caves are formed within deep artesian aquifers. Ford and Ewers claim that the features of these caves are sufficiently different from those of bathyphreatic caves that a separate category is necessary. The type examples in the United States are the gigantic three-dimensional network caves in the Pahasapa limestone in the Black Hills of South Dakota (see, e.g., Howard 1964).

The Ford–Ewers model integrates many of the concepts and models that have gone before and nicely resolves the water table controversy. By emphasizing the geologic setting, it allows one to integrate many of the European ideas, mostly based on observations in alpine caves, with North American ideas, mostly based on observations of low-relief karst in nearly flat-lying limestones.

Unlike most theoretical models, the Ford–Ewers model works best when applied to dipping limestones. At low fissure frequency, the distinction between dip tubes and tubes cutting across the bedding is more clear. Passages in flat-bedded limestones tend to follow bedding plane partings, and although the passages often cut the bedding, the lift tubes may be very short and difficult to identify in the field.

The phreatic loops predicted for states 2 and 3 of the phreatic caves are difficult to observe in the field because later processes of vadose modification, sediment in-filling and breakdown obscure the passage relations particularly when the loops are long and of low amplitude. Dry lift tubes were identified in Castleguard Cave (Ford et al., 1983) where a single conduit represents the paleodrain from beneath a presently existing glacier. More direct evidence for lift tubes is provided by divers who are beginning to survey caves below base levels. Cueva de Peña Colorado in the Huautla Area of Mexico quite clearly shows a series of lift tubes, some flooded and some air-filled, with reliefs of up to 100 m (Stone, 1984).

Development of the master conduit at any position with respect to the original water table provides a master drain of low hydraulic resistance. Its existence is frequently revealed by a pronounced groundwater trough, as mapped by Quinlan and Ray (1981) on the Sinkhole Plain aquifer of southcentral Kentucky. The master conduit then becomes a local base level toward which other parts of the aquifer and other inputs begin to drain. Hydraulic gradients are rearranged so that these secondary gradients now point toward the drain rather than toward the original base level outlet. Ewers (1978) and Ewers and Quinlan (1981) have developed this idea in some detail and show that the rearrangement of drainage migrates headward as the system develops.

9.3.2 Relief, Recharge, and Drainage Basin: Where Does the Water Come From?

The large-scale structural elements—folds and faults—also determine the placement of karst rocks with respect to nonsoluble rocks. Perhaps the most important distinction is between regions where such elements are present (i.e., folded rock terrains) and where they are absent (i.e., terrains with nearly horizontal bedrock).

Where the carbonate bedrock is horizontal, or nearly so (plateau karst and platform karst), lithologic horizons favorable to conduit development extend horizontally in two dimensions. It is possible for complex drainage systems to develop, because it is possible for many tributaries to aggregate and still remain within the same stratigraphic horizon. For this reason, most of the large cave systems of the eastern United States—the southern Indiana karst, the Mammoth Cave area of Kentucky, the margins of the Cumberland Escarpment and the Highland Rim of Kentucky, Tennessee, and Alabama, and the eastern West Virginia karst—are formed in nearly horizontal massive limestones.

In strongly folded rocks, in low-relief terrains, the caves tend to be distributed more or less along the axis of folding, and large, well-integrated conduit drainage systems are less common. Folded or faulted rocks in terrains of high relief often contain conduit systems of great complexity. Caves occur with strong vertical elements, where flow is along steeply dipping beds, major fractures, or active faults. Conduits often develop as a result of ascending waters driven upward under hydrostatic head.

The opportunities for cavern development are controlled by the sources of recharge, particularly the balance between allogenic recharge, internal runoff, and diffuse recharge (see Fig. 6.7). Palmer (1984) suggests that the development of single-conduit caves is determined by the dominance of allogenic recharge, and the development of branchwork caves is determined by the dominance of internal runoff (called authigenic recharge by Palmer).

Authigenic recharge comes from precipitation directly on a karst surface, where it is aggregated into small catchment cells, which in turn drain into closed depressions. The sinkhole drains form the extreme upstream tributaries of a branchwork cave system. Each point source feeds a distinct tributary of the system and these tributaries generally merge in the downstream direction to form cave passages of increasing size and discharge (Fig. 9.7). If bedding plane partings are the primary permeability the branchwork passages are

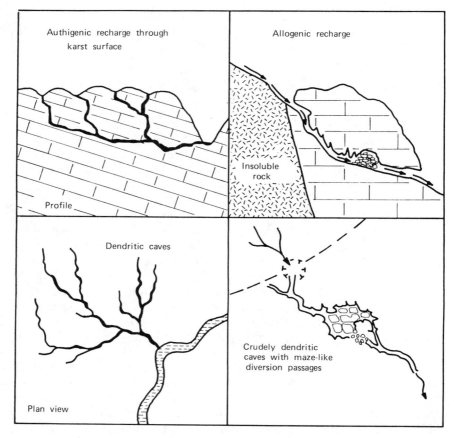

Figure 9.7 Patterns of cave development in relation to sources of recharge. [Adapted from Palmer (1984).]

likely to be sinuous. Predominantly joint control may produce an angulate branchwork pattern. In regions of strongly folded limestones, outcrop bands of limestone tend to be elongate and a more trellislike cave pattern develops.

A variant on this arrangement occurs when the aquifer is capped by noncarbonate rocks, particularly when the capping beds are partially dissected, as they are in the Mammoth Cave Plateau and parts of the Cumberland Mountains of Tennessee and Alabama. Catchment cells on the plateau drain from the edge of the clastic rocks to form vertical shafts at the limestone contact. The shaft drains become small tributaries that feed into a dendritic drainage system of branchwork cave passages.

Catchment areas on nonkarstic borderlands can be quite large.

When the allogenic streams from these basins flow across carbonate rocks, they provide a very large point source of recharge. In the initial stages, infiltration from the stream builds a groundwater mound, and thus a large hydrostatic head across the carbonate aquifer. The resulting system consists of single-conduit caves fed by a sinking-surface stream. Later processes of breakdown and sediment in-filling may fragment the conduit, so that the entire drainway is only rarely intact. Secondary passages may form diversion routes around local blockages, and alternative routes form as the conduit cuts to deeper levels. Both tubular and canyon cross-sectional passages are possible, depending on whether the master conduit develops under pipe-full conditions or as an open-channel cave stream.

Palmer (1975) pointed out that allogenic recharge may be subject to large seasonal variations. If the borderland catchment is in a region of high relief, the cave system may completely flood during periods of high recharge; this flooding creates a high hydrostatic head across all possible joints, fractures, and bedding plane partings. The result is a floodwater maze that usually appears as a tangle of anastomotic tubes. Sharp-edged residual blades, rock spans, and scoured pockets, all covered with small scallops, are suggestive of flood spate flow velocities. There seems to be little relation between the anastomotic maze pattern developed by floodwaters and the bedding plane anastomotic mazes associated with many low gradient branchwork caves. The latter are not scalloped and have a more uniform cross section.

Allogenic recharge may play a key role in the development of large cave systems in tropical karst. In the Caribbean karst, the large river caves, such as Rio Camuy System in Puerto Rico and the Cave River and Quashies River systems in Jamaica, are fed by allogenic runoff from large catchment areas on the noncarbonate mountains in islands' interiors. Because of the rapid reaction of diffuse infiltration and internal runoff at the base of the soil, the allogenic catchments provide the most efficient source of large volumes of undersaturated water for the excavation of the cave system.

Allogenic recharge and internal runoff are both generally undersaturated with respect to carbonate minerals and can initiate the dissolution process in the carbonate aquifer. Diffuse recharge through a soil mantle may react to saturation at the soil–bedrock interface, so that its role in cave development is limited to travertine deposition or as part of a mixing solution process. Some aquifers have only a diffuse flow component and no conduit development at all. In others, diffuse recharge moves laterally to the conduit drain, with little solutional modification of the bedrock. The most interesting case is that in which diffuse recharge takes place through overlying porous clastic rocks, particularly when the underlying limestones are thin

282 / *Geomorphology and hydrology of karst terrains*

and sandwiched between nonsoluble layers. Each joint and fracture in the carbonate unit receives recharge regardless of its size, and all fractures can enlarge at more or less the same rate.

Limestones with the characteristic sandwich features are typically less than 15 m thick (Fig. 9.8). Flow is retarded by lack of concentrated recharge from overlying beds, and thus channelization does not take place. Solution takes place along many available joints, generating a very dense network pattern (Palmer, 1975). An extreme example is Anvil Cave in Alabama, in which 20 km of passages have been surveyed on one plane in an area of 0.5 km^2. Caves with this pattern have been discovered most frequently near major streams, and base-level backflooding may also contribute to their development.

9.3.3 *Stratigraphic and Lithologic Controls*

The most extensive and best-developed karst occurs on pure limestones, but limestones are rarely pure (Fig. 9.9). Karst development

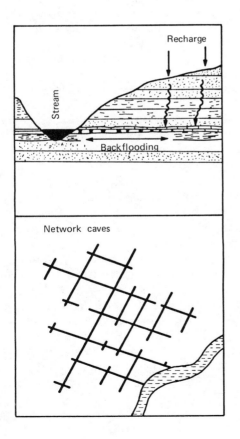

Figure 9.8 Network maze caves developed by a combination of diffuse recharge through clastic overburden and base level backflooding from nearby surface streams. [Adapted from drawings by Palmer (1984) and White (1969).]

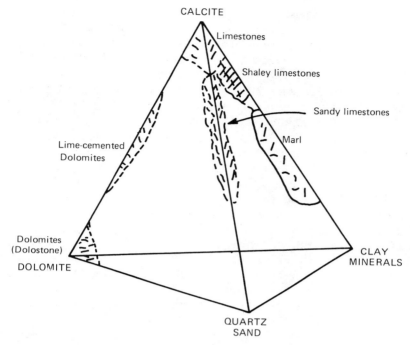

Figure 9.9 Compositional tetrahedron showing the principal components of sedimentary rocks as they relate to limestones and dolomites.

on pure massive dolomite is generally less extensive, all other things being equal. High-magnesium rocks sometimes take the form of dolomite sand cemented by calcite. Such rocks, of which the Gasconade dolomite in Missouri is an example, may have extensive cave development because the cement can be dissolved, leaving the less readily dissolved dolomite sand to be removed by mechanical transport. The same can be said for calcite-cemented sandstones and sandy limestones. The Loyalhanna limestone of western Pennsylvania, which contains 50 percent quartz sand, has some of the largest caves in the state. Percolating groundwater in the well-jointed formation dissolves the calcite cement, leaving a highly porous sand matrix behind (Schmidt, 1974). The quartz sand is removed by mechanical transport. Along some joints where the sand was not removed, ghost passages can be seen filled with uncemented sand that preserves all of the primary bedding features. In contrast, clay-rich rocks seldom develop either cave systems or good surface karst. The clay minerals released by dissolution of the bedrock are not easily removed and accumulate along joints to form impermeable barriers to further groundwater flow. Marls and chalks are rarely cavernous.

284 / Geomorphology and hydrology of karst terrains

Although all limestones are equally soluble with respect to equilibrium solubility, it has long been recognized that the rate of limestone solution varies considerably among various limestone rock types (Sweeting and Sweeting, 1969; Rauch and White, 1970; Rauch and White, 1977; Dreiss, 1974). A field test of the relative ease of solution of various rock types was made in the folded carbonate rocks of central Pennsylvania, where the upper 400 m of the Ordovician carbonate section contains a variety of rock types (Rauch and White, 1970) (Fig. 9.10). Because the limestones steeply dip along the flanks of the anticlinal valleys, the combination of structural and base-level controls allows the conduits to chose the most favorable lithology, since all lithologies are equally accessible. It is dramatically clear that some rock units are much more cavernous than others.

The Salona and Coburn formations in the upper part of the section are shaley units with interbeds of black shale near the top. As expected, groundwater circulation is greatly retarded in such units and no caves were found in these rocks. The Milroy formation at the bottom of the section is transitional with the underlying Beekmantown dolomite. It is a dolomite-rich rock and has little cave development. There are 2000 m of carbonate rock exposed in the limestone valleys of central Pennsylvania, but the known caves are concentrated in roughly 150 m of the section. Figure 9.11 gives some additional correlations. Most of the cave volume is concentrated in rock with low magnesium concentration, which may be expected from the absence of caves in dolomite. Low silica and low alumina favor conduit development, a result that may mean a low clay content. Rocks with both sparry calcite and micritic cements occur in the section. Cavern development favors the rocks with smaller grain size, a result that is in agreement with kinetic study results. Smaller grain size means larger surface area, and therefore faster dissolution rates.

9.3.4 Summary: Hydrogeologic Setting

Seen in its true context, the theory of cave origins is nothing more than the theory for the development of the conduit permeability of a carbonate aquifer. The concept of "diffuse" and "conduit" flow aquifers described in Chapter 6 can be further elaborated in terms of the type of conduit system (Table 9.1). Aquifers are placed into three main categories: diffuse flow, free flow, and confined flow. Further subdivision depends on the thickness of carbonate rock units and their relation to overlying and underlying impermeable formations and to base level (White, 1969, 1977b). Relief also plays a role in that it determines the thickness of the unsaturated zone where vadose caves can develop.

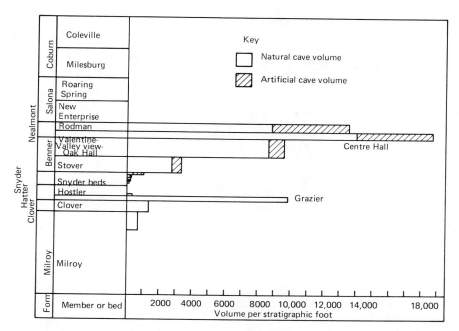

Figure 9.10 Distribution of cave volume within the Champlainian limestone sequence in central Pennsylvania. [After Rauch and White (1970).]

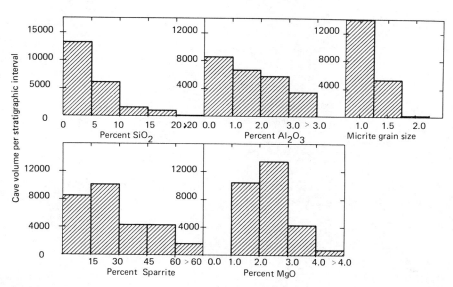

Figure 9.11 Bar graphs showing influence of chemical and lithologic variables on the development of cave volume. [Adapted from Rauch and White (1970).]

Table 9.1
Hydrologic Classification of Carbonate Aquifers

Flow type	Hydrologic control	Associated cave type
I. Diffuse flow	Gross lithology Shaley limestones; crystalline dolomites; high primary porosity	Caves rare, small, have irregular patterns
II. Free flow	Thick, massive soluble rocks	Integrated conduit cave systems
A. Perched	Karst systems underlain by impervious rocks near or above base level	Cave streams perched, often have free-air surfaces
1. Open	Soluble rocks extend upward to land surface	Sinkhole inputs; heavy sediment load; short fragments of conduit caves
2. Capped	Aquifer overlain by impervious rocks	Vertical shaft inputs; lateral flow under capping beds; long integrated caves
B. Deep	Karst system extends to considerable depth below base level	Flow is through flooded conduits
1. Open	Soluble rocks extend to land surface	Short tubular abandoned caves likely to be sediment-choked
2. Capped	Aquifer overlain by impervious rocks	Long, integrated cave, systems under caprock, active level of system likely to be inundated
III. Confined flow	Structural and stratigraphic controls	
A. Artesian	Impervious beds, which force flows below regional base level	Inclined 3-D network caves
B. Sandwich	Thin beds of soluble rock between impervious beds	Horizontal 2-D network caves

Source: After White (1969).

9.4 ANALYSIS OF CAVE DEVELOPMENT IN TERMS OF CHEMICAL AND PHYSICAL VARIABLES

Returning yet again to the coordinate system of Fig. 4.1, we now turn our attention to the physical and chemical variables in the solution process. If one thinks of the dissolution process as being described by a differential equation, then the required equation would include a fluid flow part and a reaction kinetics part. The physical and chemical factors become the parameters of this (hypothetical) equation and the geological setting provides the boundary equations that permit one to select the correct solution for the cave system of interest.

9.4.1 *The Life History of a Conduit*

Individual conduits have a discrete "life history" (Fig. 9.12). The history begins with mechanical fracturing of the bedrock: a sequence of joints, bedding plane partings, or fractures that provide a complete pathway from some source of recharge to a point of discharge. The initial fracture permeability has an aperture on the order of 10 to 25 μm. With increasing solution the fracture widens and the permeability increases, as does the flow velocity. Usually there are many fracture pathways providing alternate routes for groundwater flow. It is an observed fact that many conduit systems are made up of a few large openings. Some mechanism must be responsible for selecting one pathway out of the large number available and allowing it to grow to the observed cave dimensions while the growth of the alternative paths is suppressed. New phenomena appear when the aperture of the fracture or the diameter of the protoconduit reaches 5 to 10 mm. This critical threshold aperture also defines a boundary between a fracture aquifer and a conduit aquifer. The process of initial enlargement up to the critical size is termed initiation, and the growth of the protoconduit to full conduit size is termed enlargement.

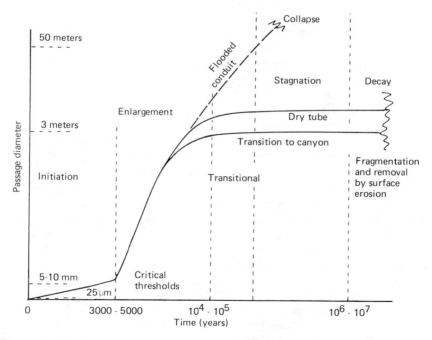

Figure 9.12 Schematic rate curve for the evolutionary history of a single conduit. The time scales are drawn from a combination of field evidence and geochemical calculation.

The succeeding history of the conduit depends on its geologic setting and the development of the drainage basin in which it is embedded. If the conduit remains flooded, it will continue to enlarge until it exceeds the critical width for mechanical stability. In most situations, however, dissection of the surface basin lowers the base level and drains the conduit. If base level lowering is slow and continuous and the source of recharge continues to flow through the conduit, downcutting gives the conduit a canyon cross section. The stream in a regime of open-channel flow occupies the canyon floor. If base level lowering is episodic or the conduit has received an insulating veneer of clastic sediment, new conduits are created at lower levels, forming tiered caves, and the high-level conduits are abandoned. The original conduit enters a stagnation phase, which can last a long time, particularly if the conduit is beneath a protecting caprock or lies deep within a mountain massif. Surface erosion continues and deepening surface valleys truncate and dissect the once continuous conduits. Eventually the lowering land surface will sweep across the cave passage and the last traces of the conduit system will be destroyed.

9.4.2 *The Initiation Phase*

Explaining the enlargement of a joint or bedding plane parting into a conduit is almost the exact opposite of describing the denudation of the limestone surface. In the latter case, completely unsaturated water from the soil contacts the underlying carbonate bedrock. Chemical reaction is prompt and rapid, and a good approximation can be obtained by simply assuming that the infiltration water reaches saturation at the interface. It is the infiltration and internal runoff water from the karst surface that is responsible for the excavation of caves. If the water were indeed saturated, no further reaction would occur and no solution conduits would develop. Although one can invoke both chemical solution and mechanical erosion by underground streams, once an open, high-flow-velocity conduit has been developed, it is difficult to account for the solutional modification of the fracture pathways in the first place. Indeed, one of the first calculations done on the rate of solution of carbonate rocks (Weyl, 1958) was a proof that caves could not exist. Most analyses of well waters and of diffuse flow spring waters show them to be slightly undersaturated. The problem is to enlarge the mechanical openings in the bedrock with a solvent that is already close to saturation when it enters the fracture.

Initial flow along a horizontal bedding plane parting has been modeled using blocks of salt (Ewers, 1972). The initial solution pathways are tendril-like distributaries with multiple branches in the downstream direction (Fig. 9.13). When one of the tendrils reaches the discharge point, it begins to grow at the expense of other tendrils

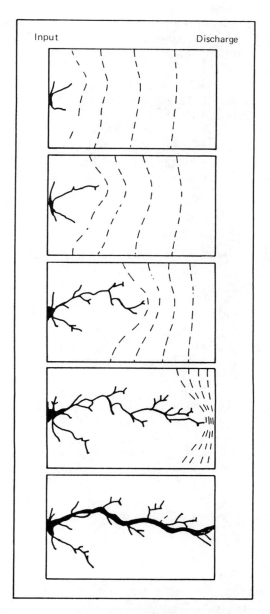

Figure 9.13 Fingering pattern of fluid penetrating a bedding plane parting or joint plane. Schematic isopotential lines are shown as dashes. [Adapted from Ewers and Quinlan (1981).]

290 / Geomorphology and hydrology of karst terrains

and soon becomes the dominant flow path. Details may not be the same in limestone aquifers as they are in the laboratory simulations, because the dissolution kinetics of salt is diffusion controlled whereas calcite dissolution is thought to be surface-reaction controlled (see Section 5.5).

Calculations on initial fracture enlargement are generally made under the assumption that the fracture enlarges uniformly and maintains its cross-sectional geometry. The calculation also depends on the choice of rate equation. Palmer (1981c) used the Plummer–Wigley second-order rate Eq. 5.54 in connection with an assumed circular opening containing water in laminar flow. By combining mass balance with the Plummer–Wigley rate equation, Palmer derived an expression for the rate of increase of passage radius:

$$\frac{dr}{dt} = \frac{Q(C - C_0)}{2\pi r \rho L} 10^{-6} \tag{9.1}$$

in which the instantaneous concentration is given by

$$C = C_s \left[1 - \left(\frac{0.002 r L k C_s}{Q} + \frac{C_s}{C_s - C_0} \right)^{-1} \right] \tag{9.2}$$

where r = passage radius in centimeters, Q = discharge in cubic centimeters per second, ρ = density of bedrock in grams per cubic centimeter, L = distance along the tube in centimeters, and C, C_s, and C_0 are instantaneous concentration, saturation concentration, and initial concentration, respectively, all in milligrams per liter. The results (Fig. 9.14) show that for various choices of discharge, passage length, and hydraulic gradient, the rate of dissolution increases with discharge up to a limit where unsaturated water is moving through the protoconduit; after this the dissolution rate is independent of discharge. Palmer's calculations are in good agreement with experimentally determined rates in laboratory modeled "fractures" (Howard and Howard, 1967). The maximum dissolution rate was calculated to be 0.14 cm/year. However, most of the initiation period of some thousands of years is a time of slow dissolution in nearly saturated water.

9.4.3 Critical Thresholds

The transition from a system in which water moves more or less uniformly through a three-dimensional network of joints, fractures, and bedding plane partings to one in which flow is localized in a well-defined conduit is a perplexing problem. If dissolution takes place uniformly along all pathways in a fracture aquifer, the evolving cave should have the pattern of a three-dimensional maze. Such patterns are rarely observed, and it appears that some sort of singularity pro-

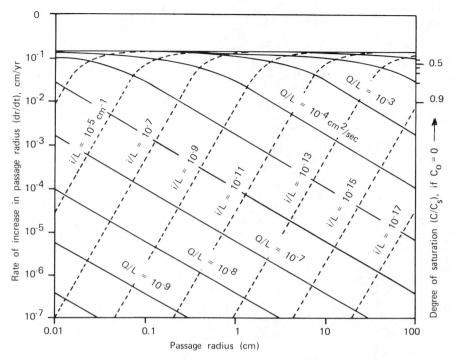

Figure 9.14 Increase in passage radius (dr/dt, in centimeters per year) for $(C_s - C) = 200$ mg/L, at different values of discharge (Q), passage length (L), and hydraulic gradient (i). [From Palmer (1981c).]

duces the shift in the cave pattern. Singularities arise because of nonlinearity in the transport processes, and these may be described as critical thresholds that mark the transition from fracture flow to conduit flow. There are thresholds in fluid flow, dissolution rate, and sediment transport rate (Fig. 9.15).

The *hydraulic threshold* is the onset of turbulence in the evolving conduit system. Flow in the fracture system is laminar under most hydraulic gradients. As the fracture aperture increases (see Fig. 6.10), velocity increases and the Reynolds number increases rapidly because it is the product of aperture and velocity. The onset of turbulence appears at Reynolds numbers between 5 and 100, depending on details of wall roughness and fracture geometry. For commonly observed hydraulic gradients, the transition to fully developed turbulent flow occurs at an aperture of 0.5 to 5 cm. Observations of solutional morphology also suggest that a change from planar to tubular forms often occurs at an aperture of about 1 cm.

The *transport threshold* is that combination of aperture and

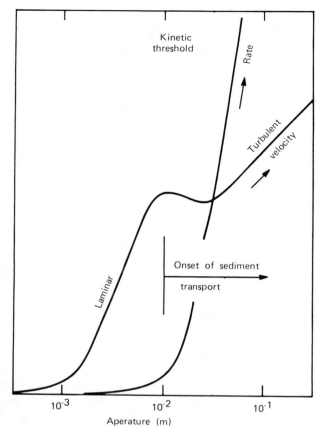

Figure 9.15 Controlled sketch showing the three critical thresholds.

velocity at which the flow system first becomes competent to carry a sediment load. The threshold for bedload transport (critical boundary shear) occurs at velocities that are reached at about the same aperture at which turbulence begins. This is fortuitous because turbulent flow is not a necessary condition for bed-load transport. In general, clastic sediments can be transported in conduit aquifers but not in fracture aquifers. The requirement that a critical aperture be reached before waters will aggregate into localized conduits is an explanation for the absence of conduits in shaley limestones. Locally derived weathering debris clogs the protoconduits before velocities become high enough to flush them out.

The *kinetic threshold* is the shift in dissolution rate that takes place when the aperture reaches centimeter sizes. None of the various models for calcite dissolution kinetics work well near equilibrium, but the experimental curve (see Fig. 5.5) shows that dissolution rates increase by orders of magnitude between saturation and a saturation index of about −0.4. As the various pathways through the aquifer enlarge, velocities increase, and the water within the fractures becomes less and less saturated. At some point water with an undersaturation level of −0.4 to −0.2 is able to penetrate the aquifer. This passage enlarges rapidly and becomes the dominant path (White, 1977c).

Buhmann and Dreybrodt (1985a,b) also show an order of magnitude increase in dissolution rate when the fluid flow becomes turbulent. This has to do with the more efficient mixing that enhances the hydration rate of aqueous carbon dioxide, which becomes an important part of the rate-controlling system at higher flow velocities.

9.4.4 Passage Enlargement and Competition

Once solutional enlargement has progressed past the critical threshold the passage will enlarge from apertures on the order of 1 cm to apertures in the range of 1 to 30 m. The hydrogeologic model is that of the free-flow aquifer without geologic barriers or path diversions. As pointed out by Palmer (1984), two recharge conditions must be recognized:

1. *The constant head case.* It is assumed that the pathway is fed by water entering at constant elevation difference with respect to the outlet. This case is relevant to caves that developed beneath an underdrained valley, where the swallow hole provided a constant head but stream flow was in surface channels, so that the developing underground pathways were always flooded. In this case, the discharge and velocity through the conduit increase as the passage diameter increases. Likewise, active conduits perched by clastic sediments provide constant head for the development of new, lower-level tiers as base levels are lowered.
2. *The constant discharge case.* If the source of recharge of a conduit is limited, such as the catchment cell for a closed depression that feeds a tributary of a branchwork cave, the flow velocity must decrease as the passage enlarges. If the conduit is above the local water table or its enlargement is accompanied by rapid base level lowering, there will be a transition from a conduit to a canyon and from pipe flow to open-channel flow.

It is difficult to model a generic conduit because of the great variety of hydrogeologic settings. Chemical data for conduit flow springs show that most conduit flows are undersaturated at the level of $SI_c = -0.3$ to $SI_c = -1$. Flows are turbulent, and the rate of dissolution is not strongly dependent on the choice of kinetic equations. Rates depend on initial CO_2 concentrations, on whether the conduit is a closed pipe or maintains communication with a fixed CO_2 reservoir, and on the kinetics of the CO_2 to H_2CO_3 hydration reaction. Under optimum conditions, the conduit could enlarge from the centimeter size to a diameter of 10 m in from 5000 to 20,000 years. There is evidence from caves at the glacial margin that complete conduits have developed since the retreat of the Wisconsin ice sheet 10,000 to 15,000 years ago. On the other hand, this rate seems too fast for many geologic settings, in which times on the order of 100,000 years are more appropriate. Inhibitors must be present to slow down the natural rate from the rate measured in laboratory experiments. One obvious candidate is a thin layer of clay and silica, the insoluble residue from dissolution of the limestone, which coats the dissolving surface and provides a barrier to migrating ions.

9.4.5 Relation of Caves to the Water Table and Base Level

An amazing amount of writings on caves is concerned with relationships of caves to the water table. Some writers have asked, "Does a water table exist in karst?" In Section 6.3.4 the answer was, "in most regions, yes, but with some rather special dynamics regarding rapid changes in precipitation." In the classical theorizing about cave origins, the central question was, "Do caves form above, at, or below the water table?" To all of these possibilities, the Ford–Ewers model answers, "yes!" Whether caves form above, at, or below the water table depends on the hydrogeologic setting: any location of cave development is possible if the geology is right. The question is now raised again—for the last time! What is the interrelationship between caves and the water table from the perspective of the hydraulics of individual conduit systems?

The elevation of the water table in a porous-medium aquifer depends on the balance between recharge rate and flow rate of water through the aquifer. Because of low permeability, a substantial hydrostatic head is developed in the recharge area until the flow through the aquifer balances the input. The water table elevation is a function of the hydraulic resistance of the porous medium.

The hydraulic resistance of a conduit falls rapidly as the conduit diameter increases (Fig. 9.16). By the time the conduit reaches a size of interest to human explorers, the supportable head is so small that the conduit acts as a short circuit to the groundwater flow system. Any previously existing water table will fall to the elevation of the conduit (or to the base level discharge point if the conduit has developed below the surface base level). The water table in the bedrock adjacent to the conduit will have a gradient oriented toward the conduit, as Ewers and Quinlan (1981) have proposed. In effect, the position of the surface base level or of the conduit determines the posi-

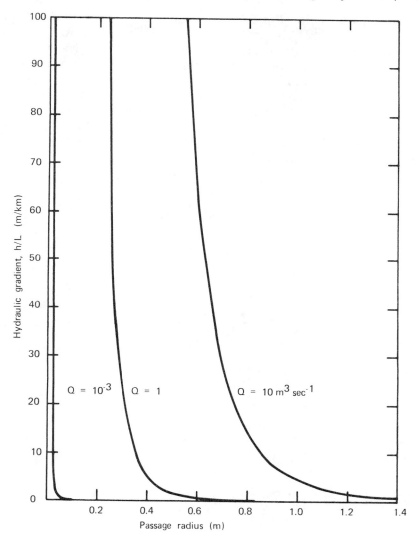

Figure 9.16 Supportable hydraulic head (maximum elevation of the water table) as a function of conduit radius for various assumed total recharge values, Q.

tion of the water table, rather than the water table determining the position of conduit development.

9.4.6 Vertical Shafts and the Development of Vadose Caves

Vertical solution is accomplished by the action of thin films of water sliding down joints and fractures but attached to the walls by surface tension. These waters, unlike seepage waters, are highly undersaturated with calcite (Fig. 9.17) and can dissolve back the walls of the joint or fracture. The carbon dioxide pressure with which such waters would be in equilibrium is in the order of $10^{-2.5}$ atm, which is high compared to the atmospheric value but is still not particularly rich in carbon dioxide.

Measurements taken at the top and bottom of vertical shafts indicate a small uptake of calcite as the moisture film moves down the shaft. The waters become somewhat more saturated and there is somewhat less carbon dioxide, although they reach the bottom of the shaft highly undersaturated.

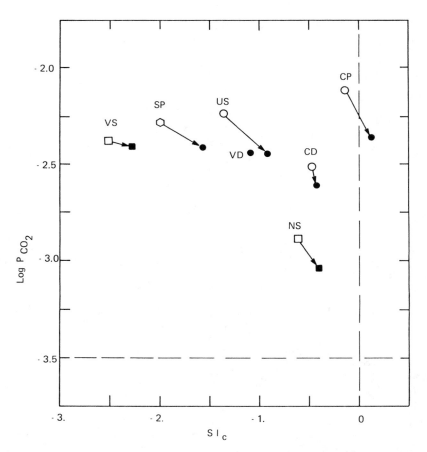

Figure 9.17 Chemical data for some vertical shafts plotted from Brucker et al. (1972). Open figure = top of shaft; solid figure = bottom of shaft; circles = Mammoth Cave System: VD = Vaughns Dome, CD = Colossal Dome, CP = Cascade Pit, US = unnamed shaft; hexagon = Swago Pit, West Virginia; squares = Alabama: VS = Valhalla Shaft, NS = Neversink Shaft.

Theories, models, and mechanisms for the origin of caves / 297

The vertical shaft can be considered as an open channel standing on end (Brucker et al., 1972). The channel width is the shaft circumference and the depth is the thickness of the moisture film. With the assumption of a wide-open channel of uniform depth, the Froude numbers and the Reynolds numbers can be calculated. Measurements of velocities for water streaming down the sides of the shafts follow the equation for films of fluid in pipes

$$\bar{v} = \frac{\rho g \delta^2}{\eta} \tag{9.3}$$

where δ = film thickness, ρ = fluid density, g = gravitational acceleration, and η = viscosity. The surprising result is that many data fall within the regime of supercritical-laminar flow, a regime seldom observed in nature. The water streams are quite shallow, typically on the order of fractions of a millimeter, and the velocities are very high, ranging from 0.3 to 2 m/sec, as measured by dye streamers.

Supercritical flow was confirmed by introducing into the streamers small gravity waves, which are swept away downstream. The flow of water in shafts increases during periods of high precipitation. Both thickness of the flowing sheet of water and the wetted fraction of the shaft circumference increase. If the discharge through the shaft increases sufficiently, the sheets will break free of the wall and become a spray or small waterfall. Increased sheet thickness with the accompanying increase in velocity drive the flow regime into a supercritical-turbulent regime.

The supercritical flow may be primarily responsible for the near-perfect vertical walls that characterize the shafts. Projecting beds, ledges, and other protuberances retard the flow of the water films. With increased water depth above the projection, the flow regime goes subcritical with a hydraulic jump at the transition point ($N_F = 1$). The accelerated erosion at the jump trims off the projections until only a vertical wall remains.

9.5 ALTERNATIVE CHEMICAL MECHANISMS FOR CAVE DEVELOPMENT

Although most solution caves in carbonate rocks are dissolved out by the action of carbonic acid-rich groundwater, it is starting to be recognized that in certain instances a different dissolution chemistry is at work. Dissolution of solution cavities occurs at depths of thousands of meters below base level, and intensive dissolution occurs in coastal karst as a result of mixing solution. In other caves sulfuric acid rather than carbonic acid seems to play an important role.

9.5.1 Sulfuric Acid Solution

Quite early, it was suggested that sulfuric acid derived from the oxidation of pyrite, FeS_2, which occurs dispersed in some limestones, might be responsible for cave enlargement (Durov, 1956; Morehouse, 1968). Morehouse was concerned with small caves in the Galena dolomite in eastern Iowa, where sulfide mineralization is common and some caves contain minable ore deposits. Depending on the mineral, 1 mole of sulfide is needed to dissolve 1 mole of calcite. From mass balance considerations it does not seem likely that sulfide-mineral-derived sulfuric acid plays an important role in most cave systems.

Egemeier (1973, 1981) proposed a different mechanism to account for the Kane Caves in northcentral Wyoming. Lower Kane Cave discharges a warm spring, and the passage is lined with fine-grained gypsum resembling moonmilk. Egemeier argued that hydrogen sulfide, seeping up from the Bighorn Basin, oxidized as it came into contact with the atmosphere in fractures at the rear of the caves, forming sulfuric acid, which gradually dissolved away the limestone:

$$H_2S(aq) + 2\ O_2(gas) \rightleftharpoons 2H^+ + SO_4^{2-}$$
$$2\ H^+ + SO_4^{2-} + CaCO_3 + H_2O \rightleftharpoons CaSO_4 \cdot 2H_2O + CO_2$$
(9.4)

A similar mechanism on a grand scale has been proposed to explain Carlsbad Caverns and other caves of the Guadalupe Mountains of New Mexico. Carlsbad Caverns contains huge rooms and passages of greatly varying cross section. It has more vertical relief in relation to its length than most caves. Passages descend steeply and then flatten out. The sculpturing of the walls is appropriate to slow percolating flow: passage walls are cut in broad, sweeping curves and there are regions of spongework maze (boneyard).

The proposed mechanism (Davis, 1980; Hill, 1981d) is that hydrogen-sulfide-bearing solutions derived from petroleum-rich sediments in the Delaware Basin to the east ascend along joints, where they meet a zone of oxygen-bearing water. In the mixing zone, hydrogen sulfide is oxidized to sulfuric acid, which then dissolves the limestone. The waste products are swept down a flow field in the freshwater zone (Fig. 9.18). As additional evidence for the mechanism, Hill (1981d) cites the occurrence of native sulfur in the Guadalupe caves, massive gypsum beds in Carlsbad Caverns with ^{34}S ratios much lighter than that of nearby gypsum, and the presence of endellite clays.

9.5.2 Mixing Solution

The saturation curve for calcite described by Eq. 5.31 shows that dissolved calcite varies with the cube root of the CO_2 pressure. If $CaCO_3$ concentration is plotted on a linear scale, the curve is concave

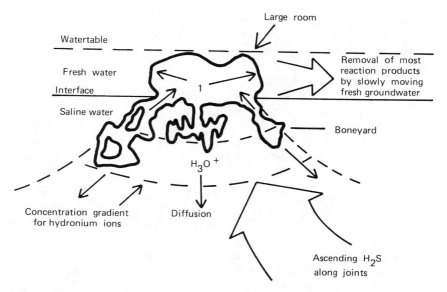

Figure 9.18 Conceptual scheme for regional flow system resulting in the dissolution of Carlsbad Caverns in an oxidizing mixing zone. [Courtesy Donald W. Ash.]

downward. It was pointed out by Bögli (1964, 1965) that a mixture of two saturated waters could produce an undersaturated water because of this curvature. Suppose a laterally moving groundwater is derived primarily from sinking streams. This water is saturated, but at the low CO_2 pressure characteristic of sinking streams. It is mixed with a vertically moving diffuse infiltration water percolating down from the limestone soil. The infiltration water is also saturated, but at a high CO_2 pressure. The mixing zone contains an undersaturated water that is capable of dissolving more carbonate (Dreybrodt, 1981).

Mixing corrosion provides one possible mechanism to explain the high intensity of solution that seems to occur near the water table. It also explains joint-controlled wall and ceiling pockets. Infiltration water seeping from joint planes meets the nearly saturated channel water and a pocket is dissolved. The mechanism is in agreement with Bretz's original contention that wall and ceiling pockets form under water-filled conditions.

In addition to mixing solution because of the nonlinear relation between calcite solubility and CO_2 concentration, other nonlinear effects lead to changes in the state of saturation of mixed waters. These include the following:

1. Purely algebraic effects, which mostly produce supersaturation.

2. Ionic strength. Mixing waters with different ionic strengths—for example, fresh water and seawater or fresh water and deep-seated brines—can lead to undersaturation or supersaturation, depending on relative proportions mixed.
3. Temperature. Mixing solutions of different temperatures changes the state of saturation because of the nonlinear dependence of the equilibrium constants on temperature.
4. Magnesium concentration. Mixing solutions of different magnesium concentration can change the state of saturation of complexes and ion pairs resulting from changes in water chemistry (Picknett, 1972).

Wigley and Plummer (1976) reduced the various nonlinear mixing phenomena to a computer program. The degree of undersaturation depends on the mixing ratio, on the particular combination of mixing phenomena involved, and on the concentrations of dissolved carbonate in the unmixed but saturated waters. The largest effect results from the CO_2 differential (Bögli's original *mischungskorro-*

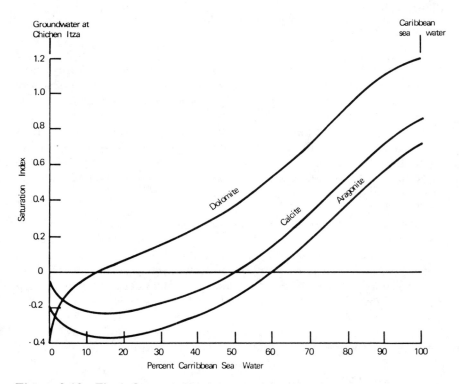

Figure 9.19 The influence of ionic strength on saturation levels of carbonate minerals in freshwater–saltwater mixtures. [Courtesy Dr. W. Back.]

sion), while the algebraic effects give nearly as large an effect in the direction of supersaturation. However, the maximum undersaturation from CO_2 mixing and the maximum supersaturation from the algebra of the mixing equations do not occur at the same mixing fraction. Ionic strength contrasts produce undersaturation at about one-third of the maximum value seen for CO_2 mixing. The other effects are relatively small.

The ionic strength effect is important in interpreting coastal karst. The deep lagoons formed on the east coast of the Yucatan were interpreted by Back et al. (1979) as solution features formed by the mixing of saturated fresh water from the interior with seawater along the coast. Figure 9.19 shows the change in saturation index for calcite, aragonite, the dolomite in the mixing zone. Shallow caves that form in the mixing zone enlarge and collapse. The collapse blocks dissolve to form the inlets along the coast.

10

Karst Evolution and Pleistocene History

The landscape as we see it is a snapshot. The time scale of landscape evolution is orders of magnitude slower than that of human observers. Some landforms are young, having evolved since the last ice cover was removed less than 20,000 years ago. Others are vastly older, formed by the slow sculpturing of rocks since early Tertiary time, more than 40 million years ago. The processes are also something like snapshots. Today's climate is not the climate of the past. Temperature, rainfall, and carbon dioxide generation have varied widely during the time over which the landscape evolved.

Caves and surface landforms record some of this history; the difficulty is in unraveling the record. The cave passages themselves record paleodischarges. The levels and sequences of caves in vertical profile tell something about pauses and changes in regional base level. However, the bulk of the historical information is encoded in speleothems and clastic sediments.

In this chapter we break away from the slice-of-life approach implicit in previous discussions of forms and processes and put together the time sequence of karstic events as read from the landforms themselves. One can go about this in either of two ways. All of the information about a particular region and drainage basin can be brought to bear on the question of the historical evolution of a particular karst drainage system. Alternatively, one can use the information recorded in caves and their deposits to learn something new about the historical evolution of the overall regions.

10.1 THE HISTORICAL MILIEU

10.1.1 *The Time Scale for Karst Processes*

The time scale for the development of karst features cannot be longer than that of the rocks on which they form. For example, shallow marine platforms were elevated above sea level at the close of Mississippian time, and extensive karst landscapes appeared along much of the southern margin of the North American Continent. These karst landscapes are recorded as paleokarst features on the unconformable boundary between the Mississippian carbonate rocks and overlying Pennsylvanian clastic rocks in many areas of the western United States. At the other end of the scale, caves and karst landforms on Bermuda and in the Bahamas occur in rocks that themselves are less than 100,000 years old.

The interest of this book is in karst that is part of the contemporary landscape. Most of the karst of the continental United States occurs in Paleozoic and, to a lesser extent, Mesozoic limestones. These rocks have been buried under younger sediments, and karst processes have been initiated after erosion exposed the rocks to surface weathering. One can get some notion of the time scale by examining the age of karst erosion surfaces.

In the Nittany Valley of central Pennsylvania is a karst surface developed on Ordovician and Cambrian dolomites on which residual soils occur with thicknesses ranging from 30 to 100 m. Taking a representative soil thickness of 50 m, with a known insoluble residue in the Gatesburg dolomite of 7.27 percent, a rock density of 2.85 g/cm^3, and a soil bulk density of 1.76 g/cm^3, simple mass balance demands that roughly 425 m of dolomite be removed to produce the observed thickness of residual soil. Using a denudation rate of 30 mm/ka, the original creation of the erosion surface is placed at 14 Ma, mid-Miocene time (Parizek and White, 1985). This figure represents a minimum age, because slope wash and leaching of silicate material could have removed some of the residual soil pile. The present-day relief between the valley upland surface and base level is about 100 m. Comparing this relief with the estimated carbonate rock removed suggests that the oldest cave fragments preserved in the residual uplands would be 2 to 3 Ma, which is consistent with paleomagnetic measurements done elsewhere in the eastern United States.

The valley upland surface in central Pennsylvania is part of an extensive erosion surface that has been traced through the Appalachian Mountains and was identified by the older school of geomorphologists as the Harrisburg surface (Fenneman, 1938). Many regions of doline karst in the eastern United States are related to the Harrisburg surface, including the Shenandoah Valley of Virginia, the

Highland Rim of Tennessee, the "Great Savannah" karst of West Virginia, and parts of east Tennessee and the dissected margins of the Cumberland Mountains. It is not unreasonable to propose that the most important karst landscapes in eastern and central United States were well established by mid-Tertiary time. That is, the protective overlying rocks had been at least partly removed, drainage systems were established, and there was internal drainage into the carbonate rocks in at least portions of the ancestral drainage basins.

It is more difficult to assign a beginning for karst landscapes in the western United States. Mountain building has been active throughout the Cenozoic period, and although the karst is mainly developed on Paleozoic limestones, the exposures and their associated drainage basins have been rearranged from time to time.

A different situation exists in the Coastal Plain, the Caribbean Islands, and parts of Mexico and Central America. Here rocks of Tertiary and Pleistocene age were subject to karstification since they were uplifted from the sea. Karstification has been an ongoing process from the beginning, although over the years, younger strata have been removed and present karst landscapes are developed on rocks that had been buried under younger carbonates. On the north coast of Puerto Rico, the Dry Harbour Mountains of Jamaica, the Yucatan, and the Maya Mountains of Belize, karst landscapes are all there ever were.

Considering all of North and Central America, it seems reasonable to set the time scale for contemporary karst processes from the beginning of Cenozoic time (Fig. 10.1). Much of the present-day karst was recognizable by the mid-Tertiary and there would have been ancestral landscapes and some fragments of caves that related to the Schooley erosion surface, the highest recognized erosion feature in the Appalachians. Although karst landscapes may have been evolving throughout the Tertiary period, most existing landforms, caves, and active underground drainage systems are rarely older than late Pliocene.

10.1.2 *Tertiary and Pleistocene Climates*

Of the seven variables of karst development (see Fig. 4.1), precipitation, temperature, and carbon dioxide production are aspects of "climate." Not only do these variables undergo seasonal cycles, but there are systematic variations on various longer time scales. The characteristic time scale for the development of a surface landform or a conduit is 10,000 to 100,000 years. Shifts in climatic regime that occur on these time scales are particularly important.

It is generally agreed that the early Tertiary climate was more or less a continuation of the Cretaceous climate and that tropical conditions extended to the northern boundaries of what is now the

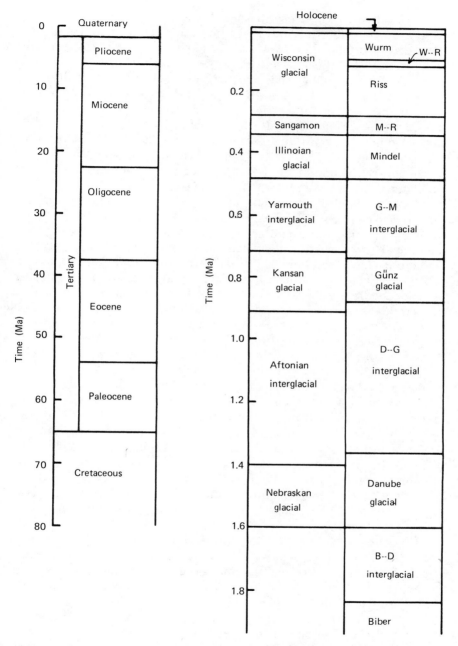

Figure 10.1 Time scales for the Tertiary and Quaternary periods. Subdivisions of the Quaternary are given for North America and for northern Europe. [Extracted from Van Eysinga (1983).]

United States. At the close of the Eocene age, the Northern Hemisphere underwent a period of cooling, and mean annual temperatures dropped from 25 down to 10°C. The temperate climate continued through the Miocene and Pliocene until the onset of the more extreme climatic swings that marked the onset of the Pleistocene.

Most of the karst lands of the United States were exposed to erosion during the Tertiary period, and few climatic records are preserved. Observations of fossil forest populations (Wolfe, 1978, 1979) suggest climatic cycles during the Tertiary spaced at about 10-million-year intervals (Fig. 10.2). There were cycles during the tropical phase before the end of the Eocene, and smaller temperature excursions in the temperate phase that followed.

Much more detail is available on glacial and climatic events of the Pleistocene (see, e.g., Wright and Frey, 1965). The onset of the Pleistocene age is marked by an abrupt lowering of temperature that is most clearly recorded in deep-sea sediments (Ericson et al., 1963). On land, the lowered temperature allowed the formation of massive continental glaciers, which flowed southward across North America and Europe. It is generally agreed that four major glacial advances occurred, separated by warmer periods of glacial retreat (see Fig. 10.1). These also are most clearly recorded in the fossils of deep-sea cores (Ericson and Wollin, 1968). On land, the retreating glaciers left moraines and till plains, and the warmer interglacial periods left paleosols, which have been the objects of a great deal of research as well as a good deal of argument about which soil was associated with which phase of which glaciation. The most recent, the Wisconsin glaciation, advanced and retreated in at least four substages, and it is likely that the earlier glaciations were also broken into periods of advance and retreat. These effects were most noticeable near the glacial margins. It remains an intriguing mystery whether the retreat of the Wisconsin ice sheet only 10,000 years ago marked the end of the ice ages with their severe climatic fluctuations or whether the present period, termed the holocene in Fig. 10.1., is merely one more interglacial period.

The effects of the Pleistocene glaciations on karst processes extended far south of the glacial margins. Large-scale effects in both North America and Europe reach into the tropical latitudes.

North of the glacial margins, any karst features that had formed during the interglacials were buried under advancing ice sheets. Alpine caves are frequently choked with glacial sediments and alpine karst is often modified by glacial processes. Scouring of soils in northern England and Europe has led to widespread pavement karst in contrast to other karst landscapes. Accessible caves north of the glacial margins tend to be young, many having formed in post-Wisconsin time.

For a wide band extending several hundred kilometers south of

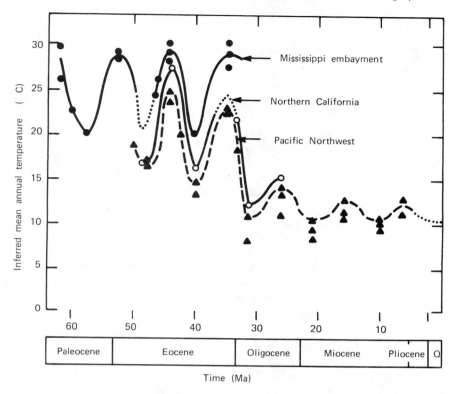

Figure 10.2 Mean annual temperature for the Tertiary, from analysis of fossil forest populations. [Adapted from Wolfe (1978).]

the glacial margins, the ice sheets had a pronounced impact on climate during the glacial advances. Winters tended to be extremely cold, and summers were cold and wet. Deep frost action on clastic sediments on mountain slopes provided the sheets of rock scree and a source of clastic sediment that is still being drained off through the cave systems today. Many of the present-day cave systems formed under climatic conditions quite different from those of the present.

Glacial damming during the ice advances caused surface drainage to be rearranged, with corresponding influence on any karst drainage within the larger basins. Release of tremendous sediment loads during periods of glacial retreat filled river valleys, raised base levels, and caused readjustment of underground drainage. This effect has been examined in greatest detail in the Mammoth Cave area, where 8 to 10 m of Wisconsin outwash material fill the deep channel of Green River, and have caused the flooding of the lowest tier of conduits in the cave system.

The influence of the glaciers in lower latitudes seems to have been to make the climate much wetter than it is at present. There is evidence (see, e.g., Flint, 1963) of pluvial lakes in what is now the arid Southwest. Massive travertine deposition in Carlsbad Caverns and other caves of the Guadalupe Mountains of New Mexico is thought to have taken place during the Pleistocene pluvials. Spectacular karst development in North Africa and the Arabian Peninsula, in what are now extremely arid areas, is likely a result of Pleistocene pluvial climates.

During the glacial maxima a substantial volume of the water of the world's oceans was tied up in the ice sheets, resulting in many fluctuations of sea level (Haq et al., 1987). Sea levels were lowered by 85 to 100 m, and the great limestone platforms of the Caribbean and the Gulf of Mexico were exposed to karstic processes. Large conduit systems formed, graded to the low stands of sea level, and extensive doline karst formed on the platform surfaces. With the melting of the ice, the sea levels rose, and many of the karst features were drowned. Locations of old spring outlets form the "blue holes" found in the Bahamas and elsewhere. Cave passages explored by divers are found to contain speleothems characteristic of air-filled caves (see, e.g., Palmer, 1985).

10.2 CLOCKS IN THE KARST

Many physical techniques suitable for dating either clastic sediments or travertine deposits have been applied to caves since the mid-1970s (Fig. 10.3). All suffer from two common defects. First, they do not span the necessary time scale—that is, from the present back to mid-Pliocene time. Second, they measure the ages of deposits in the caves rather than the caves themselves. The temporal relationships between the caves and the depositional features in them must be determined from field evidence.

10.2.1 *Carbon-14 Dating*

Cosmic rays striking the upper atmosphere convert ^{14}N to ^{14}C at a rate that can be assumed to have been constant over time. Atmospheric mixing brings the ^{14}C, with a half-life of 5730 years, into the lower atmosphere, where it forms a constant mixture with the stable ^{12}C and ^{13}C isotopes. A tree, for example, draws CO_2 from the atmosphere as it grows, and the living wood contains the background concentration of ^{14}C. When the tree dies or is burned, leaving residual charcoal, the exchange of carbon with the atmospheric reservoir is terminated and the incorporated ^{14}C begins to decay. By measuring the ^{14}C content of an old carbon, one can calculate the time elapsed since the cutoff from the atmospheric reservoir occurred. In this manner, ages of carbon-containing specimens can be determined up to a limit of about 50,000 years.

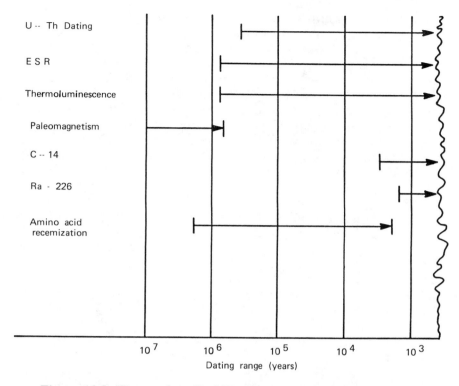

Figure 10.3 Range of applicability for various dating techniques.

In principle, ^{14}C dating methods can be applied to young speleothems, providing the source of carbon in the speleothem can be identified. The main source of dissolved carbonate in the seepage waters that deposit travertine is the dissolution of limestone bedrock at the soil/bedrock interface by carbonic acid derived mainly from the soil atmosphere. From the chemical reaction

$$CaCO_3 + CO_2 + H_2O \rightleftharpoons Ca^{2+} + 2\ HCO_3^- \tag{10.1}$$

one would expect the calcite-depositing waters to be composed half of recent carbon from the soil and half of "dead" carbon from the limestone. Actual measurements of freshly deposited cave calcites (Broecker and Olson, 1959) give an isotope ratio with 70 percent ± 20 percent modern carbon. This circumstance extends the potential dating period back to about 30,000 years, but it also requires that one know the fraction of modern carbon in the original deposit. Isotope exchange during dissolution and redeposition appears to upset the carbon balance implied by Eq. 10.1.

The best approach is to select calcite depositional sequences such as stalagmites or cores through flowstone cascades, where the "top" of the sequence is still growing. From analysis of the zero-age material, the ^{14}C deficit, and thus an age correction, can be determined which can then be applied to the lower parts of the section. The only assumption is that the fractionation processes have been constant over the period of deposition. Latham et al. (1986) found a correction term of 1000 years for a Mexican stalagmite. Dating of travertine formed over a segment of bone gave good results (Broecker et al., 1960). Extending the method to old deposits requires careful calibration, but successful dating was achieved for pluvial lakes Lahontan and Bonneville in Utah and Nevada using freshwater carbonates from lake and cave sediments (Broecker and Orr, 1958).

10.2.2 *Uranium-Series Dating*

The most abundant isotope of uranium, ^{238}U undergoes a long decay scheme (Fig. 10.4) ultimately to the stable ^{206}Pb. An intermediate step in the decay scheme is ^{234}U with a half-life of 2.45×10^5 years before it decays to ^{230}Th. Likewise, the minor isotope ^{235}U undergoes decay through ^{231}Th and ^{231}Pa on its way down the decay chain. To the convenient half-lives of ^{234}U and ^{231}Pa is added a quirk of geochemistry that turns these decay sequences into a useful dating method for speleothems. Thorium occurs in nature only as the Th^{4+} state, which forms oxides, silicates, and phosphates that are extremely insoluble. Although Th-containing minerals such as zircon and monazite are found in clastic sediments in caves, there is essentially no thorium in groundwater. Uranium, in the oxidizing environment of karst groundwater appears as the six-valent UO_2^{2+} ion, which is further solubilized by complexing with carbonate. Uranium is readily transported as $UO_2(CO_3)_2^{2-}$ and $UO_2(CO_3)_3^{4-}$ species and secondary calcite deposits commonly contain a few parts per million of uranium.

For a speleothem to be datable, it must

1. Contain some uranium (0.1 ppm or more).
2. Be free of detrital material containing thorium.
3. Be nonporous and have not undergone recrystallization at any time since the calcite was deposited.

The technique now widely used for speleothem age dating (P. Thompson et al., 1975; Harmon et al., 1975a; G. M. Thompson et al., 1975; Gascoyne, 1984; Schwarcz and Gascoyne, 1984; Gascoyne, 1985) consists of the following steps: The speleothem is dissolved, spiked with a known quantity of ^{232}U and ^{228}Th to act as an internal standard, purified, and the uranium and thorium are separated by ion exchange. These elements are then extracted and plated out on steel disks, and the alpha-spectrum is determined. Integrated areas under the alpha-peaks are directly proportional to the activity of each isotope.

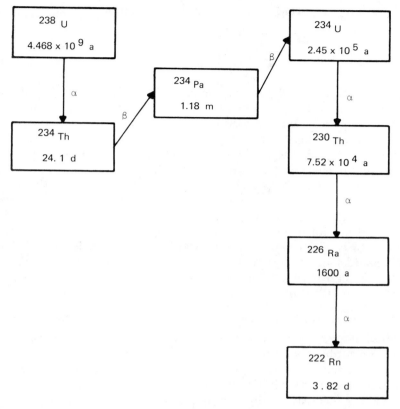

Figure 10.4 Decay scheme for ^{238}U, giving the half-lives of the intermediate isotopes. See Gascoyne (1985) for further detail.

Three different dating calculations are possible: ^{230}Th/^{234}U, ^{234}U/^{238}U disequilibrium, and ^{231}Pa/^{235}U combinations. Because thorium is insoluble and not transported into the growing speleothem, it is assumed that any thorium that appears in the analysis is a result of the decay of ^{234}U. Because of the sequence of somewhat similar half-lives, the age relationship is

$$\frac{[^{230}\text{Th}]}{[^{234}\text{U}]} t = \frac{1 - e^{-\lambda_0 t}}{[^{234}\text{U}/^{238}\text{U}]} + \frac{\lambda_0}{(\lambda_0 - \lambda_4)} \left(1 - \frac{1}{[^{234}\text{U}/^{238}\text{U}]}\right) (1 - e^{-(\lambda_0 - \lambda_4)t}) \quad (10.2)$$

where λ_0, λ_1, and λ_4 are the decay constants of ^{230}Th, ^{231}Pa, and ^{234}U, respectively. It is found that karst seepage waters are enriched in ^{234}U over the expected concentration from the equilibrium decay of ^{238}U. Therefore, the recovery of the ^{234}U/^{238}U ratio would be a method of age dating. The method has not been widely applied because of the necessity of knowing the initial ratio. The measured concentration of ^{231}Pa in the speleothem is also related to the time available for the decay of ^{235}U, according to

$$\frac{[^{231}\text{Pa}]}{[^{235}\text{U}]} = 1 - e^{-\lambda_1 t} \qquad (10.3)$$

Most reported dates have been measured by the ^{230}Th/^{234}U method, which has an effective range of 350,000 years. The protoactinium method requires a more difficult chemical separation step and has an age range going back 200,000 years. See Gascoyne (1985) for a critical comparison of the methods and Harmon et al. (1975a) for details of the experimental methods.

One other dating procedure can be extracted from the decay scheme of ^{238}U. One of the daughter products is ^{226}Ra, which decays with a half-life of 1600 years into ^{222}Rn, which is now widely observed as a hazard in caves (see Chapter 14). Radium is chemically similar to calcium and coprecipitates with it. ^{226}Ra is highly radioactive and can be detected in extremely low concentrations, thus ^{226}Ra/^{234}U can be used for dating in the time span of 300 to 7000 years, as an alternative to ^{14}C method (Latham et al., 1986).

10.2.3 *Electron Spin Resonance and Thermoluminescence Dating*

Calcite, like most other crystalline solids, is subject to radiation damage. The alpha decay of included uranium, thorium, and daughter isotopes creates defect centers in the calcite that may contain trapped electrons or holes. It takes a certain quantity of energy to release the electron or the hole from the trap—the "trap depth" on an energy scale. If the trap depth is large compared with the thermal energy of the environment, the charges remain trapped for long periods of time and the crystal becomes a dosimeter, accumulating trapped charges in proportion to the length of time it is exposed to its own internal radiation field. If the quantity of trapped charges can be measured, and the radiation field that produced them estimated, the elapsed time can be calculated. The experimental methods used to measure the time record in speleothems are electron spin resonance (ESR) spectroscopy and thermoluminescence spectroscopy.

Trapped charges in defect centers have unpaired spins, as do the electrons of certain of the transition metal ions that may also be present. ESR spectroscopy is a means of measuring these unpaired spins by tuning an applied magnetic field against a fixed-frequency microwave signal and looking at the power loss when these resonate with spin flips in the sample. Several defect centers in calcite fall in the midst of the six-line spectrum of Mn^{2+} (Fig. 10.5). The peak positions in the spectrum depend on experimental choices of magnetic field and microwave frequency and are usually described by a fixed

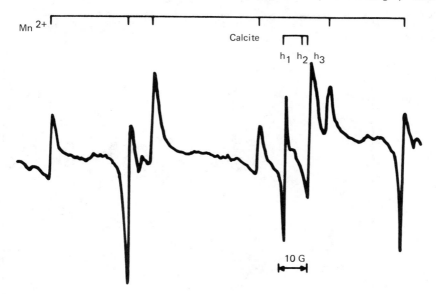

Figure 10.5 ESR spectrum of calcite showing peaks caused by trapped electrons superimposed on the six-band spectrum of Mn^{2+} impurity. [Adapted from Skinner (1983).]

parameter known as the Landé g-factor. Landé g = 2.007 for h_1, 2.004 for h_2, and 2.001 for h_3. Peak h_3 is the one usually used for ESR dating. Other impurities may also produce peaks but the great advantage of ESR spectroscopy is its sensitivity and spectral resolution, which allow discrimination against most interferences.

Since the ESR dating of cave travertines was proposed (Ikeya, 1975) a great deal of effort has been expended on refining the sample preparation and experimental methods, and evaluating the results (Hennig and Grün, 1983; Skinner, 1983; Goede and Bada, 1985; Smith et al., 1985; Grün, 1985; Grün, 1986). In principle, one need measure only the intensity of the h_3 ESR peak, calibrate this against the available alpha-dose based on the uranium content of the sample, and calculate the age. In practice, account must be taken of the additional gamma-dose received from the background and other possible interferences. Some sense of the reliability of state-of-the-art dating methods of the mid-1980s may be gained by comparing ESR dates with thorium–uranium dates on the same samples (Fig. 10.6). One great advantage of ESR dates is that they can be extended past 730,000 years, at which point it is possible to use the Brunhes/Matuyama magnetic reversal as a calibration point for the chronology.

Thermoluminescence spectroscopy is an alternative method for

314 / *Geomorphology and hydrology of karst terrains*

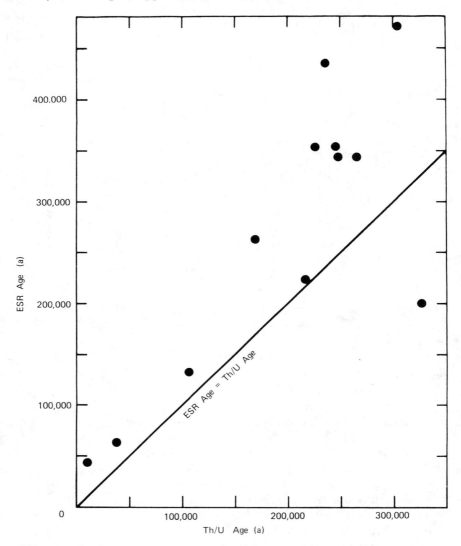

Figure 10.6 Comparison of ESR ages with ^{230}Th/^{234}U ages measured on the same specimens. Data are drawn from compilations in Grün (1985, 1986).

measuring radiation dose in calcite crystals. The decay of electron traps depends exponentially on temperature. As the sample is heated, nothing much happens until the thermal energy gets close to the trap depth energy. At this temperature, the trapped charges decay quickly to the ground state, releasing their energy as a pulse of light known as thermoluminescence. The light pulse can be measured and, with some difficulty, can be related to the age of the specimen. Thermoluminescence peaks are broad and have less discrimination against interferences than ESR. However, thermoluminescence appears to be a usable speleothem dating technique (Debenham, 1983).

10.2.4 *Magnetic Records*

The earth's magnetic pole, which is the reference point for all of the magnetic compasses used in cave mapping, is not a fixed target. The position of the pole is given by its declination and inclination (the orientation of the magnetic pole with respect to geographic north and the dip of the field with respect to horizontal). The past apparent position of the magnetic pole is affected by

1. Secular variation in polar position caused by wandering of the dipole field and particularly its nondipole component. The time scale is tens to hundreds of years, and there has been extensive drift of the field over the past few thousand years.
2. Continental drift causes the apparent position of the pole to shift because the rocks that record the information have drifted from their original positions. Because the rates of drift are on the order of centimeters per year, the time scale for substantial shifts in apparent polar position is millions of years.
3. The earth's magnetic field reverses polarity at intervals of a few hundred thousand to a million years. The magnetic reversals have been carefully documented from age-dated basaltic rocks, which have a strong magnetic signature and a magnetic time scale has been established well back into the Tertiary period.

Caves record a weak magnetic signature in the clastic sediments and in the calcite speleothems. In both cases, the signature is a result of tiny quantities of magnetic minerals such as magnetite, hematite, and other iron oxide minerals. When clastic sediments settle into place, the magnetic grains orient themselves in the direction of the magnetic pole. Once settled and buried under later sediments, the magnetic grains record the polar position as it was at the time of deposition. The magnetic signal from both clastic cave deposits and calcite speleothems is very weak but can be measured with contemporary high-sensitivity magnetometers if due care is taken in sample collection and magnetic "cleaning."

Individual measurements of polar direction are not very useful, but if a substantial sedimentary section that spans some thousands of years of deposition can be established, the polar wandering curve can be measured directly from the sediments and matched against the known secular variation curve for the region, thus establishing a chronology for the sediments. The best samples are thin-bedded silts and clays that have deposited slowly under quiet water conditions. Such magnetic fabrics have been established for Climax Cave in Georgia, where a sedimentation event was observed beginning about 700 A.D. and terminating at 1500 A.D. (Ellwood, 1971). Magnetic fabrics were established for Agen Allwedd and Pwll y Gwynt caves in Wales (Noel, 1983, 1986a) and in Peak Cavern, where 7000 B.P. was established as the date for the sedimentation event (Noel, 1986b).

Latham et al. (1979) argue that the magnetic signatures from dripstone are better than those from clastic sediments in that there is no danger of later processes disturbing the signal. Stalagmites are good stratigraphic sequences and have the advantage that they can be independently dated by uranium/thorium or carbon-14 methods. A good secular variation curve for Mexico over the past 1200 years was obtained by this method (Latham et al., 1986).

Magnetic reversals are recorded in cave sediments and provide fixed points on the time scale. Discovery of a single reversed sediment sample at least establishes an age of greater than 730,000 years, the time of the last recorded reversal. Whereas measurement of secular variation curves are done with single beds of clastic sediments or single stalagmites, magnetic reversal stratigraphy is best applied to entire sedimentary sequences in caves. The record for Mammoth Cave (Schmidt, 1982) (Fig. 10.7) shows normal sediments in the lowest levels with the first reversals appearing in the middle-level passages. Normal sediments just above these may record the Jaramillo normal, a period of normal polarity lasting only 70,000 years. Reversed sediments dominate the sequence through the level of the Main Cave in Mammoth, but higher-level passages such as Collins Avenue in the Flint Ridge portion of the system are a poorly resolved mixture of normal and reversed samples. The preferred interpretation dates the clastic sediments of the highest levels in the cave system as late Pliocene. Flowstone with a reversed magnetic signature has been found in British Columbia (Latham et al., 1982).

10.2.5 *Amino Acid Racemization*

The amino acids are asymmetric molecules with an intrinsic right-handedness or left-handedness. Living organisms are composed of left-handed amino acids. Solutions of these molecules have the property of rotating the plane of polarization of polarized light, and thus the right- or left-handed character can be determined. When an organism dies, the broken-down protein material is slowly converted to a mixture of L-amino acids and D-amino acids, a process called

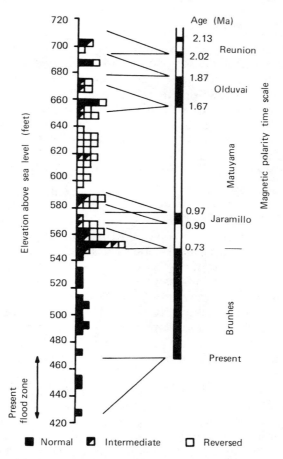

Figure 10.7 Paleomagnetic record of the clastic sediments in Mammoth Cave in relation to established magnetic polarity time scale. [From Schmidt (1982).]

racemization. Each amino acid has a characteristic conversion kinetics, all of which are strongly dependent on temperature. Given these constraints, however, measurement of the D/L ratios for specific amino acids is a means of dating in the range of a few tens of thousands of years. Goede and Bada (1985) found good agreement between aspartic acid racemization dating and radiocarbon and ESR dating for fossil bone material from Tasmanian caves.

The application to cave sediments is limited because of the scarcity of amino acid–bearing deposits. However, the method has been useful in recent coastal areas such as Bermuda and the Bahamas, where some organic material remains in the cave sediments.

10.3 TECTONIC RECORDS

The elevation difference between the sea and the uplands of the continental interior is the energy source for geomorphic processes. The energy must be renewed periodically or the land surface would long since have been graded down to sea level. Land elevations change because of movement of continental and oceanic plates, isostatic depression and rebound during and after glaciation, and rising and falling sea levels. Many of these influences begin at the sea and migrate upstream as the rivers deepen or in-fill their channels. By the time these conflicting signals reach the continental interiors, the message is considerably garbled.

10.3.1 *Cave Levels and River Terraces*

Cave levels have been correlated with nearby river terraces but rarely over entire drainage basins. Data for the main stem of the Potomac River (White and White, 1974) show that caves are clustered in well-defined levels beneath the valley uplands and beneath the accordant ridge crests in the Great Valley and Eastern Valley and Ridge portions of the basin (Fig. 10.8). The vadose zone between the land surface in the valley uplands and the base-level stream is relatively thin, about 200 m in the example given. One must take care that apparent clusterings of passages do not merely prove that caves are underground.

Correlation of cave levels with terrace levels over the scale of an entire drainage basin must take into account the exponential form of the river profile. For the river to maintain its profile as downcutting proceeds, the upstream reaches will cut down much more than the downstream reaches. Terrace levels will be more widely spaced in the upstream reaches and thus will provide more vertical separation of cave levels, making these levels easier to recognize. In the downstream reaches, levels may be stacked so closely together that vertical solution, breakdown, and other processes disguise the tiered pattern of the cave system.

The most intensively studied tiered cave in the United States is the Mammoth Cave System. Deike (1967), Miotke and Palmer (1972), and Palmer (1981a) laid out a geomorphic development for the cave system, placing the highest-lying passages in late Tertiary time and the larger upper level trunks in the early Pleistocene. Base-level trunks, in a recent history complicated by in-filling of the Green River channel with glacial outwash material, could still be related to the Illinoian and Wisconsinian glacials and the Sangamon interglacial. These interpretations were based on considerations of passage evolution and of the evolution of the Green River Valley. They turned out to be in remarkably good agreement with the Paleomag-

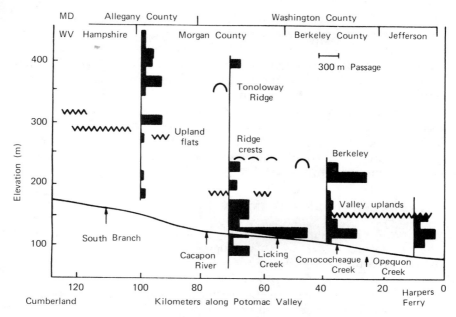

Figure 10.8 Distribution of cave length along the main stem of the Potomac River. The bar graphs give accumulated passage length within the elevation range at various segments along the river. The Potomac, here, flows eastward between Maryland and West Virginia. The counties bordering the river are shown at the top of the figure. Major tributaries entering the Potomac are indicated by arrows.

netic time scale measured in the cave by Schmidt (1982) (see Fig. 10.7).

It appears that mere correlation of cave passages with river terraces is not a useful exercise once one has accepted the Swinnerton–Davies–Ford–Ewers idea that many caves are water table caves. It would be more productive to use the caves, which are easily identifiable and easily mappable, rather than river terraces to disentangle the evolution of river basins.

10.3.2 Karst Plains

The earlier generation of geomorphologists had a great interest in peneplains and devoted a good deal of effort to identifying them, usually in the form of valley uplands or accordant ridge summits. It has never been clear how much of the observed surfaces should be ascribed to long pauses in base level lowering and what portion was the result of differential erosion of rocks of very different resistance, a process Hack (1960b) termed *dynamic equilibrium*.

Some of the best-preserved erosion surfaces in the eastern United

States are karst surfaces. Doline karsts, or sinkhole plains, extend for distances of tens of kilometers and often cut across the geologic structure. Unlike other valley upland surfaces, the karst plains are not dissected by entrenching surface streams as base levels are further lowered. Because of internal runoff, the planar surface is preserved, although it must be gradually lowered by denudation.

The elevational relations of the karst plains, and their relations to each other across the Appalachian Mountains or the Interior Lowlands—especially those obtained by reconstructing the surface through time—are some of the best means for interpreting these long puzzling remnants of erosion levels.

10.4 THE CLIMATE RECORD IN CAVES

The cave environment is remarkably constant on a human time scale. The temperature is close to the mean annual temperature of the region. Near base level, the water levels and discharge fluctuate with storm events and surface runoff. Storage and discharge of sediments may vary over the year. In the higher levels of caves, events move more slowly. However, events on the surface do influence the cave through clastic sedimentary in-fillings and speleothem deposition. The cave is a storehouse of information about the surface events of the past.

10.4.1 *The Speleothem Record*

Secondary calcite deposits contain the following records:

1. Trace quantities of uranium, which permit the deposits to be dated.
2. The bulk of the speleothems themselves, which, combined with dating, tell something about the distribution of calcite deposition over time.
3. Isotropic variations in ^{18}O, which correlate with temperature.
4. Fluid inclusions that preserve a sample of the original solution from which the speleothems precipitated. These are a source of deuterium/hydrogen ratios.
5. The organic stains, which give the speleothems their ringed or layered appearance. The organic fraction is a potential source of information on the type of plant cover at the surface.

Speleothems, particularly stalagmites and flowstone, grow very slowly to considerable masses. Cylindrical stalagmites are the best candidates for intraspeleothem dating because the sequence of superimposed growth caps provides an easily observed "stratigraphy" if the stalagmite is sawed longitudinally. Mass transfer arguments (see Section 8.5.3) show that cylindrical stalagmites grow at a constant rate and height is proportional to time of growth, a conclusion that is born out by age dating from top to bottom (Fig. 10.9).

Much additional information in the stalagmite stratigraphy has not been exploited. Fabrics vary from massive, densely crystalline layers, to thin banded layers, to breaks in the section containing silt. The organic staining is also banded, but the bands have not been interpreted. The banding appears to be too thick to represent annual cycles, and the stalagmites may record a cycle of wet and dry periods on a time scale of a few tens to a few hundreds of years.

Speleothems, occupying different levels in tiered caves, can provide minimum ages for the passages from which they are sampled. Unfortunately, the time scale for cavern development is about an

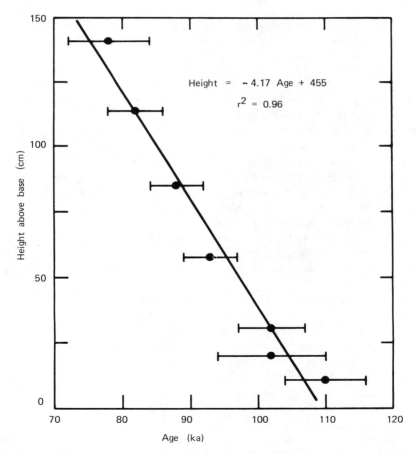

Figure 10.9 Growth rate curve for a stalagmite in Little Trimmer Cave in Mole Creek Karst, Tasmania. Data are from Goede et al. (1986). Linear least-squares regression gives a growth rate of 4.17 cm/ka for this deposit.

order of magnitude larger than the 350,000-year limit for Th/U dating, the technique most commonly used. Thus, one obtains information only about the most recent events in the evolution of a cave system and its drainage basin. Speleothem ages show that some Norwegian caves formed before the Weichselian (Wurm) glaciation (Lauritzen and Gascoyne, 1980). The relict passages of caves in the Mendips and in Yorkshire are older than 350,000 years (Atkinson et al., 1978; Atkinson et al., 1984; Gascoyne et al., 1983b). Precise age determination of tiered water table caves allows one to estimate rates of valley deepening. Calculated rates in Yorkshire were 50 to 200 mm/ka (Gascoyne et al., 1983a). Extrapolated backward, the age of the presently surviving dales and caves of Yorkshire are in the range of 1 to 2 Ma, similar to areas of comparable relief in the United States.

On a still larger scale, frequency plots of speleothem dates for all measured deposits in all caves at all levels give some indication of when calcite was deposited and when it was not. Such records have been compiled for the north of England and for western Canada (Gascoyne et al., 1983a) (Fig. 10.10). There are clearly defined periods of massive speleothem growth and periods when the growth process is shut down. In England and Scotland, Atkinson et al. (1986) found that there was continuous but sparse speleothem growth from 40,000 to 26,000 B.P., the absence of speleothems from 26,000 to 15,000 B.P. and abundant speleothem growth from 15,000 B.P. to the present. In the Mackenzie Mountains of western Canada, Harmon et al. (1977) found five distinct periods of speleothem deposition; the present to 15 ka, 90 to 150 ka, 185 to 235 ka, 275 to 320 ka, and greater than 350 ka. In these northern climates, speleothem deposition ceases during periods of intense glaciation, probably because of the development of permafrost. Extensive calcite deposition coincides with the interglacial periods of mild climate and abundant rainfall.

The opposite is true in arid regions at lower latitudes. Deposition of speleothems in the arid southwest of the United States coincides with the Pleistocene pluvials. Radiocarbon dating of massive stalagmite in the Rössing Cave in the Namib Desert of Africa assigns the time of deposition to the Weichselian (Wurm) pluvial (Heine and Geyh, 1983).

Speleothem age dating can be used to identify times of low sea level stand. The drowned caves of Bermuda and the Bahamas contain stalactites and stalagmites that formed when the caves were air-filled. A stalagmite on Andros Island in the Bahamas grew between 160,000 and 139,000 B.P., thus putting a date on the Illinoian glacial advance (Gascoyne, et al., 1979). Data from Bermuda show sequential periods of rise and fall of the sea-level (Harmon et al., 1978b).

The isotopic composition of groundwaters and precipitates from

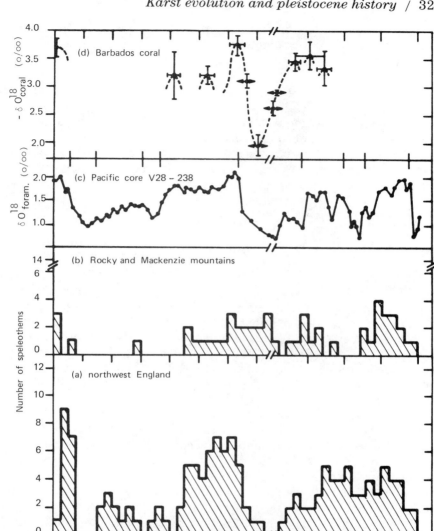

Figure 10.10 Comparison of other paleoclimatic records with speleothem records from the north of England and from the Rocky and Mackenzie Mountains of western Canada. [Adapted from Gascoyne et al. (1983b).]

groundwaters varies depending on the processes of dissolution and precipitation and the temperatures at which these processes occur. Making use of the temperature dependence, isotopic compositions, particularly $^{18}O/^{16}O$ and $^{2}H/^{1}H$, are widely used as paleothermometers. Because the changes in isotope ratio are very small, they are generally expressed in delta units as parts per thousand with respect to an arbitrary standard known as standard mean ocean water (SMOW)

$$\delta(o/oo) = 1000 \frac{R - R_{SMOW}}{R_{SMOW}} \qquad (10.4)$$

Measurements using modern mass spectrometers are accurate to ± 0.2 percent for ^{18}O and to ±2 percent for ^{2}H.

Use of oxygen isotope ratios as a paleothermometer depends on isotopic equilibrium between the seepage water and the precipitating calcite:

$$\frac{(^{18}O/^{16}O)_c}{(^{18}O/^{16}O)_w} = K_{cw}(T) \qquad (10.5)$$

The temperature is then calculated from the temperature dependence of K (Harmon et al., 1977):

$$10^3 \ln K_{cw} = 2.78 \times 10^6 \, T^{-2} - 2.89 \qquad (10.6)$$

If precipitation is slow, Eq. 10.5 should be satisfied. The test (Hendy, 1971) is to compare variations in ^{18}O with variations in ^{13}C. Their correlation is evidence for rapid precipitation out of isotope equilibrium as has been claimed (Fantidis and Ehhalt, 1970). However, most recent studies do not find this correlation, and the authors have proceeded with temperature calculations (Thompson et al., 1976; Harmon et al., 1977; Harmon, 1979).

If the speleothem oxygen isotope ratio is to record cave temperature at the time of deposition, which in turn is to measure mean annual temperature at the land surface, it is necessary that the isotopic ratio not depend on annual fluctuations in rainfall and surface temperatures. Yonge et al. (1985) show that the isotopic composition of seepage waters is constant and approximately equal to the mean annual precipitation of the area.

To solve Eq. 10.5, one needs isotope ratios for calcite and for the seepage water. The calcite can be measured but the preserved specimens of seepage water in the fluid inclusions may have undergone oxygen isotope exchange with the enclosing calcite, invalidating the measurement. A linear relation, known as the meteoric water line, exists between oxygen and hydrogen isotopes for surface waters. Comparison of the ^{18}O determined from the calcite with deuterium determined from the fluid inclusions (Fig. 10.11) shows that the seepage waters follow the meteoric water line closely. Thus, deuterium from the fluid inclusions, which should not have isotope exchanged with the enclosing calcite, can be used to calculate the corresponding ^{18}O of the seepage water; this allows paleotemperatures to be calculated. The ^{18}O ratios in present-day cave seepage waters correlate directly with cave temperatures (Fig. 10.12). Cave

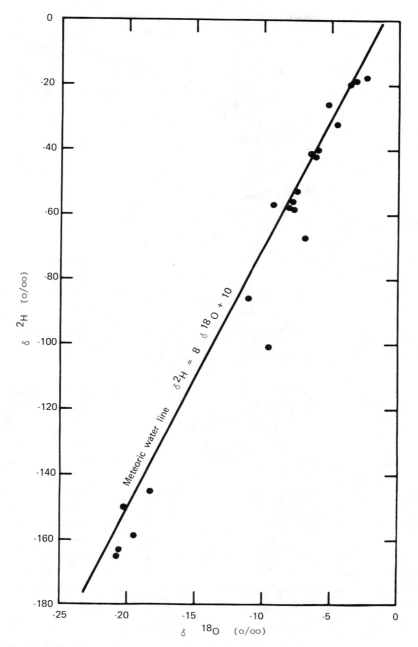

Figure 10.11 Data for various cave-seepage waters (Yonge, et al., 1985) compared with the meteoric water line. Both ^{18}O and ^2H shifts were measured with respect to SMOW.

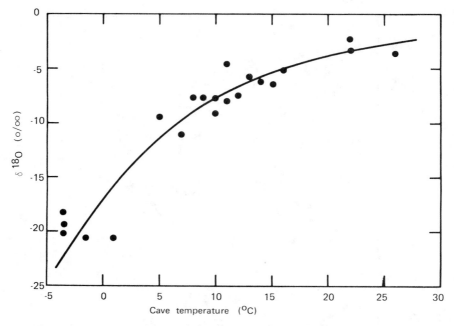

Figure 10.12 Dependence of $\delta^{18}O$ on temperature for seepage waters in various caves in North America. Data are from Yonge et al. (1985).

temperature is a good measure of mean annual surface temperature, and the oxygen isotope composition is a good measure of the isotope composition of rainfall.

The combination of Th/U dating of speleothems with their ^{18}O and ^{2}H compositions as a paleothermometer allows the construction of temperature–time curves for regions where appropriate quantities of cave deposits exist. Speleothems become the continental equivalent of the deep-sea cores that have established much of the Pleistocene climatic record (Gascoyne et al., 1980; Gascoyne et al., 1981; Harmon et al., 1978a; Harmon et al., 1979; Schwarcz et al., 1976). The complete establishment of a North American climatic record is underway, but because of the large range of space and time, many more data points are needed before the climatic picture at various times in the late Pleistocene comes into focus.

10.4.2 Clastic Sediment Records and Paleohydrology

The in-fillings of clastic sediments depend on how cave passages are disposed with respect to the sediment sources, but they also depend on high-velocity flow to transport clastic material. The required flows, interpreted in terms of the hydrogeologic setting of the cave, tell something about climatic conditions at the surface, particularly about total precipitation and how precipitation is distributed. Long periods of continuous but low-intensity rainfall produces a response in the cave system quite different from that produced by the same total precipitation appearing as episodic but very intense storms.

Among sedimentary deposits that have been described are the following:

1. Cobble and boulder fills, unbedded and unsorted, that require occasional intense flood flows to transport and emplace.
2. Bedded silts, sands, and fine gravels representative of high-velocity stream flow.
3. Massive unctuous clays that may have settled under static water conditions with the cave acting as a "settling tank."
4. Fine-grained laminated clays that look much like varved clays found in lakes and that may represent pulses of sediment-laden water moving into flooded passages (Bull, 1981; Gillieson, 1986).

All cave sediments in the active portions of caves are only in temporary residence. What appear to be passive sediment beds can be removed en mass in episodic floods, only to be replaced with other sedimentary material as part of the continuous sediment flux through the karst drainage system. Those sediments that happen to be in residence when the passages are drained for the last time are preserved when the passage moves from the enlargement to the stagnation stage. There has been little success (see Section 8.4) in discovering any sort of continuous beds that can be traced laterally for long distance. However, the form of the deposits gives some clues about the flow conditions that existed when the passage was last abandoned.

If passages with measurable cross sections (not too much of the passage obscured by sediments) and uniform scalloping can be found, then the paleodischarge can be calculated. If the catchment area can be identified and its boundaries are reasonably certain, then a paleorunoff intensity can be calculated and compared with present-day conditions in the same basin.

11
Evaporite Karst

Highly soluble rocks such as gypsum and salt can be sculpted into karst landscapes. The higher solubilities mean that the landscapes develop at faster rates and also that the landforms tend to be more transient than those formed in the less-soluble carbonate rocks.

11.1 KARST PROCESSES IN EVAPORITE ROCKS

11.1.1 Types of Evaporite Karst

Evaporite rocks are widely distributed over the earth's surface, many in extensive areas of the United States (Fig. 11.1). However, most of these rocks occur at depth because rapid dissolution quickly destroys surface outcrops. Surface expressions of evaporite karst are restricted to arid regions. In the United States, surface karst occurs in western Oklahoma (Myers, 1960a), northwest Texas (Gustavson et al., 1982), and eastern New Mexico. Other regions with extensive evaporite karst include Nova Scotia and British Columbia in Canada (Wigley et al., 1973), the Harz Mountains of Germany (Pfeiffer and Hahn, 1972; Bradt et al., 1976), various areas in the eastern Mediterranean, and very extensive areas in the Soviet Union (Shelley, 1954; Popov et al., 1972).

Quinlan et al. (1986) point out that, of the various types of karst, exposed, soil-covered, and interstratal karsts are most common in evaporite rocks. Much interstratal karst is formed by the dissolution of salt rather than gypsum because of the complex interbedding between salt, anhydrite, gypsum, carbonates, and silicoclastic rocks in evaporite sequences.

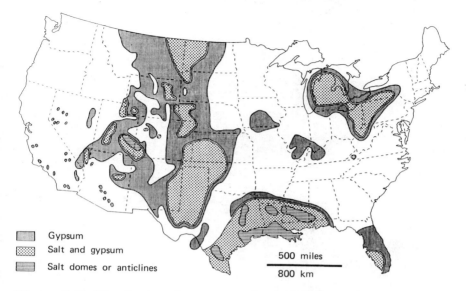

Figure 11.1 Distribution of evaporite beds in the United States. From Quinlan et al. (1986). [Courtesy Dr. Quinlan.]

11.1.2 Interstratal Karst

The term interstratal karst is applied to solution processes that take place beneath a covering layer of younger nonkarstic rocks. The term is applied to a generalized two-dimensional or three-dimensional areal dissolution, not to cave development beneath a protective caprock. This type of karst is particularly important in evaporite rocks because of the possibility of fresh water infiltrating through the cover rocks and dissolving salt or gypsum at depth. Figure 11.2 illustrates the principal features. Because the chemistry of evaporite dissolution is simple dissociation with no chemical reaction, fresh water is injected on the up-gradient side of the system, and saturated gypsum waters or salt brines are discharged on the down-gradient side. The only constraint on the amount of materials removed is the equilibrium solubility of the dissolving minerals.

Removal of substantial volumes of material at depth provides the conditions necessary for subsidence. Breccia pipes and subsidence sinkholes are characteristic of interstratal karstification areas.

11.1.3 Solubility of Evaporite Rocks

Gypsum dissolves by a simple dissociation reaction:

$$CaSO_4 \cdot 2H_2O \rightleftharpoons Ca^{2+} + SO_4^{2-} + 2\,H_2O \tag{11.1}$$

330 / Geomorphology and hydrology of karst terrains

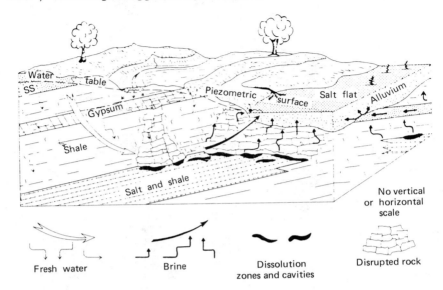

Figure 11.2 Typical interstratal karst in western Oklahoma. The horizontal dimension is 1 to 15 km; the vertical dimension is 30 to 300 m. [From Johnson (1981).]

The solubility of gypsum as a function of temperature, pressure, and concentration of dissolved salts has been measured many times. The most recent data, those of Hardie (1967) and Blount and Dickson (1973), are in agreement (Fig. 11.3). Gypsum is moderately soluble in water with a saturation concentration of roughly 2500 ppm as $CaSO_4 \cdot 2H_2O$. A rather unusual feature is the maximum in the solubility curve at 40°C. At 58°C at 1 atm, gypsum loses its water of crystallization to become anhydrite. The gypsum-anhydrite reaction can be driven at lower temperatures by lowering the activity of water; as a result, bedded evaporite deposits often contain anhydrite because of high salinities (low water activities) in the fluids from which the rocks were precipitated.

Halite is extremely soluble in groundwater. The solubility is 35.5 percent by weight at 25°C and increases rapidly with temperature. High sodium chloride concentrations influence the activity coefficients of other ions and thereby make the salinity of groundwater an important variable in both gypsum dissolution and carbonate chemistry.

It is possible to define a saturation index for gypsum analogous to the saturation indices for the carbonate minerals.

$$\mathrm{SI}_g = \frac{a_{Ca^{2+}}\, a_{SO_4^{2-}}}{K_{\mathrm{gyp}}} \qquad (11.2)$$

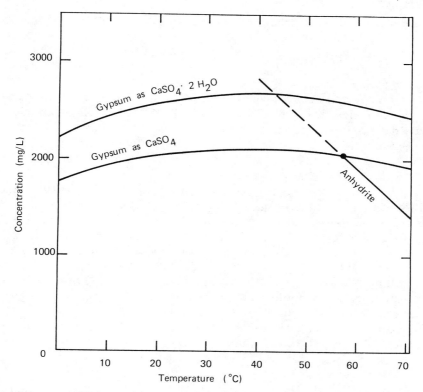

Figure 11.3 Solubility curves for gypsum and anhydrite based on the experimental data of Blount and Dickson (1973). The upper curve is calculated as gypsum and pertains to the mass loss of gypsum rock in solution. The lower curve is calculated as $CaSO_4$ and displays the invariant point at 58°C, where gypsum, anhydrite, and liquid coexist.

Wigley (1971) has fitted the available experimental data and gives the solubility product, K_{gyp}, as a function of temperature:

$$pK_{gyp} = 4.641 - 2.49 \times 10^{-3}\, T + 5.92 \times 10^{-5}\, T^2 \qquad (11.3)$$

where T is temperature in degrees Celsius.

11.2 GYPSUM KARST

11.2.1 Surface Landforms in Gypsum

Exposed karst or soil-covered karst in gypsum exhibits landforms similar to those seen on carbonate rocks. Because surface outcrops of gypsum survive only in arid climates, sinking-stream patterns tend to be small dry arroyos that end either in swallow holes or in

332 / Geomorphology and hydrology of karst terrains

short open cave passages. Closed depression features are common. They tend to be broad and shallow (Fig. 11.4). Because of the intense solution at depth in the bedrock, collapse and subsidence sinks are also common. Some of these are quite large, such as the Bottomless Lakes near Roswell, New Mexico.

An enigmatic feature that seems to be unique to evaporite karst is the subsidence trough described by Olive (1957). These are elongate depressions that range from 100 to 1000 m in width and from 1 to 15 km in length. Generally they are only a few meters deep. They are thought to result from subsidence as a result of solution at depth along faults or fracture zones.

Bare bedrock exposures on gypsum are sculpted into the same suite of minor landforms as are found on pavement karst in carbonate rocks. Rillenkarren are common (Fig. 11.5) because of rapid dissolution of exposed gypsum in sheet wash during rare rainstorms. Rinnenkarren and various small solution basins (tinajitas or kaminitza), as well as kluftkarren are found. Although no numerical measures have been obtained, there is no visually striking difference, in either scale or morphology, between solutional sculpturing on gypsum and solutional sculpturing on carbonate rocks.

Using a mass balance argument and an assumption of saturation at the soil/bedrock interface similar to the derivation of Eq. 7.18,

Figure 11.4 Shallow doline in gypsum near Vaughn, New Mexico.

Figure 11.5 Rillenkarren on gypsum. Bottomless Lakes area near Roswell, New Mexico.

one can calculate an expression for the denudation rate in gypsum karst.

$$D_n \text{ (mm/ka)} = \frac{10^{-3}C_s}{\rho_{\text{gyp}}}(P - E) \qquad (11.4)$$

$$D_n \text{ (mm/ka)} = 1.05\,(P - E)\,(\text{mm/a}) \qquad (11.5)$$

where C_s is the solubility of gypsum, ρ_{gyp} is the density of gypsum (in grams per cubic centimeter), P is precipitation, and E is evapotranspiration (both in millimeters per year). Gypsum karst is denudated rapidly. What saves it is that, in most of the semiarid regions where gypsum karst occurs, evapotranspiration is greater than precipitation, and bedrock is lost only during the infrequent storms of sufficient intensity to produce surface or internal runoff.

11.2.2 Gypsum Caves

Gypsum caves follow two main patterns. Some are fragments of conduit with ground plans and cross sections much like those of single-conduit caves in carbonate rocks. The caves in northwest Texas (McGregor et al., 1963) and Alabaster Cave in western Oklahoma (Bretz, 1952; Myers, 1960b), which is one of the few show caves in gypsum, are sinuous single-conduit caves. Fragments of conduits

334 / *Geomorphology and hydrology of karst terrains*

have also been described in Nova Scotia (Moseley, 1975) and the Harz Mountains (Hensler, 1968). In contrast, many gypsum caves are very densely packed network mazes. The most extensive of these are the large gypsum caves in the Ukraine, USSR (Dublyanskii, 1979) (Fig. 11.6).

The conduit type of cave tends to be shallow. Entrances are located at sinking streams or at the upstream ends of arroyos. Passages may be canyon-shaped, elliptical, or nearly circular tubes (Fig. 11.7). Many gypsum caves are dry most of the year, but bear the evidence of storm water runoff in the form of scoured floors with little sediment and small scallops. Breakdown occurs in gypsum caves, and the solutional sculpturing of the cave walls is much like that found in caves in carbonate rocks.

The gypsum maze caves are usually found beneath some protective caprock. They appear to form by a mechanism similar to that proposed by Palmer (1975) for tight network mazes in limestone. Water percolates slowly into the cavernous horizon from overlying permeable but insoluble rocks. Geologic circumstances prevent the accelerated enlargement of single conduits. All available joints are enlarged and if joints are closely spaced, a pattern similar to that shown in Fig. 11.6 evolves. Although Quinlan (1978), in his descrip-

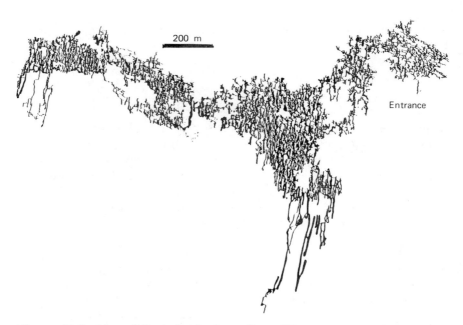

Figure 11.6 Map of Optimistcheskaya Cave, Ukraine. [Adapted from Dublyanskii (1979).]

Figure 11.7 Passage cross section in Skylight Cave in Chosa Draw Karst, Delaware Basin, New Mexico.

tion of the types of karst, clearly distinguishes between the development of caverns and interstratal karst, cave patterns such as that shown in Fig. 11.6 may be what interstratal karst would look like if it were conveniently exposed.

11.2.3 Mechanisms for Gypsum Cave Development

Gypsum caves pose an enigma for the cave theorist. Morphologically, gypsum caves are little different from limestone caves. The chemistry, however, is much simpler. There is no dependence on pH or carbon dioxide partial pressure. The mechanism of dissolution is also different. Rates of gypsum dissolution determined in the laboratory (Liu and Nancollas, 1971) depend strongly on the rate of stirring of the solutions, indicating that the transport of dissolution products across the surface boundary layer is rate controlling. Liu and Nancollas and James and Lupton (1978) give rate equations that can be written in the general form

$$\frac{dC}{dt} = \frac{A}{V} k_T (C_s - C) \qquad (11.6)$$

where C is concentration of gypsum in solution, k_T is a mass transfer rate constant ($k_T = D/\delta$, where D is the diffusivity of ions in the boundary layer and δ is its thickness), A is the surface area, and V is the volume of solution.

Equation 11.6 can be integrated to give a concentration/time relation of the form

$$C = C_s(1 - e^{-A/V\,k_T t}) \qquad (11.7)$$

This plots as an S-shaped curve that rises steeply and then curves over to approach the saturation concentration C_s asymtotically. Based on Liu and Nancollas's rate constants, gypsum solutions reach a few percentage points of saturation within a few minutes and almost complete saturation within an hour. The time scale for gypsum dissolution is a factor of 100 shorter than that for limestone dissolution. Groundwater in gypsum reaches saturation in much shorter distances than groundwater in limestone, and the initiation of solutionally widened joints and bedding plane partings at depth in the bedrock is correspondingly slower.

Gypsum caves, like limestone caves, have an initiation phase and an enlargement phase. The rapid dissolution kinetics imply that the initiation phase is longer, because the initial pathway along joints and bedding plane partings is opened by water that is very close to saturation. The initial mechanical pathway is more important, and open fractures should be important sites for the initiation of gypsum caves. At the critical thresholds, the onset of turbulence should play a more important role in gypsum dissolution because of the transport control of dissolution rate. Once initiated, however, gypsum caves should enlarge much faster than limestone caves because of the higher solubility and the more rapid dissolution kinetics.

The very long gypsum maze caves are compatible with the dissolution chemistry. Vertical seepage from overlying beds provides a supply, although perhaps a limited one, of unsaturated water at each joint in the gypsum beds. This avoids the problem of transmitting water long distances from the source without it becoming completely saturated.

11.2.4 *Blister Caves*

Some bedded $CaSO_4$ occurs as anhydrite, which is unstable in freshwater environments. Hydration of anhydrite to gypsum results in a change in volume

$$CaSO_4 + 2H_2O \rightleftharpoons CaSO_4 \cdot 2H_2O \qquad \Delta V = 39\% \qquad (11.8)$$

The pressure developed by volume expansion is sufficient to buckle overlying beds (Tsui and Curden, 1984). Sometimes buckling caused by anhydrite hydration creates small voids, which may be called blister caves.

On South Bass Island in Lake Erie are small caves in dolomite, some lined with celestite crystals, that are thought to have formed from anhydrite conversion to gypsum. According to Verber and Stansbury (1958), expansion forces caused the rock to buckle, after which dissolution of the gypsum produced the caves.

11.3 SALT KARST

11.3.1 *Landforms and Caves*

Although bedded salt formations are moderately common in many parts of the world, the solubility of salt is so great that salt survives at the land surface only in extremely arid localities. Where salt is exposed, sinkholes, shafts, and caves are found, as in any other karst.

An example of this phenomenon is Mount Sedom in the Dead Sea Depression in Israel (Donini et al., 1985). Here a massive salt bed, bounded by faults and protected above by a caprock, is exposed near the elevation of the Dead Sea, nearly 400 m below the world sea level. Caves are dissolved from the salt. Most have the pattern of meandering tubes and canyons. Downcutting of the canyons with concurrent migration of meander bends leads to complex cross sec-

Figure 11.8 Lower entrance to Forrat Mico, a salt cave near the village of Cardona in eastern Spain.

338 / *Geomorphology and hydrology of karst terrains*

tions with many offset ledges. The profiles of the caves are nearly horizontal and appear to be graded to former levels of the Dead Sea. Although most of the caves are quite small, Sedom Cave, the largest, has an aggregate length between two subparallel conduits of more than 1000 m. Feed water for the caves enters from the surface where the salt is not protected by caprock. Solution has carved out vertical shafts that extend down from the land surface to the cave level. Profiles consist of nearly horizontal components with vertical shafts here and there. The caves are decorated with halite, gypsum, and anhydrite speleothems. Halite is simply dissolved in seeping water and redeposited by evaporation of the solutions. Anhydrite occurs rather than gypsum because of the high salinity of the seepage waters.

Figure 11.9 Rillenkarren and spitzkarren on salt. Forrat Mico, Cardona, in eastern Spain.

Exposed salt domes, or diapirs, also support local salt karst. Forrat Mico, near the village of Cardona in eastern Spain, is an example. Cut through the salt dome, which also supports a surface karst, is 600 m of cave consisting of 200 m of a larger upper-level passage, 2 to 4 m in diameter, decorated with halite stalactites (Fig. 11.8), and 450 m of a lower passage of highly sinuous canyon. Although the sculpting of the passage walls differs in detail from that of carbonate rock caves, the overall ground plan and passage cross sections are very similar.

Salt has little mechanical strength and it does not support a well-developed surface karst. Pinnacles with sharp points, that give a form of spitzkarren, and a form of rillenkarren also exist (Fig. 11.9). The main surface features are vertical shafts and sinkholes, which may mark the location of collapsed shafts.

12

Karst and Karst-like Features in Slightly Soluble Rocks

12.1 SOLUTIONAL LANDFORMS ON SLIGHTLY SOLUBLE ROCKS

12.1.1 *Igneous Rock Karsts*

Igneous rocks, especially granite, contain minerals that are easily attacked by solutional weathering, and in certain instances small surficial karst features are formed. Best known of these are solution basins known as opferkessel (Hedges, 1969). These are the silicate rock analogs of the tinajitas or kaminitza found in carbonate rock terrain. Typically, opferkessel occur on exposed masses of granite (Fig. 12.1). They are roughly circular and wider than they are deep. In some cases the sides curve over, so that the basin appears as a bowl, wider inside than it is at the mouth. Opferkessel have been found in a variety of silicate rocks but usually are on exposed masses without soil cover. The basins frequently contain standing water and are sometimes floored with loose quartz sand, the only mineral to survive the weathering process.

Small caves occur in granite but are usually of tectonic origin. However, speleothems of allophane, a noncrystalline aluminosilicate, show that some chemical transport occurs (Webb and Finlayson, 1984; Finlayson and Webb, 1985).

Solutional grooving resembling rinnenkarren occurs in massive basalts, as do solution pits and other cusp-shaped forms (Bartrum and Mason, 1948). The olivine, pyroxene, and amphiboles that occur in basalt are subject to easy weathering. The process that leads to the development of karren on basalt is likely a combination of breakdown of glass and ferromagnesian minerals, solution, and mechanical transport of residual particles under the same sorts of flow regimes responsible for rinnenkarren sculpturing on soluble rocks.

Karst and karst-like features in slightly soluble rocks / 341

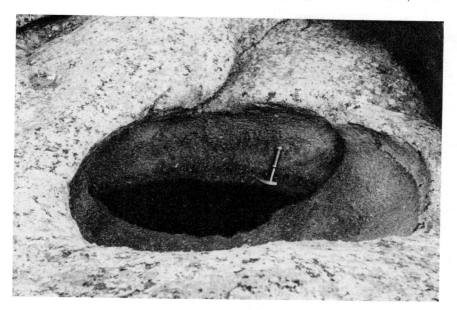

Figure 12.1 Solution basin (opferkessel) in granite. Vedauwoo Peaks, Laramie Range, in southeastern Wyoming.

Other types of karren are less common. Deep weathering along joints gives rise to forms resembling kluftkarren, but there is no reason to assign solution any special role. Broad closed depressions 100 m in diameter and 1 m deep have been found in diorite in the Appalachian Piedmont of North Carolina (LeGrand, 1952). A bibliography of the rather extensive literature on igneous rock karst landforms can be found in Hedges's (1969) paper. Karst landforms on granite in eastern Europe, including caves, have been described by Klaer (1957), Rasmusson (1959) and Wojcik (1961), among others; karst landforms in Australia have been described by Dragovich (1969), and those in Malaya by Tschang (1962).

12.1.2 Quartzite Karst

Rocks composed primarily of quartz, particularly the very pure orthoquartzites and high silica arkoses, might be expected to develop some sort of karst topography, especially under tropical weathering conditions. Quartz is slightly soluble even in cold water (Fig. 12.2). Quartz is certainly a dominant mineral in low-temperature vein deposits and moves easily when temperatures reach the hydrothermal range ($\cong 373°C$). Amorphous silica is much more soluble than quartz. Its solubility is in excess of 100 ppm even at low tempera-

342 / *Geomorphology and hydrology of karst terrains*

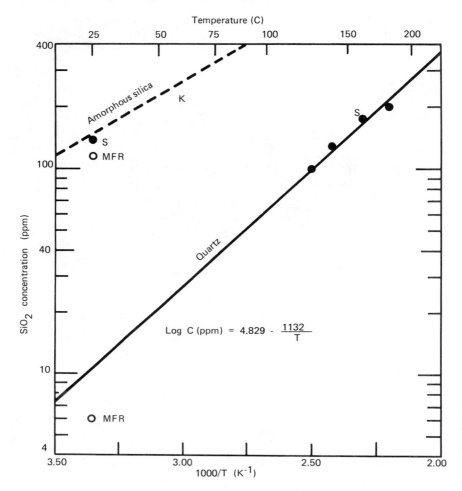

Figure 12.2 Solubility of silica as a function of temperature. Solid circles (S) are experimental points of Siever (1962); open circles (MFR) are experimental points of Morey et al. (1962, 1964). The empirical equation for the quartz solubility curve was derived by Siever from his own and other experimental data. The solubility curve for amorphous silica (K) was compiled by Krauskopf (1956) from various sources.

tures and, like that of quartz, increases rapidly with increasing temperature. Amorphous silica is apparently a distinct phase and the measured solubilities are not very sensitive to the source or form of the material. Likewise, the solubilities of both amorphous and crystalline silica are not pH sensitive over the range of pH 2 to 9. The data in Fig. 12.2 therefore could indicate that some solutional activity of quartz-bearing rock might be expected under tropical weathering conditions or in situations in which crystalline quartz could be amorphotized. Solutional landforms, particularly caves, have been found in quartzites only rarely because of the kinetics of the solution reaction, not the equilibrium solubility. Solution and reprecipitation of silica are extremely sluggish at any of the temperatures shown. Some of the experimental points of Fig. 12.2 were obtained by soaking quartz grains in water for more than a year. With such slow reactions, the channelization process outlined for carbonate and evaporite rocks is unlikely to occur in quartzite.

Minor surface forms occur on the Roraima quartzite in southeastern Venezuela (White et al., 1966). Exposed bedrock surfaces are sculpted into distinct rinnenkarren forms. Deep weathering along joints and fractures on mesa tops, where intense rain falls onto dark surfaces heated to high temperatures by the tropical sun, produce kluftkarren and cutters, whose origins are at least partly solutional. Individual cutters are up to several meters deep and extend horizontally for many meters. Similar landforms have been identified on quartzites in South Africa (Marker, 1976). A scanning electron microscope study of weathered Roraima quartzite (Chalcraft and Pye, 1984) suggests direct dissolution of crystalline quartz over very long periods.

Solution caves in quartzites have not been identified with certainty, although a large cave in the Roraima quartzite (Colveé, 1973) is suggestive. Large shafts up to 370 m deep have also been found here (Szczerban and Urbani, 1974; Urbani and Szczerban, 1974). Evidence for solutional transport of silica is found in a thin opal-like flowstone in a tectonic cave in the Roraima quartzite (White et al., 1966) and in opal-containing dripstone in an Argentine limestone cave (Siegel et al., 1968)

12.2 KARST-LIKE FEATURES FORMED BY MELTING ICE

The melting of semipermanent ice masses, glaciers, and permafrost involves processes of heat and mass transport that are mathematically similar to the solution processes of rock. Although landforms in ice are not quite identical to solutional karst features, karst-like appears to be a proper description.

12.2.1 *Glacier Caves*

The surface of a glacier is sculpted mostly by the ablating action of the wind, and few features are formed that would be considered analogous to karst. The drainage from glaciers, however, generates a series of caves and related features that have very close similarities to features formed by the solution of limestone.

Glacier caves form by one of two mechanisms: by melting of the ice through the action of invading surface water, or by a dragging action as the glacier flows over a bedrock projection in its path. The latter mechanism might be considered more analogous to the formation of tectonic caves in other types of rock. Glacier caves of the first type are typically elliptical conduits with broad, smoothly arched ceilings. The floor is usually the bedrock below the glacier and is often armored with very coarse cobbles and boulders, a testimony to the rock-moving capacity of the glacial stream. The ice walls and ceilings are frequently covered with scallops, usually 10 to 100 cm in length (Fig. 12.3). Their size suggests ablation by wind action rather than by water flow. Glacier caves are explored when the water levels are low. When melting takes place rapidly on the ice field above, the caves may flood to the ceilings. The caves alternate from a vadose to a phreatic regime on a diurnal basis. Some glacier caves are single fragments of elliptical conduits. Others, of which the Paradise Ice Caves in the Stevens Glacier of Mount Rainier are an example, have more complex branchwork patterns of interconnecting tubes (Fig. 12.4) (Anderson and Halliday, 1969). Because glacier caves form near the melting snouts of glaciers, major realignments, collapses, and other modifications of the passages are common. Glacier caves are much more ephemeral than other types of caves.

Figure 12.3 Unroofed segment of glacier cave in Paradise Valley, Mount Rainier, Washington. Note the large scallops and cobble-armored stream bed.

Karst and karst-like features in slightly soluble rocks / 345

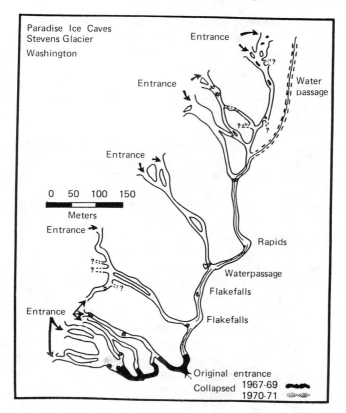

Figure 12.4 Map of the Paradise Ice Caves in Stevens Glacier, Mount Rainier National Park. [Adapted from Larson, Halliday, and Nieland (1972); see also Anderson and Halliday (1969).]

Water feeding the glacier caves is derived from ice melting on the surface of the glacier. One example is the melt water from a small catchment area on the Solheim lobe of the Myrdalsjokull, a fragment of continental glacier in southern Iceland, which drains along a distinct stream channel cut into the ice in a beautiful meandering pattern. Some hundreds of meters below the catchment area, the stream abruptly disappears into the ice through a vertical shaft at least 30 m deep. The fluting and sheer vertical walls of these moulins and their elliptical cross sections are similar in all detail to the vertical shafts in carbonate karsts.

One of the few published descriptions of the second kind of glacier cave is that of Casement Cave in the Casement Glacier of Glacier Bay National Monument, southeastern Alaska (Peterson and McKenzie, 1968). The cave tends to be angulate in pattern, presumably reflecting the tension cracks in the ice as the glacier flows over the bedrock obstruction. In this cave, as in the first type of glacier cave, speleothems of ice in the form of stalactites, stalagmites, columns, and a variety of eccentric forms are observed. Freezing of water vapor also produces feathery platelets and hairlike crystals of great delicacy.

12.2.2 Geothermal Ice Caves

Caves can be produced by melting ice or firn with geothermal heat. The best-documented example is the group of caves that occur on the summit crater of Mount Rainier in western Washington (Kiver and Steele, 1975), although similar caves have been reported on several other volcanos, including Mount Erebus in Antarctica (Giggenbach, 1976).

The Mount Rainier firn caves consist of an annular horizontal passage that curves around the crater and a radiating series of sloping passages that extend upward from the annular passage to a series of entrances around the crater. The passages are some meters in diameter, with ice (or firn) walls and ceiling and a floor of loose rocks. Kiver and Steele argue that these caves are in a state of dynamic equilibrium. The openings are maintained by continuous melting of the ice from geothermal heat and steam released from fumaroles in the volcano summit crater. The ice is replaced yearly by snowfalls on the summit, which are compacted into the firn density range and gradually subside into the crater. Because the cave pattern changes little over long periods of time, its shape must be dictated by the distribution of heat sources in the crater. In the caves are found dead birds, gloves, cans, and other litter that apparently were carried down from the surface by the subsiding firn. Kiver and Steele estimate that subsidence takes place at a rate of 3 m per year.

The Mount Rainier firn caves appear to drain internally. The considerable volume of water generated each year sinks through the permeable rock of the volcano. In the Mount Erebus example, some of the water is released as vapor through the cave entrance where it has refrozen to form ice pinnacles 4 to 5 m high.

12.2.3 Thermokarst

Thermokarst is composed mostly of closed depressions that occur in regions with permanently frozen ground (permafrost). Closed depressions form by melting of ground ice and subsequent settling of the ground. The size of thermokarst depressions varies from a few meters in diameter to a few kilometers. Generally the depressions are shallow and may be water-filled, forming thermokarst lakes.

Thermokarst can appear only in permafrost regions and so is restricted to arctic climates. Considerably large areas of thermokarst exist in the USSR, and it is not surprising that most of the literature on the subject is written by Soviets (Shelley, 1954). Thermokarst occurs in Alaska, where the thermokarst lakes have been thought to arise from changes in albedo resulting from ice-rafted sheets of mud (Ugolini, 1975). Thermokarst in Canada is described by French (1974).

12.3 PSEUDOKARST

The characteristic features of a karst landscape are closed depressions, integrated underground drainage with disappearing surface streams, and caves. All of these characteristics can arise from other processes, which may create a landscape that superficially resembles a karst landscape. The assemblage of features so produced is sometimes referred to as pseudokarst, although some object to the term (Halliday, 1960; Otvos, 1976). Closed depressions in glaciated terrain may be glacial potholes, caused by slumping of morainic material when underlying isolated blocks of ice melt away. More complete analogies to karst are found in volcanic terrains where craters, lava tubes, collapse sinks, and extensive underground drainage are all present. In terrains underlain by poorly consolidated sediments, soil piping and lateral transport by flashy runoff generates small sinks, caves, and underground drainage by mechanical transport of unconsolidated materials, in addition to badland topography.

The justification for continuing to include volcanic features with karst landforms is that speleologists have an interest in lava tubes as a type of cave; therefore, any work on underground landforms must give these features at least some attention (Halliday, 1976).

12.3.1 *Volcanic Terrains and Lava Tubes*

A variety of volcanic processes form caves, including the following types:

1. Blister caves
2. Fracture caves
3. Lava tubes

Blister caves are small pockets formed in lava flows by pockets of gas that have pushed their way to the surface leaving behind a small hollow. Fracture caves are formed mainly by mechanical processes in the nearly congealed lava; these include openings formed by pressure ridges, push-ups, and other processes that cause the lava to separate. Fracture caves are usually small—less than 100 m in length—and irregular in pattern, usually with planar, blocky surfaces. They can form in any type of lava flow. Lava tubes are elongated conduits formed by the draining of highly fluid lava from the center of a cooling lava flow. Thus, lava tubes are usually found only in pahoehoe basalt flows and not in more viscous aa flows or in andesite flows.

Lava tubes occur in great numbers in the Pacific Northwest of the United States, in Hawaii, and in recent basalt flows elsewhere. Detailed lists and descriptions of caves are available for northern California (Halliday, 1962; Knox, 1959), Washington (Halliday, 1963), Idaho (Ross, 1969), Oregon (Greeley, 1971a), and Hawaii (Greeley, 1971b). Lava tubes occur on Iceland, although descriptions are available only for Raufarholshellir and Surtshellir (Prior et al., 1971). Lava tubes and other volcanic caves are associated with the recent volcanic eruptions of the East African rift zone, including exceptionally well-developed blister caves on the sides of Mount Fantale in Ethiopia (Sutcliffe, 1973).

Lava tubes typically have an elliptical or circular cross section and reach diameters of up to 10 m or more (Fig. 12.5). They tend to fork in the downstream direction, as the tube approaches the toe of the flow, and thus a reverse branchwork or distributary pattern is common. Because the tubes form at shallow depths below the surface of the flow, roof collapse (which provides most of the known entrances) occurs easily, and these collapses dissect the tubes into shorter fragments. The tubes range in length from tens of meters to a few kilometers. Ape Cave in Washington, at 3.9 km is one of the longest lava tubes known in the United States. The tubes are sinuous in ground plan because of the meandering of the lava stream. The width–meander bend spacing follows a power law similar to that followed by surface rivers. The average orientation of the tubes parallels the primary direction of the lava flow. Indeed, lava tube densities in some low-viscosity basalts are so high that the tubes appear to have acted as primary distribution paths for the advancing flows. The total length of tube systems may amount to many kilometers, as illustrated in Fig. 12.6 for the Cave Basalt flow southeast of Mount St. Helens in Washington (Greeley and Hyde, 1972).

In regions where there have been successive periods of eruptive activity, lava tube caves can become quite complex (Wood, 1974). Old tubes can act as flow paths for younger flows, which may enlarge them, modify their courses, or choke them up. Successive tiers of tubes in layered basalts may collapse into each other to form complex passage cross sections. Continuous deep canyon cross-sectional passages are also known. The formation of the primary tube is, however, a relatively simple matter. The rapidly flowing lava becomes concentrated in a channel because of the cooling and eventual freezing of the material along the sides of the flow. The surface of the channel then freezes over to form a solid roof. Photographs of lava streams from some of the Kiluea eruptions in Hawaii show that roof formation is initially discontinuous, with windows through which

Figure 12.5 Cross section of a lava tube in the Saddle Butte System of southeastern Oregon. [Courtesy Charlie and Jo Larson.]

the rapidly flowing lava river can be observed. Eventually the entire roof becomes solid, and the lava flowing beneath feeds the advancing front of the flow. If the vent is closed off, the tube drains by means of gravity, resulting in an open cave.

The hydrology of volcanic terrains has much in common with karst hydrology. Recent lava flows, particularly those of the aa or pyroclastic type, are extremely permeable and transmit groundwater under negligible hydraulic gradient (Stearns, 1942). Integrated underground drainage systems occur where younger, highly permeable flows overlie less permeable older flows or basement rocks. In certain volcanic areas, surface streams are completely absent, as they are in karst, and all streams flowing in from borderlands sink at the edges of the lava flows. Groundwater from these aquifers

350 / *Geomorphology and hydrology of karst terrains*

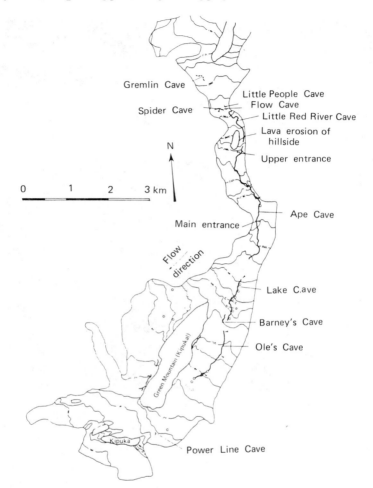

Figure 12.6 Map showing distribution of lava tubes oriented parallel to the flow direction of the cave basalt in Mount Saint Helens Volcano, Cascade Mountains, Washington. [After Greeley and Hyde (1972).]

reemerge as large springs, usually at the point of contact between the permeable and less permeable layers. The streams beneath the Snake River Plain in Idaho are some of the longest underground rivers in the world—and they are in basalt, not limestone.

The geomorphology of volcanic and impact landscapes has taken on a new importance in the space age. There is good evidence that sinuous features on the moon are collapsed lava tubes (Halliday, 1966; Greeley, 1971c), and the Hadley Rille, visited by the Apollo 15 astronauts, is of similar origin (Greeley, 1971d). The connection of lava tubes with karst becomes progressively less tenable, however. If this grouping continues, one would have to say that much of the landscape of the moon, Mars, even tiny Phobos and Deimos, and probably most other satellites and asteroids, with their profusion of closed depressions, are forms of pseudokarst. It would make as much sense to refer to the dolines of West Virginia, Indiana, and Kentucky as "pseudocraters"!

12.3.2 Soil Piping and Suffosional Pseudokarst

Regions of badland topography in which sediments are poorly consolidated and vegetative cover is thin or nonexistent suffer extremely rapid weathering and mass transport under flashy runoff conditions. A complete topography in miniature has developed, including mountains, pediments, outwash plains, and a complex branchwork of stream gulleys, all on a scale of tens to hundreds of meters. Badlands National Monument in western South Dakota (Smith, 1958) has particularly well-developed landforms, and many others are known in the American West.

Mass wasting is very rapid during the runoff periods, and the stream gulleys are cut as deep narrow canyons with steep gradients. The channels at the bottom of the canyons are often undermined to form shallow caves. Sediment walls and miniature cliffs are cut through to form tunnels and natural bridges. The vertical transport of material is by means of soil piping and it results in shallow, closed depressions or sinkholes on the upland surfaces (Fig. 12.7). The caves that form by piping and outwash are rarely more than 100 m long, but may contain roomy chambers and are irregular in ground plan and cross section. Because slumping and stream flood transport are the main means of cavern enlargement, characteristic sculptings do not occur (Clausen, 1970).

12.3.3 Pseudokarst Caves

The underground landforms called caves are primarily a human notion in the sense that they are enterable spaces. Voids that would qualify as caves form by many processes and in many types of rock. Many caves are, therefore, either a form of pseudokarst or not karst-related at all. The classification (Table 12.1) is a composite of various people's ideas, including those of Kyrle (1923), DeBellard Pietri (1956, 1967), and others. The caves are listed in two broad classes: those formed primarily by chemical processes, as already discussed, and those formed primarily by mechanical processes.

Aeolian caves are formed by the abrading action of windborne particles. They are common to desert regions where small caves are often carved in soft sandstones. Caves that were apparently caused by wind action have also been reported in loess. Aeolian caves tend

352 / *Geomorphology and hydrology of karst terrains*

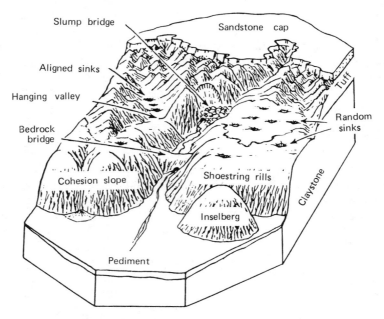

Figure 12.7 Composite block diagram showing principal landform elements of badlands and their karst-like features. Vertical relief is on the order of 30 m. [After Mears (1963).]

to consist of single elliptical chambers opening onto an exposed cliff face. Chamber sizes range up to 10 m or more, and the chambers are frequently larger than the opening through which the wind is driven. Coarse-grained, loosely cemented sandstone seems an optimal lithology. Cross-bedding seems to enhance aeolian cave formation, and many such small caves can be seen in the Navajo sandstone of the American Southwest.

Sea caves (or littoral caves) are formed by the milling of sea coasts or lake shores by wave action (Moore, 1954). They occur in many coastal areas throughout the world in many types of rock. Some of the most famous caves are sea caves, of which Fingal's Cave, on the island of Staffa in the Hebrides, and the Blue Grotto, on the Isle of Capri, are two examples. The East Coast of the United States, except for the Maine coast in the north, is unsuitable for the development of sea caves because of the low relief of the coastal plain. The West Coast in California, Oregon, and Washington has many sea caves. Descriptions can be found in the California and Washington cave surveys (Halliday, 1962, 1963).

Tectonic caves are formed by mass movement of the bedrock. They can occur in any type of rock, but are usually associated with

Table 12.1
A Classification of Caves

Type of cave	Process	Optimum host rock
Caves formed mainly by chemical processes (strongly dependent on host rock composition)		
Solution caves	Chemical solution of rock	Soluble rocks: limestone, dolomite, gypsum
Lava caves	Freezing of wall rock around channel	Pahoehoe basalts
Ice caves	Melting of ice along channel	Glacial ice
Caves formed mainly by mechanical processes (essentially independent of host rock composition)		
Aeolian caves	Scour by wind-driven sand	Massive sandstones, loosely cemented
Sea caves	Scour by wave action on sea cliffs	Massive rock with well-developed jointing
Tectonic caves	Bulk movement of rock mass	Any rock; good jointing and lubricating beds
Suffosion caves	Soil piping	Fine-grained, poorly consolidated sediments
Erosion caves	Erosion and slumping	Alternating resistant and nonresistant beds
Talus caves	Pileup of boulders	Massive rocks

massive fractured rocks. The openings of tectonic caves are formed by parting along joints with slippage along bedding planes. The passages tend to be rectangular, with matching walls. The driving force is usually gravity, although tension cracks can open in regions of intensive folding. Bedding plane slippage is enhanced by the presence of shale layers, which lubricate the planes of slippage between beds of more massive rock. Most tectonic caves consist of a single passage and their length seldom exceeds 100 m. However, some complex caves with total lengths of several hundred meters are known.

Suffosion caves are formed by soil piping and washout of soft unconsolidated sediment. They were discussed previously with other forms of suffosional pseudokarst.

The term erosion cave is used, for lack of a better name, to describe minor openings and rock shelters formed by mechanical erosion of soft rock material, usually shale. These caves develop best where a resistant sandstone layer overlies a weaker shale horizon. Often, small sandstone cliffs occur where minor streams flow down steep hillsides. The underlying shales are easily undercut, and a combination of slumping and erosion cuts the shale back, leaving a sandstone overhang. Although rock shelters, formed in this manner, usually are not considered caves, they are often referred to as caves locally, sometimes appear on maps as caves, and are often listed as cave sites by archeologists or historians who are concerned with the human usages of such shelters.

Talus caves are formed when masses of piled boulders or other rock debris arrange themselves in such a way as to leave explorable openings beneath and between the blocks. These caves are of interest only if they reach considerable size. Obviously, all boulder slopes have an indefinitely large number of small openings. In a few instances talus caves in sandstone or granite have lengthy interconnected passages. The TSOD Anorthosite Talus Cave in the Adirondack Mountains of New York perhaps holds the record, at 3.7 km of passage and an internal relief of nearly 50 m (Carroll, 1978).

13
Land Use and Land Management Problems in Karst

13.1 AGRICULTURE

It is curious that the most important problem in karst terrain can be described in a few sentences.

13.1.1 *Tillage*

The weathering of limestones and dolomites produces high-quality agricultural soils. In cooler climates, especially on terrain underlain by dolomite, the carbonate valleys provide choice farmland. Soils tend to be thick silt loams that are easy to till. In terrains underlain by soluble limestones and in warmer, wetter climates, the bedrock surface evolves into a cutter-and-pinnacle topography. Often the pinnacles emerge from the land surface to form barriers between the adjacent soil-filled cutters. Mechanical tillage of the land is difficult to impossible, and instead of croplands, this type of terrain can be used only for pasture. As the karst landforms develop, available agricultural land decreases until agriculture in many tropical karsts is limited to the bottoms of closed depressions, where thick soils can still be found (Fig. 13.1).

13.1.2 *Soil Loss*

The ability of the karst drainage system to transport soil provides an important pathway for soil erosion. Root systems from forest, brush, and grass cover retard soil loss into the solution cavities. When the protective plant cover is lost through tillage, cutting, or overgrazing, irreversible soil loss occurs.

A grim monument to the effects of overgrazing can be seen in the barren limestone mountains of the countries around the Mediterra-

356 / *Geomorphology and hydrology of karst terrains*

Figure 13.1 Agriculture in the Adriatic karst. Dolines provide traps for the only arable soils in many parts of the mountains. The stone walls prevent the soils from being washed down the drain in the middle of the doline.

nean. Out-of-control goat populations, beginning as early as biblical times, stripped trees, bushes, and grasslands. Once storm runoff had washed the soils into solution cavities, the effect was irreversible; the semiarid climate of today cannot replenish the soils faster than they are washed away. The Maya peoples of the Yucatan practiced slash-and-burn *(milpa)* farming. They burned off patches of jungle and then planted their crops. Once the tree cover had been removed, soils were quickly lost to the subsurface, and gradually the Yucatan Peninsula was transformed into a karst desert (Doehring and Butler, 1974).

13.2 FOUNDATION ENGINEERING

Karst lands are the bane of building construction (Knight, 1971). The many problems of erecting structures on karst reduce to three mechanisms: differential compaction due to the irregular bedrock surface, soil piping, and collapse of subterranean cavities.

13.2.1 *Load-Bearing of the Karstic Surface*

The conventional practice for individual homes and other small structures is to dig out a basement to the proper depth below the land surface, prepare a drain around the perimeter, and erect the basement walls, usually of cinder block. The excavation is leveled and a slab of concrete is poured to provide a floor. The above-ground portion of the structure is supported by the basement wall, so that the wall is load-bearing but the floor is not. Larger structures may have additional pillars to take up part of the load and avoid long, unsupported beams between the walls. The basement excavation may be cut into bedrock or down to bedrock if soils are thin, but with thick soils, the entire structure may be supported only by unconsolidated material.

Figure 13.2 illustrates some of the implications of placing a structure on the cutter-and-pinnacle bedrock surface. The weight of the structure causes additional compaction of the soils if it is built on unconsolidated sediment. In most terrains, the compaction would be uniform, resulting in only a slight settling of the structure. On a karst surface, the pinnacles are load-bearing but the intermediate cutters are not. Differential compaction can result in cracking of basement walls and fracturing of the concrete floor. Differential settling of the entire structure forces doorways and windows out of alignment and plaster to crack in the upper parts of the house.

Larger structures such as shopping centers, high-rise buildings, factories, and other commercial buildings are constructed with more regard for the foundation, but some of the same effects occur. Foundations for expensive structures are usually probed by test drilling before construction is started. It is important, however, that the test holes be drilled on exactly the same centers as are planned for the load-bearing pillars. A test hole can easily hit solid rock on a pinnacle beneath a thin soil while a meter away a cutter or solution cavity extends downward tens of meters. Many of the difficulties in erecting large structures on karst terrain have occurred because of inadequate test drilling before construction was started. It is also possible that what appears to be solid rock is a free-floating block isolated by solutionally widened joints and bedding planes.

13.2.2 *Infiltration and Soil Piping*

Structures with large roof areas, parking lots, streets, and driveways change the runoff and infiltration characteristics of the soils. Concentrated runoff from roof gutters, storm drains, or even the runoff from the edge of a pavement can form channelized paths through the soil to a solution opening in the underlying bedrock (Fig. 13.3). The increased hydraulic gradient and increased flow velocities speed the removal of soil from the subsurface. Because the pavement or the building itself acts as a supporting structure, the loss of soil may not be noticed until a substantial cavity has formed. At some point structural support is lost and the building collapses into the hole. This may occur gradually, with advance warning in the form of cracks in

358 / *Geomorphology and hydrology of karst terrains*

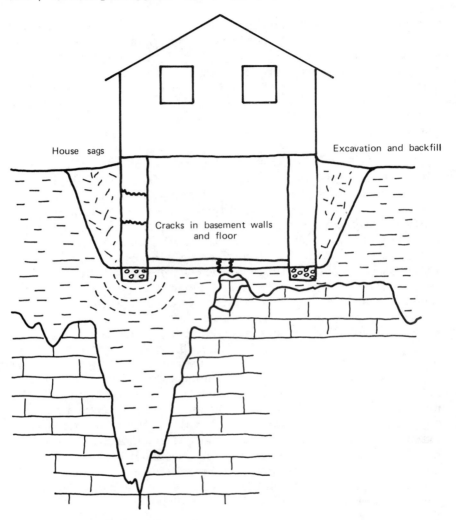

Figure 13.2 Structural damage resulting from construction on cutter-and-pinnacle karst surface.

basement walls, floors, and pavements, or it may take place abruptly. The latter is a very dangerous situation. There are horror stories of a home owner peering through a crack in his basement floor to see a cavernous void 10 m or more in diameter.

Properties can be damaged by events that occur outside their boundaries. Rerouting of storm runoff from shopping centers or streets can cause soil piping under adjacent properties. One incident involved a small housing development spread across a region of shal-

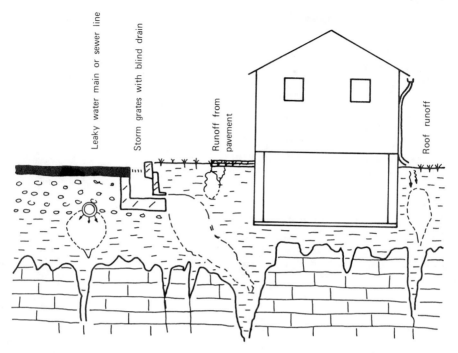

Figure 13.3 Soil piping induced by modification of natural runoff and infiltration conditions.

low dolines. A street dipped directly across one of the larger dolines and residents complained of water ponding on the street after rainstorms. The authorities constructed a storm grate into a concrete box below street level with a drain that simply terminated in the adjacent soil. Water no longer ponded in the street, but shortly thereafter an adjacent homeowner found that his basement had been completely undermined and that his house was perched on top of a large cavity. Such incidents are common in karst regions. They keep consulting hydrogeologists employed and court dockets filled with cases of assigning blame and assessing damages.

13.2.3 *Solution Cavity Collapse*

Collapse dolines form from the collapse of cave roofs. They are a natural part of the evolution of the karst landscape, the final event in the history of a cave passage as the down-wasting land surface finally passes it. Because the rate of surface down-wasting is extremely slow on a human time scale, collapse dolines are rarely observed in the act of collapsing. However, it is quite possible for construction activity to induce the collapse of near-surface solution cavities that might otherwise have remained intact for many thousands of years.

The key concept is the tension dome that develops over a void in the bedrock (see Section 8.3.3 for a discussion of cave breakdown). The boundaries of the tension dome are zones of maximum shear. The top of the dome extends upward into the bedrock for distances of some 1.5 times the cavity diameter. Any change in loading above the top of the dome is distributed over the cavern walls, and the ability of the formation to support the load is independent of the cavity's existence. If, however, the tension dome extended to the land surface or to the base of an excavation, additional loading would increase the shear along the walls, leading to collapse of the cavity and subsidence of the excavation. The relations sketched in Fig. 13.4 show the critical arrangement, where the top of the tension dome just touches the base of an excavation. The depth at which a solution cavity is not a hazard is determined by the diameter of the cavity and the shear strength of the rock formation.

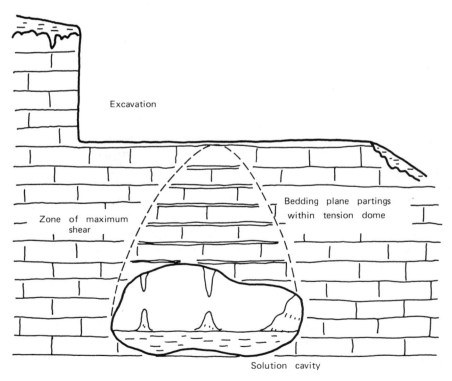

Figure 13.4 Stress region above a solution cavity, beneath a foundation excavation.

The mechanical analysis assumes that the rock units are homogeneous (although bedded) and free of fractures. Deep weathering zones with solution along joints and fractures can induce upward-stoping collapses from cavities that are much deeper than the purely mechanical analysis would imply.

Test borings are the most reliable means of detecting the presence of cavernous openings in the bedrock, but they must be both deep enough and spaced on sufficiently close centers. Because solution cavities can easily be 10 to 20 m in diameter, borings ought to be at least twice that depth and spaced on centers of less than the critical cavern diameter. Of course, the level of the final excavation should be considered in determining the depth of boring.

Cavity collapse caused one instance of major damage to a new shopping center. The rock formation was the Martinsburg shale, which contains pods of limestone known to be cavernous. Test borings were taken through the shale to a depth of 40 ft from the land surface. Construction of the shopping center called for the excavation into a hillside to prepare a flat area for the buildings and parking lot. Most of the 40 ft that had tested sound was removed in the course of construction. As it turned out, the floor of the excavation was only a few meters above a large solution cavity in a limestone pod. Shortly after excavation and before construction was completed, major collapses occurred, which left large holes in the parking lot and severly damaged the newly erected buildings.

13.2.4 Pilings, Grouting, and Other Solutions

The serious question, from the construction industry's point of view, is what can be done about cutter-and-pinnacle topography or shallow solution cavities when they are found. The easy solution is to put the structure somewhere else, but that is not always feasible.

Construction on cutter-and-pinnacle topography must provide uniform and stable support to the load-bearing portions of the structure. Since the spacing of pinnacles is unlikely to coincide with the design centers of support pillars, some of these will be over solid bedrock and some will be over cutters. The most effective, although expensive, solution is to install pilings. Large-diameter holes are bored on the same centers as the support pillars and drilled to solid bedrock. Reinforcing bars are inserted into the drill holes, which are then filled with concrete. These subterranean concrete pilings then serve as footers for the support structures of the building. In one example, an array of some 40 pilings ranging from 2 to 20 m in depth were required to support a two-story reinforced concrete frame laboratory building that was erected on cavernous limestone.

Subsurface solution cavities are often sealed by grouting. An access hole is drilled and a flowing concrete (grout) mixture is poured in. For cavities of limited lateral extent, this method works well, but when large open cave passages extend offsite, huge quantities of grout may be necessary to finally fill the portion of the cavity under the proposed structure. Sometimes one can circumvent this problem by gaining access to the cave either through natural entrances or by drilling a man-shaft. Forms can be constructed to hold the wet grout so that, instead of the entire cavity being filled, a pillar is erected in the area needing support. Most cavities contain variable thicknesses of unconsolidated clay and silt on their floors. For massive structures, this material must be removed so that the grout pillar rests on solid bedrock; otherwise the cave sediments will settle under load.

In a few instances, large solution cavities and bedrock cutters have been bridged. Usually this consists of placing steel beams across the cavity and then using either steel sheeting or reinforced concrete to provide a platform supported by the solid bedrock on both sides of the cavity. The platform is then used as a base for the structure. This technique was used to build a highway across a sinkhole that opened under the roadbed near Norristown, Pennsylvania (Steiner, 1975).

All solutions to foundation engineering problems in karst are expensive, and their justification depends on the economics of the structures being erected. For small structures such as individual homes, perhaps the only economically feasible solution (other than quietly selling the lot and moving somewhere else) is to dig the foundation well below the intended basement floor. Bedrock pinnacles should be blasted off and the extra depth backfilled to achieve a uniform load distribution. Larger structures can afford exploratory drilling and the piling and grouting correction methods. Some structures, such as nuclear power plants, require very elaborate foundation treatment (Franklin et al., 1980). Foose and Humphreville (1979) detailed the engineering solutions for construction of a medical school, a convention center, and a large motel on cavernous terrain in the Hershey Valley of Pennsylvania. Sowers (1984) summarizes many of the engineering solutions that have been investigated.

13.3 SINKHOLE COLLAPSE

13.3.1 *Soil Piping Sinkholes and Other Subsidence*

Processes of soil piping, and bedrock collapse and stoping, produce cover collapse sinks, cover subsidence sinks, and bedrock subsidence sinks as part of the natural evolution of the karst landscape. These processes can be greatly influenced by human activities, however, and in turn are among the most important land use hazards in karst terrain.

Bedrock subsidence sinks are often associated with interstratal karstification, particularly of evaporite beds at depth. Baumgardner et al. (1980) describe the formation of Wink Sink in West Texas:

> About 9 a.m. June 3, a Gulf Oil Co. maintenance crew working in the Hendrick Oil field 8.2 km southwest of Kermit, Tex., discovered a newly formed meter-wide hole in the ground. By noon it had grown to a cavity 30 m across, primarily by collapse of large blocks into the developing hole. Rapid expansion of the hole had ended by the next morning when the diameter was about 110 m. The maximum depth of the hole was 34 m, 2 days after the sinkhole began to form, and the volume of material displaced was estimated at 196,250 cubic meters.

The cavity was assumed to have originated from removal of salt in the underlying Salado bedded salt formation by groundwater, followed by upward stoping through the Rustler formation, 60 m of interbedded anhydrite and thin sandstone. Deep solution of bedded salt followed by subsidence of overlying formations is a major process in evaporite karst (Ege, 1984).

Cover collapse and cover subsidence sinkholes are more common in poorly consolidated materials. Many of the sinkholes that plague the Florida karst areas seem to be of this type (see Beck, 1984, for extensive discussions). Cover collapse sinkholes form almost instantaneously, whereas subsidence sinkholes often form over a period of hours or days. The cover subsidence sink near Winter Park, Florida, was much in the news in the spring of 1981. Over a period of several days, the advancing subsidence front swallowed a house, several cars, and half of newly rebuilt municipal swimming pool. Total damage ran into the millions of dollars. It has not been possible to reclaim the land swallowed by the sink; today, a small park has been created where the damaged structures once stood (Jammal, 1984).

The transport of sediment into the subsurface and the subsidence of cavities are rather delicately balanced. Disturbances that change the hydraulic gradients or the flux of water through the system enhance sediment transport and speed up time to failure. Thus, we have the paradoxical situation that increasing the rate of recharge and dewatering the system both lead to increased sinkhole failures.

Pavements, roofs, and storm drain systems can increase the rate of recharge in a local area, thus increasing flow velocity in solutionally widened tubes and fracture systems in the bedrock and inducing sinkhole development. A common cause of cover collapse sinks is broken water or sewer pipes. Pipelines strung through karst terrain are subject to uneven settling as soils compact or are piped into solution cavities. The result can be cracked water pipes or the separation of sections of the glazed tile frequently used for sewer lines. Utility

364 / *Geomorphology and hydrology of karst terrains*

lines are commonly strung under or along streets. The pavement acts as a protective caprock, allowing large cavities to develop before sudden failure produces a sinkhole in the middle of a street (Fig. 13.5).

If the water table is above the level of the sediment-transporting solution cavities, sediment movement is very inefficient. Further, wet sediments are more cohesive than dry sediments and have less tendency to move. If the water table is lowered because of mine or quarry pumping or from overpumping an aquifer, hydraulic gradients are increased and instead of a water-filled cave system, there may be a flowing stream, which is much more efficient at sediment transport.

The classic example of dewatering-induced sinkhole collapse is the Hershey Valley of Pennsylvania (Foose, 1953), where pumping of a deep limestone mine lowered the water table, drying up streams and inducing sinkhole formation in an adjacent golf course and park. Sinkhole development in several areas around Birmingham, Alabama, was apparently induced by dewatering of several quarries, combined with a general lowering of groundwater levels by extensive pumping of the aquifer (Powell and LaMoreaux, 1969; Newton and Hyde, 1971; Newton et al., 1973). Sinclair (1982) suggests that many of the sinkholes in Florida, a particularly sinkhole prone area,

Figure 13.5 Sinkhole collapse that truncated a country road in eastern Pennsylvania. [Courtesy of Keith Kirk.]

are a result of groundwater withdrawal and lowering of regional water tables. Here, the blanket of unconsolidated sands and silts is thick, and sinks form by the piping of these materials into deep underlying solution cavities.

Predicting incipient sinkhole collapse is difficult. Often there is no warning that a cavity in the soil has developed until the final collapse. The following quotation is from the newsletter of the National Waterwell Association:

> A giant sink hole opened up on Thursday, September 19 (1975) at a drilling site near Tampa, Florida and swallowed up a well-drilling rig, a water truck, and a trailor loaded with pipe all valued at $100,000. The well being drilled was down 200 feet when the ground began to give way to what turned out to be a limestone cavern. Within 10 minutes all the equipment was buried way out of sight in a crater measuring 300 feet deep, and 300 feet wide. Fortunately the drilling crew had time to scramble to safety and no one was hurt.

Sinclair offers the following precursors of sinkhole collapse:

1. Slumping or sagging. Canting of fenceposts or other objects from the vertical. Doors and windows that fail to open and close properly.
2. Structural failure. Cracks, however small, form along mortar joints in walls and in pavements.
3. Ponding. The ponding of rainfall may serve as the first indication of actual land subsidence.
4. Vegetative stress. One of the earliest effects at an incipient sinkhole is lowering of the water table. The lowered water table may result in visible stress in a small area of vegetation.
5. Turbidity in well water. Water sometimes becomes turbid during the early stages of development of a nearby sinkhole.

In an inventory of sink and subsidence occurrences in Missouri, Williams and Vineyard (1976) conclude that half were induced by human activity (Fig. 13.6).

13.3.2 *Catastrophic Collapse*

A family of large soil piping features that form abruptly are worth discussing separately as catastrophic collapses. Smaller cover collapse sinkholes and cover subsidence sinkholes can cause extensive property damage, but are rarely a threat to people; catastrophic collapses, however, can result in loss of life as well as property. Most are also associated with thick residual soils and dewatering of the underlying aquifer. Here is an account of a catastrophic collapse

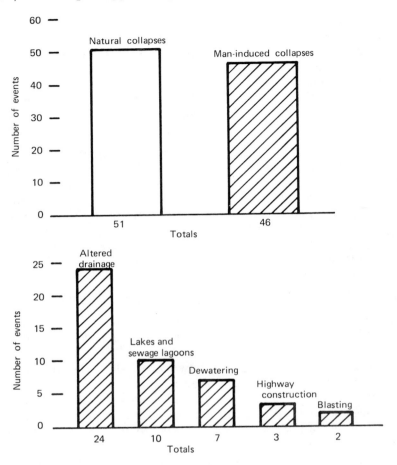

Figure 13.6 Bar graph showing distribution of sinkhole collapses in Missouri based on records kept since 1930. [From Williams and Vineyard (1976).]

some 45 km south of Birmingham, Alabama (LaMoreaux and Warren, 1973):

> Last Dec 2 (1972), Hershel Byrd, a resident of rural southern Shelby County, Alabama was startled by a rumble that shook his house, followed by the distinct sound of trees snapping and breaking. Two days later hunters in nearby woods found a crater about 140 m long, 115 m wide, and 50 m deep. These events mark the time of formation of and discovery of the largest recent sinkhole in Alabama and possibly one of the largest in the United States.

One of the most devasted areas of catastrophic sinkhole collapse is the far west Rand of South Africa. Here a very old and thick residual soil overlies the Transvaal dolomite. The hydrogeologic situation is further complicated by igneous dikes, which effectively divide the dolomite into compartments. Some compartments have been completely dewatered by deep pumping for the gold mining operations. One sinkhole collapse in December 1962 swallowed a crushing plant of the Westdriefontein Mine, with a loss of 29 lives. In another, which engulfed a housing area, one house and its five inhabitants were completely lost. The South African geologic setting is more complicated because of a chert horizon near the top of the dolomite, which necks down solution cavities, allowing large bedrock chambers to form beneath the chert into which the thick overlying soils can be easily transported. Once transport of soil through the neck is initiated, the upper soil arch can form rapidly and migrate to the surface (Fig. 13.7) (Jennings et al., 1965; J. E. Jennings, 1966; Foose, 1967).

13.3.3 Some Legal Questions

Sinkholes in their various manifestations do considerable property damage and in some cases are life-threatening. Many sinkholes result from human activity, such as lowering a water table, rearranging

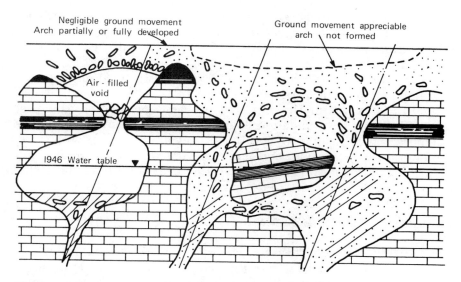

Figure 13.7 Cross-sectional view of the South African catastrophic sinkhole collapses. The situation on the left leads to instantaneous catastrophic collapse; that on the right leads to continuous ground subsidence. [Adapted from Jennings et al. (1965).]

surface drainage, and adding to recharge either directly through drains or injection wells or indirectly through leaking pipelines or other sources. Can a property owner who suffers damage or financial loss through a sinkhole of someone else's creation claim damages against the perpetrator, and compensation from his or her insurance company?

Case law concerning sinkhole collapse, sinkhole flooding, and other karst-related damages remains vague in the mid-1980s (Quinlan, 1986). Most states do not have specific statutes addressing the assignment of responsibility or the damages for which someone may be held liable. Only Florida has a specific law requiring insurance companies to cover damages resulting from sinkhole formation (Salomone, 1984). However, the Insurance Commissioner defined sinkhole collapse as follows:

> Direct loss by sinkhole collapse means actual physical damage to the property covered arising out of or caused by sudden settlement of the earth supporting such property only when such settlement or collapse results from subterranean voids created by the action of water on limestone or similar rock formations.

The definition contains ambiguous terms such as "sudden settlement," which imply, perhaps, that a cover collapse sinkhole is a sinkhole but a cover subsidence sinkhole is not. It has required an extensive series of litigations between insurance companies and insured property owners to sort out the definitions, and still an important part of case preparation is to show that ground subsidence that has caused structural damage is really caused by a sinkhole.

Development and urbanization of previously rural and agricultural karst lands is proceeding apace in the eastern and central United States. All of the expected problems of sinkhole collapse, sinkhole flooding, foundation engineering, and groundwater pollution are appearing in many locations. They are being met with suits, countersuits, legislation, and land use control through zoning and other regulations. All is in a great state of turmoil in the mid-1980s and the legal morass will take years to get itself sorted out.

13.4 RESERVOIR ENGINEERING

13.4.1 *Small Impoundments*

It is customary to construct settling ponds for quarry and industrial waste, sewage lagoons, and related small impoundments by excavating to the area and depth required, then sealing the bottom with bentonite clay or with concrete or plastic liners. The record of such impoundments on karst terrain has not been particularly good.

Lagoons and impoundments placed in unconsolidated residual soils draped over a pinnacle-and-cutter bedrock topography are subject to differential settling produced by the weight of the water when the pond is filled. Differential settling cracks linings, allowing water to drain through a small number of localized breaks into the underlying soils and from there into solution cavities in the carbonate bedrock. At best, the lagoon leaks whatever material it contains into the underlying groundwater. At worst, the localized leakages induce sinkhole collapse which destroys the impoundment.

If the lagoon is even slightly permeable, the soils beneath it will become saturated, the net effect of which is to produce a groundwater mound. The head of water produced by the mound induces a seepage pressure on the soils beneath the lagoon. This is an optimum arrangement for soil piping and sinkhole formation. Lagoon failure through catastrophic sinkhole collapse can be quite dramatic. In one case (Aley et al., 1972) a sewage lagoon at West Plains, Missouri, had an impoundment area of 49 acres (19.8 hectares) and a water depth of about 1 m. A collapse occurred in 1966 at a time when the lagoon contained about 136 acre feet (1.68×10^5 m^3) of effluent. All the water drained out within 52 hours. The collapse created a vortex of such magnitude that the manager of the treatment facility was unwilling to get close to it with a boat.

13.4.2 Dams

Large reservoirs are constructed for flood control, flat water recreation, hydroelectric power, and water supply. Those constructed in karst terrain have exhibited a distressing inability to hold water.

The usual engineering practice is to locate the dam in the narrowest part of the river channel to keep construction costs to a minimum. The regolith is excavated to bedrock in the river valley and along the flanks of the hills on both sides. The dam itself may be of the rock-and-earth-fill type, of concrete construction, or a combination of the two types. When the reservoir is filled, an additional tens to hundreds of meters of hydrostatic head is present on the floor of the reservoir, and a new water table is established within the valley walls. The principal karstic problem is deep solution cavities. If any portion of the reservoir, but particularly the dam itself, lies on carbonate rocks, there is the possibility of solution cavities, either open or clay-filled, beneath the river bed (Soderberg, 1979). Further, any dry caves that may occur in the valley walls will be flooded as the reservoir is filled, and these may provide bypass routes around the abutments of the dam. Such is apparently the case with the Gathright Dam in westcentral Virginia (Roberge, 1977).

Solution channels beneath the dam may be so severe that they threaten the mechanical integrity of the dam, but mainly they provide drainways under the dam. The increased hydrostatic head can flush sediments from solution cavities and provide open pathways for the water. There are many stories of boils appearing downstream from dams, sometimes before the reservoir is even filled. Carbonate bedrock under surface river channels is often dissolved to depths of tens of meters to more than 100 m (Grant and Schmidt, 1958).

The remedy to the problem of solution cavities used most often is to block them off. Grout curtains are used for cavities found in the bedrock below the river channels, and although expensive, these can usually be established by grouting through drill holes. Blocking open cave passages in the walls of the valley above the dam and behind the abutments is much more difficult, and no really definitive engineering practice seems to have emerged. Lateral tunnels can be excavated, from which filling and grouting can be attempted. However, if the limestone extends laterally for great distances, the caves could be anywhere, even several kilometers from the proposed abutments. Extensive lateral cavernous zones should be taken as evidence that the site is not suitable for the construction of a dam.

Subriver solution cavities were the plague of the Tennessee Valley dams (Moneymaker, 1941, 1948; TVA, 1949). It is instructive to follow the saga of the Hales Bar Dam, one of the most difficult to construct. What follows is paraphrased loosely from Burwell and Moneymaker (1950). The Hales Bar Dam is on the Tennessee River, about 20 km west of Chattanooga, Tennessee. It was built from 1905 to 1913 by a private power company and acquired by the Tennessee Valley Authority in 1939. The foundation rock is the group of Mississippian limestones now known as the St. Louis, the Monteagle, and the Bangor formations. Bedrock solution is controlled predominantly by minor faults; consequently, the largest cavities are concentrated in two zones.

The Hales Bar Dam site was selected on the basis of the narrow gorge. That the rock might be unsound or cavernous does not seem to have occurred to anyone before construction was undertaken. Hales Bar Dam was built without sufficient exploration, foundation excavation, or remedial treatment of a very cavernous foundation. Leakage through foundation rock was so great that completion was long delayed. The project, estimated at $3 million and scheduled for completion in two years, cost $11,536,889 and required eight years to be completed. Soon after its completion in November 1913, the first leak was discovered, and the first intake was discovered a few days later. Leakage increased rapidly; the number of known leakage inlets and outlets also increased. Unsuccessful attempts were made in 1913, 1914, and 1915 to cut off leakage by plugging the inlets. By 1919 there were nine small boils immediately downstream from the eastern cavernous area and eight stronger boils downstream from the western cavernous area. From 1919 to 1921 the attempt was made to stop the leakage by pumping molten asphalt into the cavities functioning as conduits. Sixty-eight holes were drilled; 78,324 ft^3 of asphalt was pumped into 45 of them. All but one of the eastern boils disappeared, but only one of the western boils was eliminated, although the flow of the others was much reduced. Soon after the asphalt grouting was discontinued, leakage began to increase. By 1930 leakage was greater than it had been in 1919, by this time amounting to at least 1200 ft^3/sec. No new attempts were made to stop the leakage until the TVA took over the project in 1939. Their investigations showed 13 large boils in the river immediately downstream from the dam and a total leakage of about 1700 ft^3/sec.

The eventual solution of the problem illustrates the heroic lengths that are necessary to salvage dam sites that were an exceedingly bad choice in the first place. Frink (1945) takes up the story.

In 1940 the Tennessee Valley Authority started a program of grouting combined with a cutoff wall to stop the leakage under the dam. The cutoff wall or grout curtain was made by drilling 18-in.-diameter holes through the rock along the upstream face of the dam. First, a row of holes spaced on 2-ft centers was drilled, lined with asbestos cement pipe, and filled with concrete. Interconnecting holes were then drilled slightly upstream, biting into adjacent concrete-filled holes. When the second set of holes was lined and concrete-filled, a continuous cutoff resulted (Fig. 13.8). At Hales Bar one dam is above and another dam is below the river.

There appears to be no real alternative to basic methods of test drilling and grouting, although the test drilling may turn up good reasons for moving the dam site somewhere else. Most of the fiascos

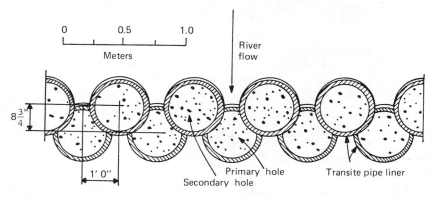

Figure 13.8 The cutoff wall or grout curtain that finally sealed the Hales Bar Dam. [Adapted from Frink (1945).]

that have made news in the past 50 years have resulted from inadequate hydrogeological characterization at the site before final designs were settled and contracts were let. Failure to spend on site characterization at the beginning results in cost overruns, long delays, and litigation with contractors.

13.4.3 Bank Stability and Bank Storage

Construction of large reservoirs in karst areas will mean that previously dry caves along the river valley will be flooded and water tables will be raised. Further, use of reservoirs for flood control or for hydroelectric power may mean that pool stage varies by 10 to 50 m over the course of a year. In karst terrain the water table rises and falls with pool stage in the reservoir. In limestone karst the fluctuating water tables may induce soil piping and the type of sinkhole collapse previously discussed. In gypsum karst it can greatly accelerate the solution process because the caverns in the walls of the reservoir are being flushed with fresh water once a year.

One of the most intensively studied areas is the gigantic Kama River Reservoir in the Soviet Union (Pechorkin, 1966, 1969). The Kama River Reservoir is constructed mainly on gypsum. Fluctuating water levels have caused collapse in the gypsum as much as 10 km from the reservoir.

13.5 GEOPHYSICAL AND REMOTE SENSING OF THE KARST

Problems with foundation engineering, property damage caused by sinkhole collapse, and reservoir engineering would all be less traumatic if the designers knew ahead of time the true character of the karst terrain. Visually, they are often presented with nothing more than weed-grown fields with no hint of what lies beneath the regolith or in the bedrock. Most characterization to date has depended on test drilling, which is both expensive and uncertain because unless one drills on very close centers it is easy to miss important features. A considerable effort has been made, therefore, to develop geophysical methods to detect cavernous terrain beneath a soil mantle.

The results of the geophysical research investigations have been ambiguous. Many of the measurements were deliberately made over known cave passages, and many of them have certainly detected something. Shallow cavities appear as anomalies on resistivity surveys, microgravity surveys, and shallow seismic measurements. See reviews by Greenfield (1979) and Kirk and Werner (1981). Much of the ambiguity seems to arise from lack of definition of the targets. Figure 13.9 illustrates the types of features one would like to map in the subsurface in a land use or foundation evaluation study. Most of the studies in the literature used a site where only one of the fea-

Land use and land management problems in karst / 373

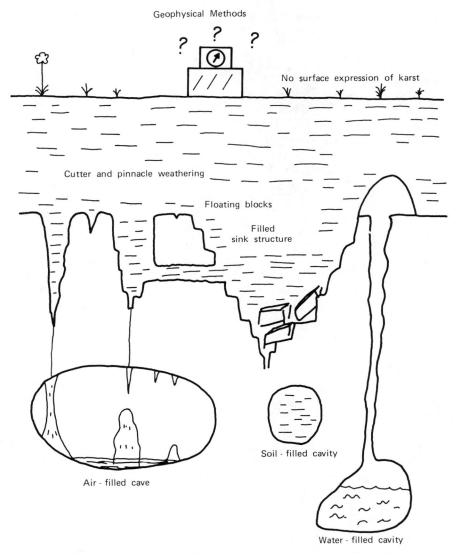

Figure 13.9 Some of the "targets" for geophysical methods of characterizing shallow karstic features in soil-mantled terrain.

tures (usually the shallow cave passage) was known to occur and then attempted to find it with various geophysical tools. Few methods available discriminate between the different kinds of subsurface features if a number of them are present.

Gravity surveys seem to offer the best prospect for detecting air-filled cavities. What one measures is the gravitational acceleration

374 / Geomorphology and hydrology of karst terrains

of the earth itself, but with a very high degree of precision. Near-surface geological information is contained in the Bouguer anomaly

$$B = g - (g_0 - cz - T) \qquad (13.1)$$

where g is the measured gravitational acceleration, g_0 is the value on the international ellipsoid, cz is an elevation or "free air" correction, and T is a terrain correction. All terms are in units of milligals (1 mgal = 10^{-5} m sec^{-2}). For regional scale gravity surveys, the corrections must be made with great care. If the object is to search for possible cavities in a small area, such as a construction site, one can look for local anomalies in the Bouguer anomaly.

What has made gravity methods for detecting cavities feasible is the invention within the past decade or so of very sensitive gravimeters. Early attempts (e.g., Chico, 1964) using gravimeters with a detectivity of 0.1 mgal were able to locate only large and shallow cavities. Chico was just able to locate the main hall of Luray Caverns, a very large cave room at a shallow depth below the surface. New gravimeters are precise to 0.01 or even 0.004 mgal, and at these levels of sensitivity, shallow cavities produce measurable anomalies.

For complete site investigations, the gravity survey should be on a rectangular grid, and the grid spacing should be kept small. Omnes (1977) used a 10-m spacing and by contouring on a 0.02- or 0.01-mgal interval was easily able to display karstic cavities as negative anomalies against the general Bouguer anomaly field. Joiner and Scarbrough's (1969) earlier gravity surveys in Alabama allowed only 0.5-mgal contours, and only general trends of cavernous zones could be detected. Kirk (1974) mapped across Fletcher's Cave with a gravimeter with 0.01-mgal sensitivy and found an anomaly that coincided with one known cave passage but not with a second. From a careful look at his data, it appears that the spacing between stations must be kept small if important features are not to be overlooked. The size of the expected anomaly is determined by the size of the cavity and its depth (Fig. 13.10).

Resistivity measurements may be the preferred way to characterize the soil–bedrock interface, to look for incipient sinkhole collapse and for cutter-and-pinnacle topography. Kirk and Werner (1981) give an extensive review of the methodology. Resistivity surveys consist of inserting electrodes into the soil and measuring the electrical resistance between them. Many arrangements have been recommended, with the Wenner array being most widely used. One passes a current between one pair of electrodes and measures the potential developed between a second pair. By using one pair of electrodes to insert the electrical current and a separate pair to measure the voltage drop, one avoids various artifacts from the electrodes and their contact with the soil. Other arrays use different numbers

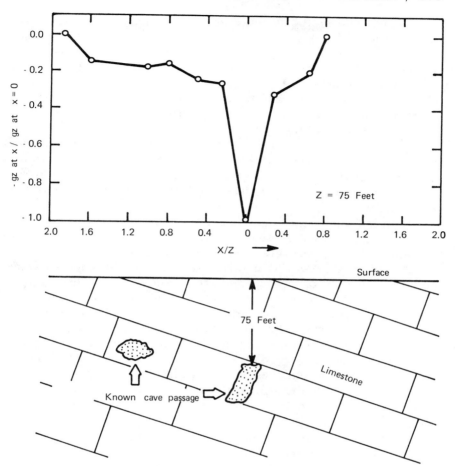

Figure 13.10 Gravity anomaly observed over Fletcher's Cave in Monroe County, West Virginia. [Adapted from Kirk (1974).]

of electrodes and electrode arrangements. A traverse across an area consists of moving the electrode array from point to point. Cavities appear as anomalies in a plot of apparent resistivity against traverse distance. Positive anomalies occur for air-filled cavities whose resistance is greater than that of the surrounding soil or rock, and negative anomalies for water or sediment-filled cavities whose resistivity is likely to be lower than that of the surrounding rock.

Shallow seismic reflection profiles have been applied to the cavity search. Inducing sound waves into the ground by striking a heavy hammer against a specially shaped metallic plate and then picking up the echos with an array of geophones, has offered some promise

376 / *Geomorphology and hydrology of karst terrains*

(Kirk and Snyder, 1977). Since seismic measurements depend on measurement of travel time, they can provide unique information on depth (given some assumptions about density and resultant sound velocity in the media), which neither gravity nor resistivity methods provide. On the other hand, seismic echos are obtained from any density discontinuity that reflects the waves, so that interpretation is more complex. Seismic profiles are an alternative way to determine thickness of soil and regolith.

Other geophysical techniques for detecting cavities include ground-penetrating radar (Kennedy, 1968; Kirk and Werner, 1981), deflections of the geothermal gradient (Ebaugh, Parizek, and Greenfield, 1976), and very low frequency electrometers (Whittemore and Parizek, 1976).

All of the geophysical methods described seem to detect cavities satisfactorily, providing that the cavities are large and shallow enough, and that their setting is not confused by bedrock contacts, joints and fractures, and multiple targets. Further, most of the devices seem to be operating on the threshold of their sensitivity, so that signal/noise ratios are unacceptably high. A fundamental problem, as can be seen from the array of targets sketched in Fig. 13.9, is that observed anomalies are interpreted on very simple models whereas the real-life situation is much more complex. Here the application of computer methods for information processing and use of computer graphics combined with new high-sensitivity devices may revolutionize our ability to "see" the karst landscape beneath the soil mantle.

Geophysical techniques share the requirement of a closely spaced network of measurement points; the spacing on the grid generally must be comparable to the size of feature one is trying to detect. The measurements, therefore, are time-consuming. They are less expensive than test drilling, and so are suitable for evaluations of small, well-defined sites such as foundations, landfills, or dam sites, but less suitable for large-area reconnaissance.

There has been interest in the remote sensing of karst as a better approach to the problem of large areas. Remote sensing uses such techniques as infrared photography, multispectral scanners, thermal imaging, and side-looking radar as means of bringing out features not observed in visible-light aerial photography. The objective is not so much to detect specific cavities or other features, as to map out sinkhole-prone areas and areas of cavernous bedrock, or to distinguish cutter-and-pinnacle bedrock topography from smooth and less difficult surfaces. Few of these efforts have appeared in print, although Warren and Wielchowsky (1973) describe some studies in Alabama. Infrared photography displayed as false color images (green = blue, red = green, and infrared = red) show tonal variations in vegetation that relate to the properties of the soils beneath. Healthy vegetation appears as red, whereas stressed vegetation—a sign of incipient sinkhole collapse—appears as green.

Thermal imaging, particularly in the 8- to 14-μm wavelength band, appears to be very promising because of the warm air that escapes from cave entrances and solution crevices in the winter (or cool air in the summer). It is a high-tech version of the old cave-hunter's trick of looking for steam on the winter hillsides. Unfortunately, the best equipment for these measurements was developed for military objectives and is rarely available for purely geological research.

13.6 GEOESTHETICS

The preceding sections have treated karstlands as a nuisance, their impact on human affairs being mainly negative. There is a positive aspect: karstlands are aesthetically pleasing. The dramatic surface landforms, sinking streams, gigantic springs, and vast cave systems have an immense human appeal. There is a wildness to the karst landscape that is attractive to most human observers. Thus, there are the "cave" parks and monuments within the national park system, a number of state parks, and many municipally owned parks and natural areas that center on big springs or caves. National scenic waterways, such as the Current and Eleven Point Rivers in Missouri, derive much of their charm from the karst landscape that forms their borders. Indeed, much of the research on caves and karst is motivated by their intrinsic aesthetic appeal rather than land use and water supply problems.

It could be argued that land use planning should take aesthetic factors into account, and there have been some attempts to do this. The difficulty is finding comparative values that are not purely economic. Leopold and O'Brien-Marchand (1968) derived a set of criteria for small drainage basins that included aesthetic considerations, and Morisawa (1971) devised a quantitative scale for beauty in river valleys. Giusti (1977) prepared geoesthetic maps of the karst lands of Puerto Rico. Those regions that rated high on the geoesthetic scale should be planned for use in recreation and tourism rather than for development or industrial use. Because of the many negative aspects of karstlands for urbanization and industrial development, more serious consideration of their very positive aesthetic character might aid greatly in the planned use of such terrains.

13.6.1 Commercial Caves

Oldest of the methods for experiencing the underground landscape is the commercial cave. Commercial caves are those that have been modified for easy access by visitors without special training or equipment. Traverse of the commercial cave is over graded walkways, up and down steps, and across bridges. Low places have been dug or blasted out so that the greatest discomfort is the need to duck one's head. The caves are sometimes lighted with great artistic skill, with white lights arranged to bring out the natural contour and highlight features of interest. Other times one must walk along an array of gaudy colored lights or a row of naked bulbs strung down the center of the passage.

Commercial cave tours are guided. The quality of the underground experience varies directly with the skills and training of the guides and these, unfortunately, are quite variable. Some are very knowledgeable of caves and can explain the strange features of the underground landscape; others tell entertaining stories and point out rows of stalagmites that look like "Snow White and the Seven Dwarfs."

Although a few show caves, such as Mammoth Cave or Luray Caverns, have a longer history, most are the product of the 1920s and the 1930s when assembly-line production of the automobile made travel possible for the average citizen. Show caves were originally opened as tourist attractions, huckstered like carnival sideshows. Today, many people have traveled widely and have seen the televised events of the world in their living rooms, yet the operation of many show caves in the 1980s has changed little since the 1920s. In the past several decades, cavers have developed a considerable number of show caves that tend to emphasize the natural features of the cave and provide something a bit closer to the caving experience. Some 200 commercial caves are open to the public in the United States (Gurnee and Gurnee, 1980), although as business enterprises, show caves are often rather marginal. New caves are opened and old ones closed every few years. Show caves are found in most developed countries, although to the author's knowledge no comprehensive world guide has been published.

13.6.2 Recreational Caving

Cave exploring is a popular recreational activity in the United States and in many other countries. Cavers in the United States are linked by a single national organization, the National Speleological Society. In many countries, caving clubs abound.

Cavers' motives are complex (see, e.g., Watson, 1966a), but at least two stand out: the desire to explore the unknown and the physical challenge. There is a great emphasis on the discovery of new caves and the exploration of known caves to their physical limits. In the underground, it is still possible, as it is nowhere else on earth, to tread ground that no one has trod before. Caving is a goal-directed activity that has much in common with back-country hiking or mountain climbing. The objective is to enter the cave, traverse it to the end and return to the surface. Very simple. Yet the accomplishment of this goal is a challenge requiring rope climbing, rock climbing, and the manipulation of one's body through low crawlways, up and down crevices, and through streams. It is strenuous activity and requires physical skill—perhaps even courage.

13.6.3 Underground Wilderness

The Wilderness Act of 1964 defines wilderness as "an area where the earth and its community of life are untrammeled by man, and where man himself is a visitor who does not remain."

It has been argued (Watson and Smith, 1971) that, although they may occur in highly populated or urbanized areas, caves can be wilderness as much as empty expanses of mountain and forest miles from human civilization. This is, in part, because the underground landscape, with its total darkness and unusual forms and shapes of rock and mineral deposits, is so alien compared with the familiar surface landscape.

Another reason is the remoteness of the cave. The farthest reaches of a large cave are seldom more than a few kilometers from the entrance, as the crow flies, and may be no more than 10 or 15 km as the caver crawls. And, of course, the outside world is usually only 100 m away vertically through solid rock. However, reaching the far corners of the cave, doing a bit of surveying, a little exploring, and then returning over the wearing, flesh-rending distance requires 24 to 36 hours.

A caver, resting a moment at the end of a cave, is as far away as if he or she were in another country or on the surface of the moon. The terrain is strange. No familiar sun or landmarks chart the way. Caving from this point of view is a genuine wilderness experience. It requires solitude, a leisurely pace, and a sense of absorption in the environment just as a wilderness experience on a mountain or forest does.

As landscapes to be modified and adapted to human purposes, karstlands present formidable challenges. As landscapes offering an intensely satisfying human experience, karstlands are valuable. The unsolved problem in this, as in many other areas of human affairs, is to balance the essentially economic focus of land use and land development against the essentially noneconomic focus of the wilderness experience.

14

Water Resources Problems in Karst

14.1 WATER SUPPLY DEVELOPMENT

An ever-present problem for the inhabitants of karstlands is procuring a reliable water supply. Absence of surface drainage and rapid infiltration and internal runoff combine to make water difficult to obtain. Further, water supplies often fail during droughts and are unusually susceptible to pollution from many sources. Water management in the karst is a special problem as is, for example, water management in deserts or on small islands.

14.1.1 Prospecting for Water Wells

Access to groundwater is obtained through water wells. The problem in the karst is knowing where to drill the wells. Karst aquifers may contain large quantities of water, but they are extremely anisotropic. Figure 14.1 illustrates the difficulties. Because the primary permeability of most carbonate rocks is very small, a well drilled into massive limestone or dolomite is likely to have only small yields. At the other extreme, wells that fortuitously intersect water-filled conduits may provide very large quantities of water with negligible drawdown.

Fractures—particularly those enlarged by solution—are zones of high permeability in conduit and fracture aquifers, and wells drilled along fractures or at their intersections are most likely to provide good yields. Where they intersect the land surface, fractures enhance weathering, and often appear on aerial photographs as linear features known as fracture traces. These traces consist of such features as aligned sinkholes, aligned tonal patterns in soil and veg-

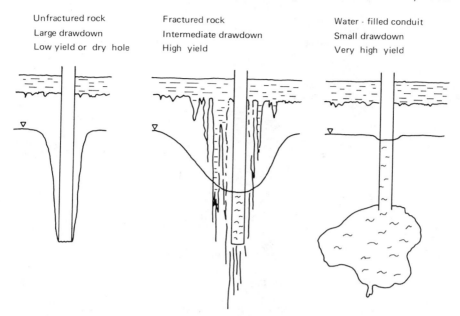

Figure 14.1 Different responses to pumping from wells in solid bedrock, wells that intersect fracture systems, and wells that intersect water-filled conduits.

etation, and linear reaches of stream channels, gulleys, and minor stream valleys.

Techniques for mapping fracture patterns were developed for the carbonate aquifers of central Pennsylvania (Lattman and Parizek, 1964; Parizek, 1976). There the spacing between fractures ranges from a few hundred meters to 1000 m, but the fracture swarms themselves—as judged from sparse outcroppings in road cuts and quarries—are only a few meters wide. Wells must hit their targets with some precision because the high concentration of solution cavities along the fractures fades into solid rock a short distance into the interfracture blocks. When municipal water supply wells were sited using fracture-trace mapping, production reached 3500 gal/min (0.22 m^3/sec), compared with a few tens of gallons per minute in nearby nonfractured rock. Figure 14.2 shows the productivity distribution of wells in fractured carbonate aquifers in central Pennsylvania. Random selection of sites produces some high-capacity wells, but the fraction of high-yield wells is much less than would exist if some guidance from the fracture pattern was used. Similar methods of well-siting have been used with some success in Alabama (Sonderegger, 1970; Moore et al., 1977). Fracture traces do not help

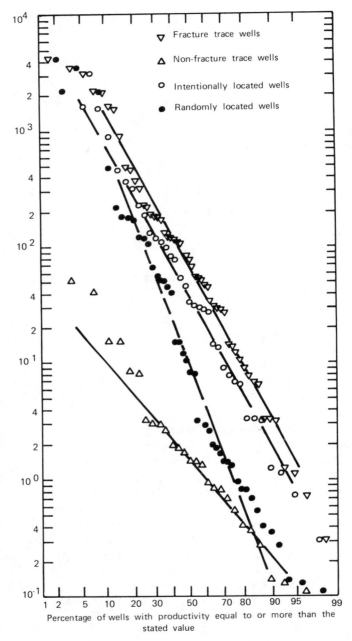

Figure 14.2 Productivity—frequency charts for wells drilled on and off fracture traces in the dolomite aquifers of Centre County, Pennsylvania. [Adapted from Siddiqui and Parizek (1971) and Parizek (1976).]

much in locating conduits, even when the conduits are joint and fracture-controlled.

14.1.2 *Water Resources Evaluation*

Any aquifer can be regarded as a renewable resource if it is developed with due regard for rates of recharge and is protected from adverse impacts of land use and pollutants in the recharge area. The information needed for resource planning includes the following:

1. Specific yield. The quantity of water that can be pumped from the aquifer at any one time.
2. Sustained yield. The amount of water that can be continuously removed from the aquifer without depleting it.
3. Locations of recharge and natural discharge.

Pump tests are the standard method for measuring specific yield. Wells are pumped at known rates for given periods of time, and water levels in nearby observation wells are measured. Established methods (e.g., Walton, 1962) permit the calculation of aquifer properties. Pumping tests are much less reliable in karst terrain. A pump test measures the specific yield of a particular well, but the results may mean little regarding the sustained yield of the aquifer. Pump tests on fracture aquifers can be made to work reasonably well by accumulating a statistically significant quantity of wells, distinguishing wells on fractures from those that are not, and taking careful account of the vertical distribution of permeability along the well bore (Parizek and Siddiqui, 1970; Eagon and Johe, 1972). Yet pump tests yield little valid information about the conduit portion of aquifers, and indeed may be very misleading.

Evaluation of aquifers with conduit permeability is best accomplished by a more comprehensive program consisting of the following steps:

1. Mapping all spring discharges and sinking stream and other recharge sources.
2. Tracing the conduit flows from cave streams and swallow holes to the springs to delineate the subbasins to which conduits and springs belong.
3. Direct mapping of accessible conduits and spillover dry cave passages.
4. Constructing water table maps from well-level data to further delineate groundwater basins and obtain information on the diffuse flow part of the system.
5. Constructing a water budget for the overall system to ensure that all inputs and outputs are accounted for.
6. Determining dynamic storage, the part above spring discharge levels, by analyzing spring discharge hydrographs.

It is difficult to obtain information about the deep storage portion of a conduit aquifer, but the shallow portion can be characterized in some detail using a comprehensive approach that may cost less in the long run than a program of test drilling. One could also attempt to reduce the characteristics of the karst aquifer to a mathematical and economic model, as suggested by Harpaz and Schwarz (1967).

To construct a complete map of losing and gaining streams and all swallow holes and rise points in karst areas, one must first find them. Thermal imaging techniques can be particularly valuable because the temperature contrast between the surface water and discharging groundwater allows rise points to be recognized (Brown, 1972b; Harvey et al., 1977). Aerial photography and LANDSAT imagery can also be useful in spotting sinking streams in rough terrain and in observing the flooding of closed depressions (Brook, 1983)

14.1.3 Use of Karst Springs

High-discharge springs provide easy access to the excess groundwater draining from carbonate aquifers, and such springs do, indeed, provide water supplies for many small communities. Diffuse-flow springs, which maintain more uniform discharge and are less threatened by pollutants, make excellent water supplies. Big Spring, which supplies the borough of Bellefonte, Pennsylvania, with more than 0.5 m^3/sec; (11 million gal/day), discharges from a fracture system in the Cambrian Gatesburg dolomite and is of sufficiently high quality that the citizens have opposed chlorination of the water. Most karst springs, however, discharge from conduit systems. These have a more variable discharge and their quality may be threatened if pollution is introduced in the drainage basin.

Other than the pollution threat, the principal disadvantage of using conduit-flow springs as water supplies is their low base flow, resulting from the ease with which the conduit system drains. This has led to attempts to use the air-filled portion of the conduit as a storage for periods of low recharge by constructing underground dams. Underground impoundments have the additional advantages of reducing land loss and evaporation losses, and offering better protection against contamination. Perič (1964), among others, has been an advocate of this use of karst waters.

Meddling with the natural flow regimes of karst conduit systems should be undertaken with some caution. Milanovič (1984) describes an experiment in damming the Obod Spring in the Fatničko polje with a concrete plug. The original purpose was water management rather than impoundment; flows were to be diverted to a lower spring for irrigation and hydroelectric power generation. Shortly after the plug was installed, heavy rains—more than 100 mm—fell in the Obod catchment. Water levels rose rapidly behind the plug and the pressure head reached 1.06 MPa (154 psi). Springs burst forth from the hillside above the spring up to elevations of 100 m, and some springs broke through in the basements of houses. There was considerable seismic activity and land slippage, including a section of highway more than 80 m above the spring, and several houses were damaged by flooding and slippage. After a few days, it was necessary to blow the plug, using 10 kg of explosive that had wisely been incorporated in it during construction. Once pressure was released, water levels dropped rapidly, the temporary springs dried up, and the hillslope quickly stabilized.

14.2 FLOOD HAZARDS IN KARST

14.2.1 *Flood Statistics and Recurrence Intervals*

River discharge at any moment depends on recent precipitation events, their distribution over the drainage basin, groundwater levels, soil characteristics and moisture conditions, the shape of the basin and tributary channel pattern, and plant cover. Daily discharge is often taken as Poisson distributed, with many days of moderate flow and a few days of flood flow. The custom has been to select the highest peak discharge in a given water year and term this the "annual flood." When annual flood discharges are examined over many years of record, they are also found to be log-normal statistically distributed. There are many years with only modest floods; larger floods occur with rapidly decreasing probability. Rather than calculate the probability of the occurrence of a flood of a particular value, we describe large floods in terms of a return interval, defined as the number of years between floods of a given magnitude.

The flood with an average return period of 2.33 years is the mean value and is the flood discharge that brings the river channel to bank-full conditions. Floods with longer return intervals spill out of the channel onto the adjacent flood plains. Floods used in classic hydrology include those with 10-year, 25-year, 100-year, and 1000-year return periods. The 100-year flood is most often used as the design criterion for bridges, flood-control impoundments, and related structures. The long-recurrence-interval floods are the most important because they cause the most damage, but their calculation is uncertain because their return period may exceed the period of record (often no more than 30–50 years). Calculation is also difficult because the correct statistical function is not certain. In the United States the double exponential distribution function introduced by Gumbel (1942) is often used along with Pearson distributions and Gamma distributions; in the Soviet Union, a three-parameter gamma distribution has been adopted (Kritsky and Menkel, 1967).

14.2.2 Flooding of Caves and Closed Depressions

There is usually a steep rising limb on a flood hydrograph. Discharges can increase by orders of magnitude in a matter of minutes or hours. Floods in some drainage basins appear as a wall of water sweeping down the channel. Because of the rapid response time of open conduit systems, flooding in caves has much the same characteristic. Rather than gradually rising in the water table, with consequent inundation of succeedingly higher cave passages, cave streams can quickly change from a trickle to a raging torrent. Rapid internal runoff through sinkholes and sinking streams fills the main conduit system much faster than the diffuse flow in smaller solution openings and fractures can accept it. High discharges are ponded behind breakdown piles and sediment in-fillings. Being trapped and drowned by flood waters is perhaps the greatest natural hazard faced by cave explorers. Exploration of large caves near base level requires careful weather checks and sometimes radio contact with the surface.

Closed depressions can fill with water either because of surface runoff that exceeds the infiltration capacity of the drain or because of rising groundwater levels. The drains reverse and water backs up into the closed depression. Except for the more permanently plugged sinkhole ponds, sinkholes flooded by storm runoff usually drain quickly. Closed depressions flooded by rising regional water levels often remain flooded for long periods of time. If the flooding occurs only during long-return-period floods, the closed depressions may be known locally as "phantom lakes," which contain water only once every few decades. The Phantom Lake depression in Centre County, Pennsylvania, is 2 km diameter but only 20 m deep, with a flat floor punctuated with smaller sinkholes. A highway crosses the depression and the bottom lands are in agricultural use. During Hurricane Agnes of 1972, the groundwater level, which had been low during the 1960s drought, rose above the bottom of the depression creating a lake 1000 m in diameter and 1 to 10 m deep, which persisted for several months (White and White, 1984).

Sinkhole flooding can be a serious problem when karst regions are urbanized. It is common to use the natural karst drainage system in such terrains as an inexpensive alternative to constructing storm drain systems. In addition to natural discharge of storm water runoff into sinkholes, drainage wells may be drilled in strategic locations. Flood flow in the shallow conduit system, ponding behind blockages in the natural channels, causes reverse flow through the drains and sinkholes, resulting in basement flooding and other problems. Bowling Green, Kentucky, is one of the best studied examples (Crawford, 1984a). Home owners may not expect karst terrains to be flood-prone and are caught unaware when their basements fill with water. However, one might reasonably expect developers and contractors to understand the landscape on which they are building (Quinlan, 1984, 1986).

Yugoslav hydrologists and civil engineers have long labored to control polje flooding. A little flooding renews soils; a large flood is a hazard to property and citizens. Flood control has been achieved in part by using debris grates to prevent the ponors that serve as drains from becoming plugged with debris, and in part by diverting some of the potential flood water.

14.2.3 *Hazards from Long-Recurrence-Interval Floods*

The most common floods—those with recurrence intervals of less than 10 to 20 years—have only modest impact on karst terrains. Dry streambeds may flow for a while, temporary ponds or lakes may form in closed depressions, caves may be flooded, and spring discharges will be high. Threats to life and property are no worse here than in drainage basins on nonkarstic rocks; indeed, the karst may be less hazardous because of the flood-damping achieved by temporary storage of water in the conduit system.

Now consider extreme events—floods with recurrence intervals of 100 to 1000 years. These improbable floods occur because of the concurrence of a series of low-probability events—for example, a hurricane-force storm that pauses and backtracks over a basin coincident with saturated ground or thick snow pack. The underground conduit system has a finite storage and water-carrying capacity. If the extreme flood completely fills the conduit system, the swallow holes will back up and divert the sinking streams into surface routes.

If all drainage is underground during normal floods, the old channels downstream from the swallow holes may become completely degraded or grown over, or all trace of the channels may disappear. In developed or urbanized areas, the abandoned water courses may not be recognized and various structures, such as houses, commercial buildings, or streets, may be constructed directly in their path. During extreme flow events, the spillover water has no available channel and must carve a new one directly through whatever has been constructed in its path. Thus, moderate floods may be less severe in karstic basins while large floods may do more damage than in nonkarstic basins. Dry valleys and doline karsts are flood-prone areas, although they may not be easily recognized as such.

14.3 POLLUTION THREATS TO CAVES AND KARST WATERS

14.3.1 *Sources and Measures of Pollution*

Pollutants are substances introduced into water by human activities that are generally dangerous to aquatic life or further human use of the water. Filtration by soils, adsorption on mineral grains, dispersion over large volumes of water, and long residence times usually ensure relatively clean groundwater. Because much of the water moves through open conduits, karst waters are no less susceptible to pollution than surface waters. Further, the injection of surface waters into the subsurface through sinking streams easily mixes polluted surface waters with groundwater.

Table 14.1 is a matrix relating various categories of pollutants to their sources. Some of these are more relevant to karst aquifers than others. Releases to the environment at specific places such as a sewage outfall are known as point sources. Leaky pipes and influent streams are line sources. Most difficult to deal with are the dispersed or nonpoint sources such as those resulting from crop spraying, fertilization, and other agricultural activities.

Although the debate continues over whether there actually is a threshold concentration below which pollutants are harmless, legally at least, maximum permissible concentrations have been established (Table 14.2). The U.S. Environmental Protection Agency (EPA) sets drinking water and other water quality standards. Drinking water standards are based on human health criteria; surface water standards are based on such criteria as toxicity to aquatic organisms and concentrations of toxic substances in the food chain. For this reason, surface-water standards may be much more stringent than drinking water standards.

Dissolved oxygen can be regarded as an "inverse pollutant" in the sense that depletion of the oxygen supply below normal levels has an adverse effect on aquatic organisms. The dissolved oxygen standard is a *minimum* level of 5 mg/L (EPA, 1976). The saturation concentration of oxygen in water is described by Henry's law, and in equilibrium with the oxygen partial pressure of the atmosphere has the solubility given by

$$C_{O_2} \text{ (ppm)} = 14.161 - 0.3943\, T + 0.007714\, T^2 - 0.0000646\, T^3 \quad (14.1)$$

where T is temperature in degrees Celsius (Truesdale et al., 1955). Reduced metal species can extract dissolved oxygen and oxidize to more stable states. Organic materials, particularly when reactions are enhanced by microorganisms, consume oxygen during decay and ultimate conversion to CO_2 and water. The concentrations of oxygen-consuming substances can be measured in terms of the amount of oxygen that is required for oxidation of inorganic constituents (chemical oxygen demand, COD) or for complete decay (biological oxygen demand, BOD). BOD, in particular, is often used as a measure of organic pollution.

14.3.2 Sinkhole Dumps and Solid Waste Injection

Some time ago, the local paper reported a small mystery: three dead sheep and five dead pigs had been found in the woods near the lime-

Table 14.1
Sources of Groundwater Pollutants

	Oxygen demand	Nitrogen Phosphates	Chlorides	Heavy metals	Hydrocarbons Organics	Bacteria Viruses
Domestic and municipal waste						
Septic tanks	XXX	XX				XXX
Outhouses	XXX	XX				XXX
Sewer lines	XX	X				X
Landfills	XXX	X	X	XX	XX	XX
Sinkhole dumps	X	X	X	XX	XX	XX
Agricultural activities						
Barnyard waste	XXX	XXX				XX
Fertilizer		XXX				
Insecticides and herbicides					XXX	
Construction and mining						
Road salt			XXX			
Parking lot runoff			XXX	X	X	
Spoil piles				XX		
Oil fields			XXX		XX	
Industrial activities						
Gasoline storage and distribution					XXX	
Waste outfalls			X	XXX	XXX	
Chemical dumps				XXX	XXX	

The number of Xs indicate, very roughly, the degree of pollution threat.

Table 14.2
Water Quality Standards[a]

	Drinking water[b]	Surface water[b,c,d]
General inorganic		
TDS	500	
Cl^-	250	
SO_4^{2-}	250	
NO_3^-	45 (as nitrate)	
Fe	0.3	1.0
Mn	0.05	0.1
F^-	1.4–2.4	
NH_3		0.02
CN^-	0.05	0.005
pH	6.5–8.5	
Selected heavy metals		
As	0.05	0.1 (irrigation water)
Ba	1.0	
Cd	0.01	0.0004–0.012
Cr (as Cr^{6+})	0.05	0.1
Cu	1.0	0.1 LC_{50}
Hg	0.002	0.00005 (aquatic life)
Pb	0.05	0.01 LC_{50}
Zn	5.0	0.01 LC_{50}
Representative organics		
Petroleum hydrocarbons		0.01 LC_{50}
Trihalomethanes[e]	0.1	
Phenol		0.001
Polychlorinated biphenyls (PCBs)		0.000001
Selected pesticides and herbicides		
2,4-D (2,4 dichlorophenoxyacetic acid)	0.1	
2,4,5-TP (Silvex)	0.01	
Endrin	0.0002	0.000004
Heptachlor		0.000001
Lindane	0.004	0.00001
Malathion		0.0001
Mirex		0.000001
Biological contaminants		
Total colliforms	1 per 100 mL	200 for bathing

[a] Drawn from the Code of Federal Regulations (40 CFR 143) and the EPA (1976).
[b] All concentrations are given in milligrams per liter.
[c] For many substances there are different standards for fresh water and for marine water. The table lists the freshwater standards.
[d] Many surface-water standards are given as a fraction, 0.1 or 0.01, of LC_{50}. LC_{50} is that concentration that is lethal to 50 percent of the population of a specified organism after a specified period of time (usually 96 hours).
[e] The standard refers to the sum of all trihalogen-substituted methanes such as chloroform ($CHCl_3$) and bromoform ($CHBr_3$). Standards are not specified for other light chlorinated hydrocarbons such as carbon tetrachloride or chlorinated ethanes, although these are known to be toxic.

stone valley farmland of central Pennsylvania. After speculating about possible origins for the dead animals, an article in the April 5, 1974 issue of *Centre Daily Times* concluded, "If a farmer loses an animal, he customarily buries it or puts it in a sinkhole. A farmer losing eight animals would not dump them in the woods."

Sinkholes are nuisances to farmers; they are acreages that cannot be farmed. Being worthless (in the farmer's view) they at least make attractive sites for trash disposal. It has been the custom to dispose of the carcasses of dead, often diseased, animals in sinkholes and open pits since time immemorial. At the turn of the century, Martel railed against the custom in France. Today, attempts to explain the relation between sinkhole dumps and water supplies still often fall on deaf ears. Sinkholes in the eastern United States are often filled with domestic trash, farm waste products, and junk automobiles, and are occasionally used as municipal dumps. Several municipal dumps in Missouri have been traced directly to springs as much as 25 km distant (Aley, 1972).

The drainways from many sinkholes connect directly with underlying conduit systems. Solid waste either slumps or is washed into the conduits, where corrosion and decay provide a source of contamination for long periods of time. Some sinkholes flood, and all take surface runoff from small catchments surrounding the sink. The pondings and injections of fresh water continually leach the accumulated debris in the sink and carry the leachate into the groundwater system.

In rural areas with sparse populations, sinkhole dumps, like other primitive means of waste disposal, may have caused few serious problems. As populations increase, both water supplies and waste product disposal must be managed more carefully. Thousands of sinkhole dumps exist. As population density increases in their vicinity, these steps are advisable (Aley, 1972):

1. Cease dumping (advice that could apply to all sinkhole dumps).
2. Trace the groundwater flow from the dumps, using trucked water to inject the tracers if necessary, to determine the direction of flow to affected springs and wells.
3. Monitor affected springs to determine their water quality during various recharge and discharge stages.
4. If possible, fill the sink and seal it with clay-rich soil to prevent as much internal runoff as possible. It may also be necessary to construct ditches to divert runoff from the sink.
5. Continuously monitor the site. Slumping, soil piping, and further collapse may re-open the sink.

14.3.3 Sanitary Landfills

Sanitary landfills were invented to replace the open dump as a means of municipal waste disposal. The basic practice is to excavate a trench on the order of 5 to 8 m deep and 8 to 15 m wide. The hauling trucks dump the waste into the trench, which is then covered each day with a meter of soil. The landfill cells are permeable, so that infiltration water soaks and leaches waste material. On some sites the base of the cells are lined with impermeable clays or with plastic or concrete liners before waste is dumped. Others use a system of drains to collect the leachate for processing or controlled disposal.

Municipal trash has a remarkably constant composition (Table 14.3). It decomposes slowly in the sealed, reducing environment of the landfill cell and produces a complex leachate that is the real contamination threat from landfill sites. Because decaying organic matter in the debris keeps oxygen potentials low, metals occur in their more soluble lower-valence states, and thus are more mobile than they would be in an oxidizing environment. The reducing capacity of the leachate is demonstrated by the BOD and iron concentrations beneath a landfill on carbonate terrain in central Pennsylvania (Fig. 14.3). BOD levels reached 10,000 mg/L and iron reached 300 mg/L—1000 times the drinking water standard. The heavy metal component of landfill leachate can be greatly increased if industrial wastes are mixed with the municipal wastes.

Landfills should be located on thick soils, above the water table, to provide the greatest possible barrier for leachate transport. Karsted carbonate terrain is one of the worst possible sites because of the generally thin soils and open solution cavities in the bedrock beneath. The uneven support in cutter-and-pinnacle bedrock topog-

Table 14.3
Composition of Municipal Waste
(National Average)

Solid waste material	Weight percent
Paper (newspaper, cardboard, misc.)	48
Food wastes	19
Metals	9
Glass and ceramics	8
Vegetation (grass, leaves, prunings)	4
Textiles	3
Rubber, leather, plastics	4
Dirt, rocks, concrete, and similar debris	3
Wood	2

Source: Data are from Hickman (1969).

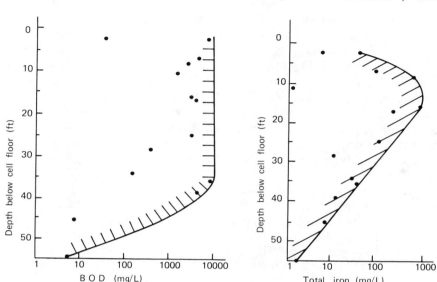

Figure 14.3 Profiles of BOD *(left)* and dissolved iron *(right)* in leachate beneath a sanitary landfill. Data were collected in 1970, for a cell emplaced in 1962; they show an advancing front of pollutants moving downward through the thick residual soils of the Gatesburg dolomite. [Adapted from Apgar and Langmuir (1971).]

raphy can cause differential settling of the landform cell and cause liners to crack. If there is a choice, location on dolomite is generally preferable to one on limestone because dolomite has less development of vertical solution openings and fewer problems with soil piping. Terrains with shallow conduit systems should be avoided because they may transport leachate directly to springs, where it can contaminate water supplies or surface streams.

A landfill was installed 1.5 km south of Kohl Spring, near Arnold, Missouri (Murray et al., 1981). The spring turned black and began to outgas hydrogen sulfide. The contaminated water killed all the fish in an adjacent pond. In addition, a small sinkhole 1 km south of Kohl Spring began to emit toxic gases, killing rodents, birds, and insects in the area. In this case landfill operations were stopped and conditions at the spring gradually improved. Clearly, leachate was being taken directly into an open conduit system.

14.3.4 *Road Salt and Parking Lot Runoff*

Snow removal in many northeastern states is accomplished by dumping copious quantities of salt (mostly NaCl, occasionally $CaCl_2$) onto the road surfaces. Salt accumulates during very cold weather and is washed off during thaws or late winter rains. The runoff in karst regions is likely to discharge into sinkholes, where it provides a source of chloride to the groundwater. Wells and springs near highways are easily contaminated (Parizek, 1971). In urban settings large areas are paved as streets, driveways, and parking lots. These impermeable surfaces accumulate hydrocarbon products, dust and dirt, small quantities of metals, and other particulate debris from automobiles. During heavy rains and snowmelt the accumulated fine-grained materials are flushed into storm drains, which in karst areas, are frequently connected to sinkholes, either deliberately or inadvertently.

The chloride and sodium ions themselves are not toxic pollutants. However, concentrations in excess of drinking water standards may give water a salty taste. High chloride levels in underground streams may adversely affect aquatic life. This is particularly significant if the limestone springs are used as water supplies for fish hatcheries (Werner, 1977). More important, the presence of high concentrations of chloride derived from road salt indicates that the other trace materials scrubbed from highways and parking lot pavements may also be present.

Chloride from pavement runoff can also affect fracture aquifers, as in the example of the Chicago area (DuPage County) of Illinois. A salt balance on the Salt Creek drainage basin (Wulkowicz and Saleem, 1974) showed that only 55 to 72 percent of the road salt was removed by surface streams. Chloride concentrations increased over time in the groundwater, showing that the aquifer system was acting as a reservoir for the salt. Chloride concentrations in the dolomite aquifer increased by more than 800 percent between the data collection in 1972–1973 and that of 12 to 40 years earlier (Long and Saleem, 1974; Saleem, 1977).

14.3.5 Septic Tanks, Disposal Wells, and Leaky Sewer Lines

High population densities require something beyond the country outhouse for disposal of human waste. In rural areas, wastes are discharged into septic tanks, where most of the material is digested by bacteria, sludge settles, and the liquid is dispersed into the soil through a tile field, all contained within the homeowner's private property. More urbanized areas, with closely spaced single-family dwellings, apartment buildings, and high-rise office and factory buildings, require piping the waste through sewer mains to a central treatment plant. In the treatment plant, wastes are digested and aerated, sludges are separated, and the effluent is discharged into surface streams—in principle without exceeding surface-water quality standards.

Properly functioning tile fields require several feet of soil because bacteriological processes in the soil are responsible for breakdown of the waste material. Thin soil in karst regions may not be adequate. Excess water from the tile field may induce soil piping, which can open direct pathways that will carry the waste to the subsurface. Groundwater contamination from septic tank effluent is a particularly common feature in karst regions.

Some residents of karst areas have found a method of sewage disposal that bypasses clogged tile fields and avoids the necessity of pumping out septic tanks: find a nearby sinkhole, dig out the soil down to the bedrock drain, run the sewer line to the bedrock drain, add some crushed stone, and backfill with soil. Cave streams beneath become running sewers, but the disposal system rarely chokes up.

Collector lines and sewage treatment plants do not necessarily end contamination problems. Most sewers are gravity drains built of ceramic tile pipe. Soil slumping into solution cavities or soil piping causes sections of pipe to pull apart, discharging the sewage into the soil, which exacerbates any piping problems that may have caused the break originally. The effluent from the treatment plant must be discharged somewhere. Normally it is discharged to a surface stream of sufficient flow to dilute the effluent. Large surface streams are rare in karstlands and effluent may be discharged into sinkholes, sinking streams, or loosing streams. Treated effluent is less of a hazard than raw sewage but is still not a desirable additive to karst groundwater and easily contaminates nearby wells and springs.

Sometimes disposal wells are used as an alternative to surface discharge of sewage effluent. This is a singularly poor practice in karst areas because of the interconnections between the disposal wells and water supply wells. A revealing case study is that of Bellevue, in the northwest corner of Ohio. The town is located on the middle Devonian Dundee limestone. In the past the town obtained its water supply from a combination of municipal and private wells drilled into the carbonate aquifer. Unfortunately, they also disposed of septic tank outflow as well as some raw sewage and industrial waste into disposal wells drilled in sinkholes or into the fractured carbonate bedrock. By the late 1960s the entire aquifer had been contaminated, as evidenced by foam, ammonia, and detergents in well water (Walker, 1961).

An attractive alternative for the disposal of sewage effluent is spray irrigation, the "living filter" concept. By spraying effluent on forest or cropland, the nutrients are extracted and used by plants, the wastewater is renovated and filtered by soil as it percolates downward, and becomes essentially freshwater recharge by the time it reaches the bedrock (Parizek et al., 1967). Spray irrigation in karstlands requires the same concern with soil conditions as any method of waste disposal. Thin or irregular soils draped over cutter-

and-pinnacle bedrock topography can provide direct nonfiltering pathways to the groundwater, bypassing the renovating plant root zone that is the key to the living filter concept.

Relatively few data document the effect of sewage contamination on cave life. Holsinger (1966) examined the cave community in Banners Corner Cave in Virginia, which had an obvious sewage contamination. Dratnal and Kasprzak (1980) examined communities in a karst stream in south Poland. In both cases, the main effect of sewage contamination was to lower dissolved oxygen levels. There were changes in the microbiological community and population differences in the larger organisms. Pollution is advantageous to some organisms that can use it as a source of nutrient, and detrimental to others.

14.3.6 Feedlot and Barnyard Runoff

Dairy herds, beef cattle, pig farms, and chicken and turkey brooders produce large quantities of organic waste. Much of this waste is concentrated in barnyards and feedlots, where large numbers of animals are confined to small areas. Runoff from these areas may be rich in organic decomposition products as well as nitrogen and phosphorus.

Manure is degraded protein material mixed with cellulose fiber and other waste products. Proteins are composed mainly of amino acids, which break down to form urea $(NH_2)_2CO$ and NH_3, as well as CO_2 and H_2O. Ammonia is highly soluble in water, where dissolved NH_3 and NH_4^+ are in equilibrium. Under highly reducing conditions such as are found in the interior of manure piles, the reduced form of nitrogen is stable. Oxidation occurring when manure is spread on fields or when leachates from manure piles moves into the soil produces nitrite (NO_2^-) or nitrate (NO_3^-) nitrogen. The extremely complex nitrogen chemistry in surface and soil environments is rarely described by equilibrium reactions. Purely inorganic reactions are extremely sluggish, and most are mediated by various microorganisms: nitrogen-fixing bacteria, bacteria that promote the decay process and breakdown of protein, and bacteria that catalyze the oxidation of nitrogen. Some nitrogen from waste products is discharged into the atmosphere as N_2 or NH_3. NH_3 and NO_2^- in leachates are usually oxidized before they reach the groundwater. The most common nitrogen-bearing species in groundwater is NO_3^-. Once NO_3^- enters the groundwater, it behaves as a conservative ion and undergoes no further reaction along the groundwater flow path. Nitrate levels are higher in karst springs than in nearby surface streams, which suggests that leachate from agricultural lands are the primary source. Kastrinos and White (1986) were able to show a linear relation between nitrate levels in springs and the fraction of the catchment area that was used for agriculture (Fig. 14.4).

The phosphorus component of animal waste appears as various

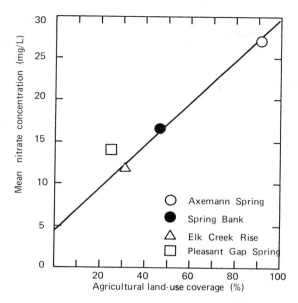

Figure 14.4 Relation of nitrate levels in karst spring waters to the fraction of agricultural land in the catchment area. [Adapted from Kastrinos and White (1986).]

phosphate species, which in near-neutral pH regimes, are mainly a mixture of $H_2PO_4^-$ and HPO_4^{2-}. These are the anions that often appear in the various phosphate minerals found in the reaction zone between guano piles and limestone wall rock (see Section 8.8.1). Phosphate rarely migrates far in surface soils because it is immobilized as hydroxyapatite $[Ca_5(PO_4)_3OH]$ and related minerals.

Barnyards and feedlots are also sources of coliform bacteria, which can be transported through solution openings into the karst drainage system. These organisms die off once they are removed from their natural habitat in the animal (or human) gut but can persist for some distances in open conduits. High nitrate levels in cave or spring waters indicate infiltration from agricultural lands or outfalls from the treatment of human waste. Concentrations of reduced nitrogen or coliform bacteria are indications of direct pollution of the water by animal manure or human sewage.

14.3.7 Gasoline and Other Petroleum Hydrocarbons

Modern civilization runs on hydrocarbon fuels. Gasoline, kerosene, diesel fuel, and home heating oil are transported in tank trucks and pipelines and stored in underground tanks. The tanks and their connecting pipes are placed underground to avoid fire hazard, but corrosion and leakage goes unnoticed until it is obvious that major losses are occurring. Light saturated hydrocarbons have low solubilities in water, on the order of 5 mg/L for gasoline. They are also substantially less dense than water. Contamination of any groundwater system by hydrocarbons is serious and difficult to clean up. Karst aquifers are a special case, with two unique differences that greatly increase the hazard. (1) Because of the open solution cavities in the bedrock, hydrocarbon spills and leakages can migrate easily to the water table. Because the petroleum products are essentially insoluble and lighter than water, the spilled material floats on the water table, where it can easily migrate along fractures and float down cave streams. (2) Low-molecular-weight hydrocarbons, particularly gasoline, are volitile even at groundwater temperatures. Hydrocarbon fumes can migrate through air-filled solution cavities into caves and upward into the basements of homes and other buildings, where they become a fire and explosion hazard.

Although small hydrocarbon leaks are likely to be extremely common, they are rarely discovered. Occasionally a large one occurs. Rhindress (1971) described a spill in a karst terrain just south of Harrisburg, Pennsylvania, from which 216,000 gal of gasoline were eventually recovered. The source of the gasoline was likely a tank farm, although which of three companies with tank farms in the area was responsible was never determined. The gasoline pool was initially discovered by a local businessman drilling an illegal drainage well. The pool was rediscovered by the Highway Department in the course of test borings for bridge foundations. At this time the oil companies drilled a series of wells in an effort to recover the gasoline. From the drilling records, the gasoline pool, floating on the water table in a solutionally modified carbonate aquifer, was a rectangular pool 500 m wide, 2000 m long, and at least 2 m deep at the deepest measured point. Under low recharge conditions, the water table was shallow but in July 1969, 230 mm of rain brought the water table near the land surface in a condition that would normally have resulted in some basement flooding. The gasoline pool floated up with the water table, and liquid gasoline was forced into some basements and gasoline fumes into others. Several houses had to be abandoned because of fire and explosion hazard. Continued pumping gradually removed the gasoline, and the intention at the time the report was published was to continue cleanup operations until all the gasoline was removed.

The explosion hazard from hydrocarbon fumes in air-filled solution cavities is very real. Gold et al. (1970) report on an exploding water well:

> During a thunderstorm of Friday, June 12, 1970, at 15:48:36 hours an explosion in a hand-dug well in the yard of a private dwelling demolished a concrete pump house and excavated a

crater 25 feet across. The explosion threw debris hundreds of feet into the air and the pump itself (200 lb) cleared the adjacent trees and a 38-foot high power line and landed in the field across the highway, 179 feet away. Most of the debris fell near the well and caused considerable damage to the house and garages, as well as to the adjacent buildings. . . .

The time of the explosion was known precisely because it was picked up by the Seismic Observatory at The Pennsylvania State University, some 20 km to the west. Analysis of the crater and the debris allowed the energy released by the explosion to be estimated at 5×10^{15} erg, but further calculations showed that this spectacular energy release could be achieved by the combustion of only 2.5 gal of gasoline! The source was apparently a storage tank used to supply gasoline for farm machinery. A leak allowed the gasoline to move along solution cavities to a cave passage that had been intersected by the hand-dug well, where a spark from the pump motor, or perhaps from lightning, ignited the explosive mixture.

Accumulation of gasoline fumes in caves can be a threat to cave explorers. Howards Waterfall Cave, in northwestern Georgia, was the scene of an explosion in 1966 that ultimately cost three people their lives, two of whom were would-be rescuers (Black, 1966). Near the back of the cave is a 6-m pit into which a small stream carried gasoline apparently leaked from a nearby filling station. When the leader of a boy scout exploring group descended into the pit, his carbide lamp ignited the fumes and produced an explosion. Amazingly, neither the leader nor the scouts were killed by the blast, but incomplete combustion of the gasoline filled the cave with carbon monoxide. Some of the scouts escaped and some took refuge on a high ledge above the fumes. The leader began to lose consciousness and fell, breaking his neck. Since the explosion was heard outside, people congregated at the scene and the first two who entered the cave were overcome by gas and died. The bodies were recovered and the three trapped scouts rescued only after intensive efforts by rescue personnel with portable breathing apparatus and after the cave had ventilated somewhat.

14.3.8 Industrial Waste: Heavy Metals and Organic Chemicals

Waste products from industry are released at outfalls where liquid waste streams are discharged and as solid wastes which may take the form of sludges, chemical wastes, and trash. The metal mining industry can also be a source of metal contamination (Procter et al., 1977). Industrial outfalls range from essentially pure water from cooling systems, with only excess heat as a "pollutant" (thermal pollution, characteristic of power plants and similar industry), to highly acid solutions rich in metals from certain metal-processing industries, to washwaters of various sorts that contain concentrations of organic chemicals and other substances. Solid chemical wastes are often packed in 55-gal steel drums and placed in shallow land burial sites. Sludges may be similarly treated. Industrial waste is sometimes treated as domestic trash and is placed in sanitary landfills with other municipal wastes.

Toxic waste sites, at least those established legally, require permits that usually specify details of site construction and post emplacement monitoring. Chemical waste sites in karst terrains are subject to the same threats as sanitary landfills and sewage lagoons. There is a possibility of soil piping beneath the site, permitting toxic substances to transport directly into the underlying conduit system. Leachate from the waste can enter the subsurface either directly through solution openings or by percolation through soils. The steel drums containing chemical wastes corrode after a few years in most environments and release their contents to whatever transport processes are available. Monitoring regulations written for sites overlying porous media aquifers prescribe observation wells, usually one well up the potential gradient from the site and three down gradient from the site. Quinlan and Ewers (1985) argue that monitoring springs and mapping the local conduit system is a more reliable way of spotting release of hazardous materials and predicting the route such materials would take through the groundwater system.

Most metals of concern—chromium, nickel, copper, zinc, cadmium, arsenic, lead, and mercury—are not highly mobile if brought into equilibrium with karst groundwater. Acid wastewaters are quickly neutralized by reaction with limestone. The buffer capacity of bicarbonate ion keeps the pH near neutral, where most of these metals form insoluble carbonates or hydroxides, or are adsorbed on iron hydroxides. However, the beneficial effect of carbonate water chemistry is offset by the open conduit flow paths. Leachates may enter cave streams where they flow on a chemically inert bed of quartz sand and clay minerals. Rapid flow in the conduit system may transport metals for long distances before reactions come to equilibrium. This is a particularly serious problem when industrial waste outfalls are discharged directly into sinkholes or sinking streams.

Quinlan and Rowe (1977) describe heavy metal releases into the carbonate aquifer near Horse Cave, Kentucky. Effluent from a metal-treating plant was allowed to enter the municipal sewer system. The sewer outfall discharges into a sinkhole, which connects to the south branch of Hidden River Cave. The ultimate discharge of the Hidden River underground catchment is on the Green River some 8 km from the town. Heavy metals arrived at an observation point near the entrance of Hidden River Cave with little evidence of precipitation or dilution. Although concentrations were much lower at the springs, levels were still anomalously high and were used to confirm the dye tracing.

The hydrochemistry of other industrial waste outfalls and hazardous chemical disposal sites in karst terrains is almost unknown. Chlorinated hydrocarbons and other more complex organic molecules are often poorly soluble in water, and so could be floated for considerable distances with little dispersion along conduit drainage paths. Adsorption on clastic sediments, oxidation, or other reactions with karst groundwater, and rates of dilution and dispersion are all of interest. Likewise there are few data on the effects of chemical contamination on aquatic organisms in underground streams. It has been argued that these organisms, adapted to the cave environment, would be more at risk than their surface relatives, but the measurements available are somewhat ambiguous (Bosnak and Morgan, 1981).

14.3.9 *Toxic and Explosive Gases*

The air-filled part of the integrated system of caves and vertical solution openings poses a poorly recognized hazard in urbanized karst terrain. Volatile substances carried into caves from leaks in pipelines and storage tanks, spills in sinkholes, and solid materials slumped into the subsurface from sinkholes or washed into sinking streams evaporate to fill the caves with fumes that may be toxic, explosive, or both. The particularly widespread problem of gasoline fumes was discussed earlier, but many other substances pose a threat. Volatile materials can be transported laterally by underground streams to places where the released fumes can migrate to the surface along solution openings. They can invade buildings through drains or cracks in floors or emerge from the ground in inhabited areas. At greatest risk are pets and children who breathe air closer to the ground.

Bowling Green, Kentucky, has been the most extensively investigated area (Elliott, 1976; Crawford, 1984b). As a moderately industrialized community with an extensive cavern system at shallow depths, it is an optimum environment for toxic and explosive gas hazard. Crawford points out the role of rising flood waters in forcing toxic substances upward into basements and sinkholes. Chemical scums are found on cave walls, and chemicals are found floating on residual pools when floodwaters recede. One such pool analyzed by Crawford was found to contain benzene and methyl chloride mixed with diesel fuel. Similar hazards can be expected in other communities built on cavernous limestone.

14.3.10 *Radon and Its Daughter Products*

One of the most recent discoveries is that caves can also possess a radiation hazard. This is a natural contaminant—radon gas and its daughter products. ^{222}Rn is produced in the decay chain of ^{238}U, which occurs widely in small concentrations in sedimentary rocks. A second decay chain from ^{230}Th produces another radon isotope, ^{220}Rn, commonly known as thoron. Because radon is a noble gas, it does not combine chemically and can migrate through cavities and fractures, apparently accumulating in caves. ^{222}Rn has a half-life of 3.83 days. It decays into polonium and other heavy metal isotopes that can accumulate in the body. Uptake of radioactive heavy metals is a substantial part of the hazard of breathing radon in the cave atmosphere.

The background radiation was first noticed in Alabama caves, where it was initially ascribed to the effects of fallout from the nuclear testing program (Reckmeyer, 1963a,b). It was eventually established that the high radiation levels resulted from natural sources, and that the radioactivity levels fluctuated with the season and the movement of air currents (Wilkening and Watkins, 1976). The National Park Service became interested in radon levels because of possible overexposure of their tour guides (Yarborough, 1976, 1977, 1978; Ahlstrand, 1976, 1980; Ahlstrand and Fry, 1977). Radiation levels in the caves examined are less than the allowable limit for continuous exposure during a normal 40-hour work week. Caving is an occasional activity for most people and the danger of excessive radiation is small. Few caves outside of the national park system have been measured, but the wide distribution of the park caves suggests that radon daughters are everywhere and it is quite possible that some caves would be unexpectedly "hot" (Table 14.4).

14.4 COASTAL KARST AQUIFERS

14.4.1 *The Ghyben–Herzberg Principle*

The water table must meet the ocean at grade when the aquifer dips under the sea. The freshwater–saltwater interface slopes back under the land surface at an angle dictated by the relative densities of salt water and fresh water. If the system is static, the water table and the freshwater–saltwater interface meet at sea level (Fig. 14.5). Under static conditions the relation between the water table level and the level of the interface is given by the Ghyben–Herzberg principle (see, e.g., Freeze and Cherry, 1979):

$$Z_s = \frac{\rho_w}{\rho_s - \rho_w} z_w \qquad (14.2)$$

Given $\rho_w = 1.00$ and $\rho_s = 1.025$

$$Z_s = 40 z_w \qquad (14.3)$$

Table 14.4
Radon Levels in Caves[a]

Cave	Radon	Thoron
Mammoth Cave, KY[b](1)		
Historic route	0.69	0.13
Scenic route	0.73	0.10
Wild cave	0.57	0.11
Lantern tour	0.43	0.10
Carlsbad Caverns, NM(2)		
Warm season	0.63	
Cool season	0.20	
Lehman Cave, NV(3)	1.12	
Crystal Cave, CA(3)	1.37	
Indian Cave, Cumberland Gap, VA(3)	1.18	
Lilburn Cave, CA(4)	3.09	0.12
Soldiers Cave, CA(4)	1.43	0.05
Castleguard Cave, Canada[c](5)	0.047	

[a]Following Park Service practice, all data are given in working levels, which relate to total human exposure over an assumed year of 40-hour work weeks. One working level is defined as the combination of radon daughters in 1 L of air that will release 1.3×10^5 MeV of alpha energy. One working level corresponds to the total decay of short-lived daughters of 100 pCi of radon per liter of air, or 8 pCi of thoron per liter. The total annual dose specified by NIOSH and OSHA must not exceed 4 working levels.
[b]Data for tourist routes. Overall cave data average 1.2 working levels.
[c]Measured values ranged from 2 to 8 pCi L^{-1}.
Sources: (1) Yarborough (1977); (2) Alhstrand (1980); (3) Yarborough (1977); (4) DesMarais (1978); DesMarais, personal communication; (5) Atkinson et al. (1983).

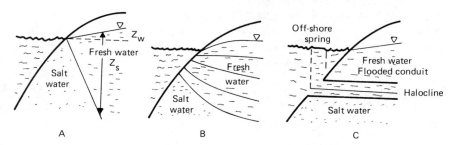

Figure 14.5 The freshwater–saltwater interface: (A) the hydrostatic (Ghyben–Herzberg) case; (B) the dynamic case in porous media; (C) a karst aquifer with drowned Pleistocene conduit that acts as a freshwater drain.

The freshwater–saltwater interface is depressed 40 m for each meter of elevation of the water table above sea level.

In a dynamic system with a continuous flow of fresh water toward the sea, the interface is depressed below sea level because of the front of fresh water moving seaward. In porous media aquifers, flow nets can be calculated using the principles set forth by Hubbert (1940). The higher the flow, the greater the depression of the saltwater interface. In peninsular and island aquifers, the salt water can extend completely under the land mass, so that the available fresh water floats entirely on the salt water, as a Ghyben–Herzberg lens with the interface curving up to the sea on all sides.

The freshwater–saltwater interface for dynamic systems is modified by diffusion and, in karstic aquifers, by direct mixing. The interface can be more accurately portrayed by calculating isochlors using some sort of mathematical model such as that proposed by Lee and Cheng (1974).

14.4.2 *Saltwater Intrusion*

When wells drilled into the freshwater lens are pumped, the drawdown releases the hydrostatic pressure and permits the saltwater interface to rise. Because of the density contrast, 1 m of drawdown results in a 40-m rise of the interface. Pumping also reduces the flow of fresh water to the sea and thus allows the interface near the coast to rise toward sea level. Overpumping of wells in peninsular Florida has caused substantial increases in salinity and has forced some wells in the Miami area to be abandoned. With sufficient overpumping, it is possible to destroy the Ghyben–Herzberg lens completely. Because saltwater intrusion greatly increases the electrical conductance of the water, resistivity measurements are a convenient way to map the intrusion front (Fretwell and Stewart, 1981).

In karstic carbonate aquifers the conduit permeability short circuits the flow net of fresh water. In many coastal karsts, conduits formed at grade with sea level during the Pleistocene sea level minima and are now drowned. Divers have observed the freshwater–saltwater interface as a planar surface through the conduit, and indeed can swim up and down through the interface. The fresh water now uses the conduits on its way to offshore springs. Disturbing the water table by pumping the system can easily reverse this flow, and salt water can be swept through the conduit system to contaminate the freshwater aquifer.

14.4.3 *Waste Injection*

Waste disposal is a problem in coastal karsts because of the ease with which the delicate water supply can be affected. Figure 14.6 shows a concept in use for Merida, the capital of the Mexican state

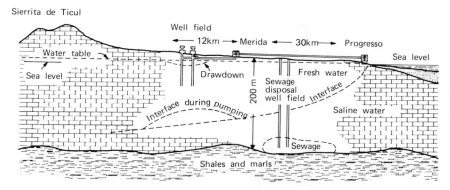

Figure 14.6 Strategy of both water supply and sewage disposal in the Yucatan Peninsula. [Adapted from Back and Hanshaw's (1974) adaptation of a drawing by H. Lesser Jones (1965).]

of Yucatan. The Yucatan Peninsula is a nearly flat karst plain with only a few tens of meters of relief; it is bounded on west, north, and east by the Gulf of Mexico and the Caribbean Sea. Freshwater is obtained from production wells south of the city. Drawdown is small in the karsted limestone and if pumping rates are carefully controlled, saltwater invasion of the producing aquifer can be avoided. Wastewater is pumped to the north and injected in cased deep wells that penetrate far below the freshwater–saltwater interface. The wastewater is injected into a relatively static zone of saline water, where it cannot interfere with the freshwater supply.

References

Ahlstrand, G. M. (1976) Alpha radiation associated studies at Carlsbad Caverns. *Proc. Natl. Cave Management Symp.* T. Aley and D. Rhodes, Eds. (Speleobooks, Albuquerque, NM), 71–74.

Ahlstrand, G. M. (1980). Alpha radiation levels in two caves related to external air temperature and atmospheric pressure. *Natl. Speleol. Soc. Bull.* 42, 39–41.

Ahlstrand, G. M. and P. L. Fry (1977) Alpha radiation project at Carlsbad Caverns: Two years and still counting. *Proc. Natl. Cave Management Symp.* R. Zuber, J. Chester, S. Gilbert, and D. Rhodes, Eds. (Adobe Press, Albuquerque, NM), 133–137.

Aley, T. (1964) Echinoliths—an important solution feature in the stream caves of Jamaica. *Cave Notes 6*, 3–5.

Aley, T. (1972) Groundwater contamination from sinkhole dumps. *Caves and Karst 14*, 17–23.

Aley, T. and M. W. Fletcher (1976) The water tracer's cookbook. *Missouri Speleol. 16*, 1–32.

Aley, T. J., J. H. Williams, and J. W. Massello (1972) *Groundwater contamination and sinkhole collapse induced by leaky impoundments in soluble rock terrain.* Eng. Geol. Series No. 5, Missouri Geological Survey.

American Public Health Association (1985) *Standard methods for the examination of water and wastewater,* 16th ed. American Public Health Association, Washington, DC.

Anderson, C. H. and W. R. Halliday (1969) The Paradise Ice Caves, Washington: an extensive glacier cave system. *Natl. Speleol. Soc. Bull. 31*, 55–72.

Andrieux, C. (1962) Étude cristallographique des édifices stalactitique. *Bull. Soc. franc. Minéral. Crist. 85*, 67–76.

Andrieux, C. (1965a) Étude des stalactites tubiformes monocristallines. Mechanisme de leur formation et conditionnement de leurs dimensions transversales. *Bull. Soc. franc. Minéral. Crist. 88*, 53–58.

Andrieux, C. (1965b) Morphogenèse des hélictites monocristallines. *Bull. Soc. franc. Minéral. Crist.* **88**, 163-171.

Apgar, M. A. and D. Langmuir (1971) Groundwater pollution potential of a landfill above the water table. *Ground Water* **9**, 76-96.

Aronis, G., D. J. Burdon, and K. Zeris (1961) Development of a karst spring in Greece. *Internatl. Assoc. Sci. Hydrol.* Special Pub. 57, 564-585.

Ashton, K. (1966) The analysis of flow data from karst drainage systems. *Trans. Cave Res. Group Great Britain* **7**, 161-203.

Atkinson, T. C. (1977a) Diffuse flow and conduit flow in limestone terrain in the Mendip Hills, Somerset (Great Britain). *J. Hydrol.* **35**, 93-110.

Atkinson, T. C. (1977b) Carbon dioxide in the atmosphere of the unsaturated zone: An important control of groundwater hardness in limestones. *J. Hydrol.* **35**, 111-125.

Atkinson, T. C., R. S. Harmon, P. L. Smart, and A. C. Waltham (1978) Palaeoclimatic and geomorphic implications of $^{230}Th/^{234}U$ dates on speleothems from Britain. *Nature* **272**, 24-28.

Atkinson, T. C., T. J. Lawson, P. L. Smart, R. S. Harmon, and J. W. Hess (1986) New data on speleothem deposition and palaeoclimate in Britain over the last forty thousand years. *J. Quaternary Sci.* **1**, 67-72.

Atkinson, T. C., P. L. Smart, and T. M. L. Wigley (1983) Climate and natural radon levels in Castleguard Cave, Columbia Icefields, Alberta, Canada. *Arctic and Alpine Res.* **15**, 487-502.

Atkinson, T. C., P. L. Smart, and J. N. Andrews (1984) Uranium-series dating of speleothems from Mendip Caves. 1: Rhino Rift, Charterhouse-on-Mendip. *Proc. Univ. Bristol Speleol. Soc.*, **17**, 55-69.

Aubert, D. (1966) The structure, activity, and evolution of a doline. *Bull. Soc. Neuchateloise Sci. Nat.* **89**, 113-120; English translation by C. and J. Gunn, *Cave Geology*, **1**, 259-266.

Back, W. and B. B. Hanshaw (1974) Hydrochemistry of the northern Yucatan Peninsula, Mexico with a section on Mayan water practices. In *Field Seminar on Water and Carbonate Rocks of the Yucatan Peninsula, Mexico.* A. E. Weidie, Ed. New Orleans Geol. Soc. 45-77.

Back, W., B. B. Hanshaw, T. E. Pyle, L. N. Plummer, and A. E. Weidie (1979) Geochemical significance of groundwater discharge and carbonate solution to the formation of Caleta Xel Ha, Quintana Roo, Mexico. *Water Resources Res.* **15**, 1521-1535.

Baes, C. F. Jr., H. E. Goeller, J. S. Olson, and R. M. Rotty (1976) *The global carbon dioxide problem. Oak Ridge Natl. Lab. Report. ORNL-5194.*

Bagnold, R. A. (1966) *An approach to the sediment transport problem from general physics. U.S. Geol. Survey Prof. Paper 422-I.*

Bagnold, R. A. (1968) Deposition in the process of hydraulic transport. *Sedimentology* **10**, 45-56.

Bagnold, R. A. (1977) Bed load transport by natural rivers. *Water Resources Res.* **13**, 303-312.

Bakalowicz, M. (1976) Geochimie des eaux karstiques. Une methode d'etude de l'organisation des ecoulements souterrains. *Ann. Scien. University of Besancon*, No. 25, 3rd Series, 49-58.

Bakalowicz, M. and A. Mangin (1980) L'aquifère karstique. Sa définition, ses caractéristiques et son identification. *Mem. h. sér. Soc. geol. France* No. 11, 71–79.

Baker, G. and A. C. Frostick (1947) Pisoliths and ooliths from some Australian caves and mines. *J. Sedimentary Petrol.* **17**, 39–67.

Baker G. and A. C. Frostick (1951) Pisoliths, ooliths and calcareous growths in limestone caves at Port Campbell, Victoria, Australia. *J. Sedimentary Petrol.* **21**, 85–104.

Baker, V. R. (1973) Geomorphology and hydrology of karst drainage basins and cave channel networks in East-Central New York. *Water Resources Res.* **9**, 695–706.

Balch, E. S. (1900) *Glacières or freezing caverns* (Allen, Lane and Scott, Philadelphia; reprinted by Johnson Reprint Corp., New York), 1970.

Balenzano, F., L. Dell'Anna, M. DiPierro, and S. Fiore (1984) Ardealite, $CaHPO_4CaSO_4 \cdot 4H_2O$; a new occurrence and new data. *Neues Jahr. Mineral. Mh.*, 1984, 461–467.

Barnes, H. H., Jr. (1967) Roughness characteristics of natural channels. *U.S. Geol. Surv. Water Supply Paper 1849.*

Baron, G., S. Caillère, R. Lagrange, and Th. Pobeguin (1959) Étude du mondmilch de la grotte de la Clamouse et de quelques carbonates et hydrocarbonates alcalino-terreux. *Bull. Soc. franc. Minéral Crist.* **82**, 150–158.

Barr, T. C., Jr. (1961) *Caves of Tennessee.* Tennessee Div. Geol. Bull. 64.

Bartrum, J. A. and A. P. Mason (1948) Lapiez and solution pits in basalts at Hokianga, New Zealand. *New Zealand J. Sci. Technology* **30B**, 165–172.

Bassett, J. L. and R. V. Ruhe (1973) Fluvial geomorphology in karst terrain. In *Fluvial Geomorphology*, Marie Morrisawa, Ed. (State University of New York at Binghamton), 75–89.

Bates, R. G. (1964) *Determination of pH. Theory and Practice* (John Wiley, New York).

Bauer, F. (1962) Nacheiszeitliche Karstformen in den österreichischen Kalkhochalpen. *Proc. 2nd Internatl. Congress Speleol., Bari,* 1, 299–328.

Baumgardner, R. W., T. C. Gustavson, and Ann D. Hoadley (1980) Salt blamed for new sink in W. Texas. *Geotimes* (Sept.), 16–17.

Bear, J. (1972) *Dynamics of flow in porous media* (Elsevier, Amsterdam).

Beck, B. F. (1984) *Sinkholes: Their geology, engineering and environmental impact* (A. A. Balkema, Rotterdam).

Berner, R. A. (1975) The role of magnesium in the crystal growth of calcite and aragonite from sea water. *Geochim. Cosmochim. Acta* **39**, 489–504.

Berner, R. A. and J. W. Morse (1974) Dissolution kinetics of calcium carbonate in sea water. IV. Theory of calcite dissolution. *Amer. J. Sci.* **274**, 108–134.

Black, D. F. (1966) Howard's Cave disaster. *Natl. Speleol. Soc. News* **24**, 242–244.

Bleahu, M. (1965) Sur les confluences souterraines. *Internatl. J. Speleol.* **1**, 441–459.

Blount, C. W. and F. W. Dickson (1973) Gypsum-anhydrite equilibria in systems $CaSO_4$-H_2O and $CaCO_3$-NaCl-H_2O. *Amer. Mineral. 58*, 323–331.

Blumberg, P. N. and R. L. Curl (1974) Experimental and theoretical studies of dissolution roughness. *J. Fluid Mech. 65*, 735–751.

Bögli, A. (1960) Kalklösung und Karrenbildung. *Zeits. Geomorph. Suppl. 2*, 4–21.

Bögli, A. (1964) Mischungskorrosion—ein Beitrag zum Verkarstungsproblem. *Erdkunde 18*, 83–92.

Bögli, A. (1965) The role of corrosion by mixed water in cave forming. In *Problems of the Speleological Research*, O. Stelcl, Ed. (Czechoslovak Academy of Science, Prague), 125–131.

Bögli, A. (1980) *Karst hydrology and physical speleology.* (Springer-Verlag, Berlin).

Bosnak, A. D. and E. L. Morgan (1981) Acute toxicity of cadmium, zinc, and total residual chlorine to epigean and hypogean isopods (Asellidae) *Natl. Speleol. Soc. Bull. 43*, 13–18.

Bradbury, J. C. (1959) Crevice lead-zinc deposits of northwestern Illinois. *Ill. Geol. Surv. Rpt. Invest. 210.*

Braker, W. L. (1981) *Soil Survey of Centre County, Pennsylvania.* U.S. Soil Conservation Service.

Brandt, A., S. Kempe, M. Seeger, and F. Vladi (1976) Geochimie, Hydrographie und Morphogenese des Gipskarstgebietes von Düna/Südharz. *Geol. Jahr. C-15*, 3–55.

Bretz, J H. (1942) Vadose and phreatic features of limestone caverns. *J. Geol. 50*, 675–811.

Bretz, J H. (1949) Carlsbad Caverns and other caves of the Guadalupe Block, New Mexico. *J. Geol. 57*, 447–463.

Bretz, J H. (1952) A solution cave in gypsum. *J. Geol. 60*, 279–283.

Bretz, J H. (1955) Cavern-making in a part of the Mexican Plateau. *J. Geol. 63*, 364–375.

Bretz, J H. (1956) *Caves of Missouri. Missouri Geol. Surv. Water Resources*, 39.

Bretz, J H. and S. E. Harris, Jr. (1961) Caves of Illinois. *Ill. Geol. Surv. Rpt. Invest. 215.*

Bridge, P. J. (1971) Analyses of altered struvite from Skipton, Victoria. *Mineral. Mag. 38*, 381–382.

Bridge, P. J. (1973a) Urea, a new mineral, a neotype phosphammite from Western Australia. *Mineral. Mag. 39*, 346–348.

Bridge, P. J. (1973b) Guano minerals from Murra-el-elevyn Cave, Western Australia. *Mineral. Mag. 39*, 467–469.

Bridge, P. J. (1974) Guanine and uricite, two new organic minerals from Peru and Western Australia. *Mineral. Mag. 39*, 889–890.

Bridge, P. J. (1975) Urea from Wilgie Mia Cave, W. A., and a note on the type locality of urea. *Western Australian Naturalist, 13*, 85–86.

Bridge, P. J. and R. M. Clarke (1983) Mundrabillaite—a new cave mineral from Western Australia. *Mineral. Mag. 47*, 80–81.

Bridge, P. J. and B. W. Robinson (1983) Niahite—a new mineral from Malaysia. *Mineral. Mag. 47*, 79–80.

Brod, L. (1962) Cave mapping: A systematic approach. *Missouri Speleol. 4*, 1–52.

Broecker, W. S. and E. A. Olson (1959) C^{14} dating of cave formations. *Natl. Speleol. Soc. Bull. 21*, 43.

Broecker, W. S., E. A. Olson, and P. C. Orr (1960) Radiocarbon measurements and annual rings in cave formations. *Nature 185*, 93–94.

Broecker, W. S. and P. C. Orr (1958) Radiocarbon chronology of Lake Lahontan and Lake Bonneville. *Geol. Soc. Amer. Bull. 69*, 1009–1032.

Brook, G. A. (1983) Application of LANDSAT imagery to flood studies in the remote Nahanni Karst, Northwest Territories, Canada. *J. Hydrol. 61*, 305–324.

Brook, G. A., D. W. Cowell, and D. C. Ford (1977) Comment on 'regional hydrochemistry of North American carbonate terrains' by Russell S. Harmon, William B. White, John J. Drake, and John W. Hess and 'The effect of climate on the chemistry of carbonate groundwater' by John J. Drake and T. M. L. Wigley. *Water Resources Res. 13*, 856–859.

Brook, G. A., M. E. Folkoff, and E. O. Box (1983) A world model of soil carbon dioxide. *Earth Surface Processes and Landforms 8*, 79–88.

Brook, G. A. and D. C. Ford (1973) The Nahanni North Karst: A questionmark on the validity of the morphoclimatic concept of karst development. *Proc. 6th Internatl. Congress Speleol., Olomouc.* 2, 43–57.

Brook, G. A. and D. C. Ford (1977) The sequential development of karst landforms in the Nahanni Region of northern Canada and a remarkable size hierarchy. *Proc. 7th Internatl. Congress Speleol., Sheffield*, 77–81.

Brook, G. A. and D. C. Ford (1982) Hydrologic and geologic control of carbonate water chemistry in the subarctic Nahanni Karst, Canada. *Earth Surface Processes and Landforms 7*, 1–16.

Brown, M. C. (1972a) Karst hydrology of the Lower Maligne Basin, Jasper, Alberta. *Cave Studies 13*.

Brown, M. C. (1972b) Karst hydrogeology and infrared imagery: An example. *Geol. Soc. Amer. Bull. 83*, 3151–3154.

Brucker, R. W. (1966) Truncated cave passages and terminal breakdown in the central Kentucky karst. *Natl. Speleol. Soc. Bull. 28*, 171–178.

Brucker, R. W., J. W. Hess, and W. B. White (1972) Role of vertical shafts in the movement of ground water in carbonate aquifers. *Ground Water 10*, 5–13.

Buhmann, D. and W. Dreybrodt (1985a) The kinetics of calcite dissolution and precipitation in geologically relevant situations of karst areas. 1. open system. *Chemical Geol. 48* 189–211.

Buhmann, D. and W. Dreybrodt (1985b) The kinetics of calcite dissolution and precipitation in geologically relevant situations of karst areas. 2. closed system. *Chemical Geol. 53*, 109–124.

Bull, P. (1981) Some fine-grained sedimentation phenomena in caves. *Earth Surface Processes and Landforms, 6*, 11–22.

Burdon, D. J. and N. Papakis (1963) Handbook of karst hydrogeology (United Nations, Athens).

Burdon, D. J. and C. Safadi (1963) Ras-el-Ain: The great karst spring of Mesopotamia. *J. Hydrol. 1*, 58–95.

Burdon, D. J. and C. Safadi (1965) The karst groundwaters of Syria. *J. Hydrol. 2*, 324–347.

Burke, A. R. and P. F. Bird (1966) A new mechanism for the formation of vertical shafts in carboniferous limestone. *Nature 210*, 831–832.

Burwell, E. B., Jr., and B. C. Moneymaker (1950) Geology in dam construction. In *Application of Geology to Engineering Practice* S. Paige, Ed. *Geol. Soc. Amer. Bull.* 11–44.

Busenberg, E. and L. N. Plummer (1982) The kinetics of dissolution of dolomite in CO_2-H_2O systems at 1.5 to 65 C and 0 to 1 atm P_{CO_2}. *Amer. J. Sci. 282*, 45–78.

Butcher, A. L. and C. L. Railton (1966) Cave surveying. *Trans. Cave Res. Group Great Britain 8*, 1–37.

Campbell, N. P. (1978) *Caves of Montana.* Montana Bur. Mines and Geol. Bull. 105.

Campbell, N. (1979) Alpine karst of the Scapegoat-Bob Marshall Wilderness and adjoining areas, north-central Montana. *Natl. Speleol. Soc. Bull. 41*, 66–69.

Carroll, D. E. and H. C. Starkey (1959) Leaching of clay minerals in a limestone environment. *Geochim. Cosmochin. Acta 16*, 83–87.

Carroll, R. W., Jr. (1978) TSOD: Adirondack Anorthosite Talus Monster. *NSS News 36*, 119–120.

Carwile, R. H. and E. F. Hawkinson (1968) Baselevel sedimentation, Flint Ridge, Kentucky. BS Thesis, Geology (Ohio State University, Columbus, OH).

Castillo, E., G. M. Karadi, and R. J. Krizek (1972) Unconfined flow through jointed rock. *Water Res. Bull. 8*, 266–281.

Cavaillé, A. (1965) Observations sur l'evolution des grottes. *Internatl. J. Speleol. 1*, 71–100.

Chabert, C. (1977) *Les Grandes Cavites Mondiales. Spelunca*, Suppl. 2, 4th Series, *17*.

Chalcraft, D. and K. Pye (1984) Humid tropical weathering of quartzite in southeastern Venezuela. *Zeits. Geomorph. 28*, 321–332.

Charlton, R. E., Jr. (1966) Cave-to-surface magnetic induction direction finding and communication. *Natl. Speleol. Soc. Bull. 28*, 70–79.

Chico, R. J. (1964) Detection of caves by gravimetry. *Internatl. J. Speleol. 1*, 101–108.

Chinese Academy of Geological Sciences (1976) *Karst in China* (Shanghai Peoples Publishing House).

Chow, V. T. (1959) *Open-Channel Hydraulics* (McGraw-Hill, New York).

Clark, W. C. (1982) *Carbon Dioxide Review: 1982* (Oxford University Press, New York).

Clausen, E. N. (1970) Badland Caves of Wyoming. *Natl. Speleol. Soc. Bull. 32*, 59–69.

Colby, B. R. (1964) Practical computations of bed-material discharge. *J. Hydraulics Div. Proc. Amer. Soc. Civil Engineers 90*, 217–246.

Cole, L. J. (1911) The caverns and people of northern Yucatan. *The Leucocyte 18*, 153–163.

Coleman, J. C. (1949) An indicator of water-flow in caves. *Proc. Univ. Bristol Speleol. Soc. 6*, 57–67.

412 / References

Colveé, P. (1973) Cueva en Cuartcitas en el Cerro Autana, Territorio Federal Amazonas. *Bol. Soc. Venezolana Espel. 4*, 5–13.

Compton, R. G. and P. J. Daly (1984) The dissolution kinetics of Iceland spar single crystals. *J. Colloid Interface Sci. 101*, 159–166.

Conn, H. and J. Conn (1977) *The Jewel Cave Adventure* (Zephyrus Press, Teaneck, NJ).

Corbel, J. (1957) Les Karsts du Nord-Ouest de L'Europe et de Quelques Regions de Comparaisons. Institut des études Rhodaniennes de l'université de Lyon Mémoires et Documents No. 12.

Corbel, J. (1959a) Erosion en terrain calcaire. *Ann. Geograph. 68*, 97–120.

Corbel, J. (1959b) Vitesse de L'erosion. *Zeits. Geomorph. 3*, 1–28.

Courbon, P. (1972) *Atlas des grands gouffres du monde* (Vioud and Coumes, Apt en Provence, France).

Courbon, P. and C. Chabert (1986) *Atlas des Grandes Cavités Mondiales* (Fédération Francaise de Spéléologie, Paris).

Crawford, N. C. (1979) *The Karst Hydrogeology of the Cumberland Plateau Escarpment of Tennessee*. Part 1. Subterranean Stream Invasion, Conduit Cavern Development, and Slope Retreat in the Lost Creek Cove Area, White County, Tennessee. Cave and Karst Studies Series, No. 1 (Western Kentucky University, Bowling Green, KY).

Crawford, N. C. (1984a) Sinkhole flooding associated with urban development upon karst terrain: Bowling Green, Kentucky. In *Sinkholes: Their Geology, Engineering, and Environmental Impact*, B. F. Beck, Ed. (A. A. Balkema, Rotterdam), 283–292.

Crawford, N. C. (1984b) Toxic and explosive fumes arising from carbonate aquifers: a hazard for residents of sinkhole plains. In *Sinkholes: Their Geology, Engineering, and Environmental Impact*, B. F. Beck, Ed. (A. A. Balkema, Rotterdam), 297–304.

Crowther, J. (1984) Soil carbon dioxide and weathering potentials in tropical karst terrain, peninsular Malaysia: A preliminary model. *Earth Surface Processes and Landforms 9*, 397–407.

Curl, R. L. (1958) A statistical theory of cave entrance evolution. *Natl. Speleol. Soc. Bull. 20*, 9–22.

Curl, R. L. (1962) The aragonite-calcite problem. *Natl. Speleol. Soc. Bull. 24*, 57–73.

Curl, R. L. (1964) On the definition of a cave. *Natl. Speleol. Soc. Bull. 26*, 1–6.

Curl, R. L. (1965) Solution kinetics of calcite. *Proc. 4th Internatl. Congress Speleol., Ljubljana, 3*, 61–66.

Curl, R. L. (1966a) Caves as a measure of karst. *J. Geol. 74*, 798–830.

Curl, R. L. (1966b) Scallops and flutes. *Trans. Cave Res. Group Great Britain 7*, 121–160.

Curl, R. L. (1972) Minimum diameter stalactites. *Natl. Speleol. Soc. Bull. 34*, 129–136.

Curl, R. L. (1973) Minimum diameter stalagmites. *Natl. Speleol. Soc. Bull. 35*, 1–9.

Curl, R. L. (1974) Deducing flow velocity in cave conduits from scallops. *Natl. Speleol. Soc. Bull. 36*, 1–5.

Curl, R. L. (1986) Fractal dimensions and geometries of caves. *Mathematical Geol.* 18, 765–783.
Cvijič, J. (1893) Das Karstphanomen. Geographische Abhandlungen herausgegeben von A. Penck, Bd. *V*, H. 3, Vienna.
Cvijič, J. (1960) La Geographie des Terrains Calcaires. *Monogr. Serbian Acad. Sci. Arts* Vol. 341.
Dalton, R. F. (1976) Caves of New Jersey. *New Jersey Geol. Surv. Bull.* 70.
Daugherty, R. L. and A. C. Ingersoll (1954) *Fluid Mechanics* (McGraw-Hill, New York).
Davies, W. E. (1949) Features of cave breakdown. *Natl. Speleol. Soc. Bull.* 11, 34–35.
Davies, W. E. (1951) Mechanics of cavern breakdown. *Natl. Speleol. Soc. Bull.* 13, 36–43.
Davies, W. E. (1952) *The Caves of Maryland.* Maryland Dept. Geol., Mines, and Water Resources. Bull. 7.
Davies, W. E. (1957) Rillenstein in northwest Greenland. *Natl. Speleol. Soc. Bull.* 19, 40–46.
Davies, W. E. (1958) Caverns of West Virginia. *West Virginia Geol. Econ. Surv.* Vol. 19A.
Davies, W. E. (1960) Origin of caves in folded limestone. *Natl. Speleol. Soc. Bull.* 22, 5–18.
Davies, W. E. and E. C. T. Chao (1959) Report on sediments in Mammoth Cave, Kentucky. Administrative report, U.S. Geol. Surv. to Natl. Park Service.
Davies, W. E. and G. W. Moore (1957) Endellite and hydromagnesite from Carlsbad Caverns. *Natl. Speleol. Soc. Bull.* 19, 24–27.
Davis, D. G. (1973) Sulfur in Cottonwood Cave, Eddy County, New Mexico. *Natl. Speleol. Soc. Bull.* 35, 89–95.
Davis, D. G. (1980) Cave development in the Guadalupe Mountains. A critical review of recent hypotheses. *Natl. Speleol. Soc. Bull.* 42, 42–48.
Davis, N. W. (1970) Optimum frequencies for underground radio communication. *Natl. Speleol. Soc. Bull.* 32, 11–26.
Davis, S. N. and R. DeWiest (1966) *Hydrogeology* (John Wiley, New York).
Davis, W. M. (1930) Origin of limestone caverns. *Geol. Soc. Amer. Bull.* 41, 475–628.
Davis, W. M. (1931) The origin of limestone caverns. *Science* 73, 327–331.
Day, M. (1976) The morphology and hydrology of some Jamaican karst depressions. *Earth Surface Processes* 1, 111–129.
Dayton, G. O., W. B. White, and E. L. White (1981) The Caves of Mifflin County, PA. *MAR Bull.* 12.
Deal, D. E. (1962) Geology of Jewel Cave National Monument, Custer County, South Dakota with special reference to cavern formation in the Black Hills. MS Thesis, Geology, University of Wyoming.
DeBellard Pietri, E. (1956) Observaciones Espeleologicas. *Acta Cien. Venezolana* 7, 122–124.
DeBellard Pietri, E. (1967) Espeleogenesis. Clasificacion de las cuevas por su origen. *Bol. Acad. Cien. Fisicas, Matematicas Naturales* (Venezuela) 26, 7–33.

Debenham, N. C. (1983) Reliability of thermoluminescence dating of stalagmitic calcite. *Nature 304*, 154–156.
Deike, G. H. III (1960a) Origin and geologic relations of Breathing Cave, Virginia. *Natl. Speleol. Soc. Bull. 22*, 30–42.
Deike, G. H. III (1960b) X-ray analysis of some Missouri cave clays. *Missouri Speleol. 2*, 9–11.
Deike, G. H. III (1967) The development of caverns in the Mammoth Cave Region. PhD Thesis, Geology, The Pennsylvania State University.
Deike, G. H. III and W. B. White (1969) Sinuosity in limestone solution conduits. *Amer. J. Sci. 267*, 230–241.
Deike, R. G. (1969) Relations of jointing to orientation of solution cavities in limestones of Central Pennsylvania. *Amer. J. Sci. 267*, 1230–1248.
Delecour, F., F. Weissen, and C. Ek (1968) An electrolytic field device for the titration of CO_2 in air. *Natl. Speleol. Soc. Bull. 30*, 131–136.
DesMarais, D. J. (1978) Radon and carbon dioxide in the air of four caves within Sequoia and Kings Canyon Parks, California. Cave Research Foundation Annual Report, S. G. and B. J. Wells, Eds., p. 13.
Dilamarter, R. C. and S. C. Csallany (1977) *Hydrologic Problems in Karst Regions* (Western Kentucky University, Bowling Green, KY).
Doehring, D. O. and J. H. Butler (1974) Hydrogeologic constraints on Yucatan's development. *Science 186*, 591–595.
Donini, G., G. Rossi, P. Forti, A. Buzio, and G. Calandri (1985) Monte Sedom. *Societa Speleologica Italiana*.
Douglas, H. H. (1964) Caves of Virginia (Privately published, Virginia Cave Survey, Falls Church, VA).
Dragovich, D. (1969) The origin of cavernous surfaces (tafoni) in granitic rocks of southern South Australia. *Zeits. Geomorph. 13*, 163–181.
Drake, J. J. (1980) The effect of soil activity on the chemistry of carbonate groundwaters. *Water Resources Res. 16*, 381–386.
Drake, J. J. (1983) The effects of geomorphology and seasonality on the chemistry of carbonate groundwater. *J. Hydrol. 61*, 223–236.
Drake, J. J. and D. C. Ford (1972) The analysis of growth patterns of two-generation populations: The example of karst sinkholes. *Canadian Geograph. 16*, 381–384.
Drake, J. J. and D. C. Ford (1973) The dissolved solids regime and hydrology of two mountain rivers. *Proc. 6th Internatl. Congress Speleol., Olomouc, 4*, 53–56.
Drake, J. J. and R. S. Harmon (1973) Hydrochemical environments of carbonate terrains. *Water Resources Res. 9*, 949–957.
Drake, J. J. and T. M. L. Wigley (1975) The effect of climate on the chemistry of carbonate groundwater. *Water Resources Res. 11*, 958–962.
Dratnal, E. and K. Kasprzak (1980) The response of the invertebrate fauna to organic pollution in a well oxygenated karst stream exemplified by the Pradnik Stream (South Poland). *Acta Hydrobiol. 22*, 263–278.
Drever, J. I. (1982) *The Geochemistry of Natural Waters* (Prentice-Hall, Englewood Cliffs, NJ).
Dreiss, S. (1974) Lithologic controls on solution of carbonate rocks in Christian County, Missouri. *Proc. 4th Conf. Karst Geol. Hydrol.*, H. W.

Rauch and E. Werner, Eds. (West Virginia Geol. Surv., Morgantown, WV), 145–152.

Dreybrodt, W. (1981) Mixing corrosion in $CaCO_3$-CO_2-H_2O systems and its role in the karstification of limestone areas. *Chemical Geol. 32* 221–236.

Droppa, A. (1957) *Demänovske Jaskyne* (Slovak Academy of Science, Bratislava).

Dublyanskii, V. N. (1979) The gypsum caves of the Ukraine. *Cave Geol. 1*, 163–183.

Dunkley, J. R. and T. M. L. Wigley (1967) *Caves of the Nullarbor* (Sydney University Speleological Society, Sydney).

Durov, S. A. (1956) On the question about the origin of the salt composition of karst water. *Ukranian Chem. J. 22*, 106–111; English trans. *Cave Geology 1*, 185–190.

Eagon, H. B., Jr. and D. E. Johe (1972) Practical solutions for pumping tests in carbonate-rock aquifers. *Ground Water 10* (4), 6–13.

Ebaugh, W. F., R. R. Parizek, and R. Greenfield (1976) Channel detection by geothermal methods. In *Karst Hydrology and Water Resources* V. Yevjevich, Ed. (Water Resources Publications, Fort Collins, CO). 648–658.

Ege, J. R. (1984) Formation of solution-subsidence sinkholes above salt beds. *U.S. Geol. Surv. Circ. 897*.

Egemeier, S. J. (1973) Cavern development by thermal waters with a possible bearing on ore deposition. PhD Thesis, Geology, Stanford University.

Egemeier, S. J. (1981) Cavern development by thermal waters. *Natl. Speleol. Soc. Bull. 43*, 31–51.

Einstein, H. A. (1964) River sedimentation. In *Handbook of Applied Hydrology*, Ven Te Chow, Ed. (McGraw-Hill, New York), Sect. 17-II, 35–67.

Ek, C., F. Delecour, and F. Weissen (1968) Teneur en CO_2 de l'air de quelques grottes Belges. Technique employée et premiers résultats. *Ann. Speleol. 23*, 243–257.

Ek, C., S. Gilewska, L. Kaszowski, A. Kobylecki, K. Oleksynowa, and B. Oleksynówna (1969) Some analyses of the CO_2 content of the air in five Polish caves. *Zeits Geomorph. 13*, 267–286.

Eller, P. G. (1981) Chemical aspects of the conversion of cave nitrates to saltpeter. *Natl. Speleol. Soc. Bull. 43*, 106–109.

Elliott, L. P. (1976) Potential gas accumulation in caves in Bowling Green, including relationship to water quality. *Natl. Speleol. Soc. Bull. 38*, 27–36.

Ellwood, B. B. (1971) An archeomagnetic measurement of the age and sedimentation rate of Climax Cave sediments, southwest Georgia. *Amer. J. Sci. 271*, 304–310

Endo, H. K., J. C. S. Long, C. R. Wilson, and P. A. Witherspoon (1984) A model for investigating mechanical transport in fracture networks. *Water Resources Res. 20*, 1390–1400.

EPA (1976) *Quality Criteria for Water* (U.S. Environmental Protection Agency, Government Printing Office).

Ericson, D. B., M. Ewing, and G. Wollin (1963) Pliocene-Pleistocene boundary in deep-sea sediments. *Science, 139,* 727–737.

Ericson, D. B. and G. Wollin (1968) Pleistocene climates and chronology in deep-sea sediments. *Science 162,* 1227–1234.

Even, H., I. Carmi, M. Magaritz, and R. Gerson (1986) Timing the transport of water through the upper vadose zone in a karstic system above a cave in Israel. *Earth Surface Processes and Landforms 11,* 181–191.

Ewers, R. O. (1966) Bedding-plane anastomoses and their relation to cavern passages. *Natl. Speleol. Soc. Bull. 28,* 133–140.

Ewers, R. O. (1972) A model for the development of subsurface drainage routes along bedding planes. MS Thesis, Geology, University of Cincinnati.

Ewers, R. O. (1978) A model for the development of broad scale networks of groundwater flow in steeply dipping carbonate aquifers. *Trans. British Cave Res. Assoc. 5,* 121–125.

Ewers, R. O. (1982) An analysis of solution cavern development in the dimensions of length and breadth. PhD. Thesis, Geography, McMaster University, Hamilton, Ontario.

Ewers, R. O. and J. F. Quinlan (1981) Cavern porosity development in limestone: A low dip model from Mammoth Cave, Kentucky. *Proc. 8th Internatl. Congress Speleol.* (Bowling Green, KY), 727–731.

Fantidis, J. and D. H. Ehhalt (1970) Variations of the carbon and oxygen isotopic composition in stalagmites and stalactites: Evidence of non-equilibrium isotropic fractionation. *Earth Planetary Sci. Lett. 10,* 136–144.

Fellows, L. D. (1965) Cutters and pinnacles in Greene County, Missouri. *Natl. Speleol. Soc. Bull. 27,* 143–150.

Fenneman, N. M. (1938) *Physiography of Eastern United States* (McGraw-Hill, New York).

Ferguson, G. E., C. W. Lingham, S. K. Love, and R. O. Vernon (1947) Springs of Florida. *Florida Geol. Surv., Geol. Bull. 31.*

Finlayson, B. L. and J. A. Webb (1985) Amorphous speleothems. *Cave Sci., Trans. British Cave Res. Assoc. 12,* 3–8.

Fischbeck, R. and G. Müller (1971) Monohydrocalcite, hydromagnesite, nesquehonite, dolomite, aragonite, and calcite in speleothems of the Fränkische Schweiz, West Germany. *Contr. Mineral. Petrol. 33,* 87–92.

Fish, J. E. (1977) Karst hydrogeology and geomorphology of the Sierra de el Abra and the Valles-San Luis Potosi Region, Mexico. PhD Thesis, Geology, McMaster University, Hamilton, Ontario.

Flint, R. F. (1963) Pleistocene climates in low latitudes. *Geograph. Rev. 53,* 123–129.

Folk, R. L., H. H. Roberts, and C. H. Moore (1973) Black phytokarst from Hell, Cayman Islands, British West Indies. *Geol. Soc. Amer. Bull. 84,* 2351–2360.

Foose, R. M. (1953) Ground water behavior in the Hershey Valley, Pennsylvania. *Geol. Soc. Amer. Bull. 64,* 623–645.

Foose, R. M. (1967) Sinkhole formation by groundwater withdrawal: Far West Rand, South Africa, *Science 157,* 1045–1048.

Foose, R. M. and J. A. Humphreville (1979) Engineering geological approaches to foundations in karst terrain of the Hershey Valley. *Bull. Assoc. Eng. Geol. 16*, 355–381.

Ford, D. C. (1965a) Stream potholes as indicators of erosion phases in limestone caves. *Natl. Speleol. Soc. Bull. 27*, 27–32.

Ford, D. C. (1965b) The origin of limestone caverns: A model from the Central Mendip Hills, England. *Natl. Speleol. Soc. Bull. 27*, 109–132.

Ford, D. C. (1968) Features of cavern development in Central Mendip. *Trans. Cave Res. Group Great Britain 10*, 11–25.

Ford, D. C. (1971) Geologic structure and a new explanation of limestone cavern genesis. *Trans. Cave Res. Group Great Britain 13*, 81–94.

Ford, D. C. (1977) Genetic classification of solutional cave systems. *Proc. 7th Internatl. Congress Speleol.*, Sheffield, 189–192.

Ford, D. C. (1979) A review of alpine karst in the southern Rocky Mountains of Canada. *Natl. Speleol. Soc. Bull. 41*, 43–65.

Ford, D. C. and R. O. Ewers (1978) The development of limestone cave systems in the dimensions of length and depth. *Canad. J. Earth Sci. 15*, 1783–1798.

Ford, D. C., P. L. Smart, and R. O. Ewers (1983) The physiography and speleogenesis of Castleguard Cave, Columbia icefields, Alberta, Canada. *Arctic Alpine Res. 15*, 437–450.

Ford, T. D. (1984) Paleokarsts in Britain. *Trans. British Cave Res. Assoc. 11*, 246–264.

Ford, T. D. and C. H. D. Cullingford (1976) *The Science of Speleology* (Academic Press, London).

Frank, R. M. (1965) Petrologic study of sediments from selected central Texas caves. MS Thesis, Geology, University of Texas, Austin.

Frank, R. M. (1969) The clastic sediments of Douglas Cave, Stuart Town, New South Wales. *Helictite 7*, 3–13.

Frank, R. M. (1975) Late Quaternary climatic change: Evidence from cave sediments in central eastern New South Wales. *Australian Geograph. Stud. 13*, 154–168.

Franke, H. W. (1965) The theory behind stalagmite shapes. *Studies in Speleol. 1*, 89–95.

Franklin, A. G., D. M. Patrick, D. K. Butler, W. E. Strohm, Jr., and M. E. Hynes-Griffin (1980) Siting of nuclear facilities in karst terrains and other areas susceptible to ground collapse. Report of U.S. Army Corps of Engineers Waterways Expt. Station. to U.S. Nuclear Regulatory Commission.

Franz, R. and D. Slifer (1971) Caves of Maryland. *Maryland Geol. Surv. Educational Ser. No. 3*.

Freeman, J. P., G. L. Smith, T. L. Poulson, P. J. Watson, and W. B. White (1973) Lee Cave, Mammoth Cave National Park, Kentucky. *Natl. Speleol. Soc. Bull. 35*, 109–125.

Freeze, R. A. and J. A. Cherry (1979) Groundwater (Prentice-Hall, Englewood Cliffs, NJ).

French, H. M. (1974) Active thermokarst processes, Eastern Banks Island, Western Canadian Arctic. *Canad. J. Earth Sci. 11*, 785–794.

Fretwell, J. D. and M. T. Stewart (1981) Resistivity study of a coastal karst terrain, Florida. *Ground Water 19*, 156–162.

Frink, J. W. (1945) Solution of limestone beneath Hales Bar Dam. *J. Geol.* 53, 137–139.

Frondel, C. (1962) *The System of Mineralogy*, Vol. III, Silica Minerals (John Wiley, New York).

Gale, S. J. (1984) The hydraulics of conduit flow in carbonate aquifers. *J. Hydrol.* 70, 309–327.

Gams, I. (1969) Some morphological characteristics of the Dinaric Karst. *Geograph. J.* 135, 563–574.

Gams, I. (1977) Towards the terminology of the polje. *Proc. 7th Internatl. Congress Speleol.*, Sheffield, 201–202.

Gams, I. (1978) The polje: The problem of definition. *Zeits. Geomorph.* 22, 170–181.

Gardner, J. H. (1935) Origin and development of limestone caverns. *Geol. Soc. Amer. Bull.* 46, 1255–1274.

Garrels, R. M. and C. L. Christ (1965) *Solutions, Minerals, and Equilibria* (Harper & Row, New York).

Garza S. (1962) Discharge and changes in ground-water storage in the Edwards and associated limestones, San Antonio area, Texas. Texas Board of Water Engineers Bull. 6201.

Gascoyne, M. (1977) Trace element geochemistry of speleothems. *Proc. 7th Internatl. Congress Speleol.*, Sheffield, 205–208.

Gascoyne, M. (1984) Twenty years of uranium-series dating of cave calcites. *Studies in Speleol.* 5, 15–30.

Gascoyne, M. (1985) Application of the $^{227}Th/^{230}Th$ method to dating Pleistocene carbonates and comparison with other dating methods. *Geochim. Cosmochim. Acta* 49, 1165–1171.

Gascoyne, M., G. J. Benjamin, H. P. Schwarcz, D. C. Ford (1979) Sea-level lowering during the Illinoian glaciation: Evidence from a Bahama "blue hole." *Science* 205, 806–808.

Gascoyne, M., H. P. Schwarcz, and D. C. Ford (1980) A palaeotemperature record for the mid-Wisconsin in Vancouver Island. *Nature* 285 474–476.

Gascoyne, M., H. P. Schwarcz and D. C. Ford (1981) Late Pleistocene chronology and paleoclimate of Vancouver Island determined from cave deposits. *Canad. J. Earth Sci.* 18, 1643–1652.

Gascoyne, M., D. C. Ford and H. P. Schwarcz (1983a) Rates of cave and landform development in the Yorkshire Dales from speleothem age data. *Earth Surface Processes and Landforms*, 8, 557–568.

Gascoyne, M., D. C. Ford and H. P. Schwarcz (1983b) Uranium-series ages of speleothem from northwest England: Correlation with Quaternary climate. *Phil. Trans. Royal Soc. London B-301*, 143–164.

George, A. I. (1974) Preliminary index of gypsum speleothems in the caves of Kentucky, Indiana, and Tennessee. *Proc. 4th Conf. on Karst Geol. Hydrol.* H. W. Rauch and E. Werner, Eds., West Virginia Geol. Surv., 169–177.

Gerstenhauer, A. (1960) Der tropische Kegelkarst in Tabasco (Mexico). *Zeits. Geomorph. Suppl.* 2, 22–48.

Géze, B. (1953) La genese des gouffres. *Proc. 1st Internatl. Congress Speleol.*, Paris, 2, 1–13.

Géze, B. (1957) *Les cristallisations excentriques de la Grotte de Moulis* (Ateliers d'Impressions d'Art Jean Brunissen, Paris).

Géze, B. (1965) Les conditions hydrogéologiques des roches calcaires. *Chronique d'Hydrogeol.* 7, 9–39.

Giddings, M. T. Jr. (1974) Hydrologic budget of Spring Creek drainage basin, Pennsylvania. PhD Thesis, Geology, The Pennsylvania State University.

Giggenbach, W. F. (1976) Geothermal Ice Caves on Mt. Erebus, Ross Island, Antarctica. *New Zealand J. Geol. Geophys.* 19, 365–372.

Gillieson, D. (1986) Cave sedimentation in the New Guinea Highlands. *Earth Surface Processes and Landforms, 11*, 533–543.

Giusti, E. V. (1977) Hydrogeology and "geoesthetics" applied to land use planning in the Puerto Rican karst. *Mem. 12, Internatl. Assoc. Hydrogeol.* 149–167.

Glazek, J. (1966) On the karst phenomena in North Vietnam. *Bull. Acad. Sci. Poland 14*, 45–51.

Glazek, J. and M. Markowicz-Lohinowicz (1973) Remarks to the use of quantitative methods to karst denudation velocity. *Proc. 6th Internatl. Congress Speleol., Olomouc, 3*, 225–230.

Glew, J. R. (1977) Simulation of rillenkarren. *Proc. 7th Internatl. Congress Speleol., Sheffield*, 218–219.

Glover, R. R. (1972) Optical brighteners—new water tracing reagent. *Trans. Cave Res. Group Great Britain 14*, 84–88.

Goede, A. and J. L. Bada (1985) Electron spin resonance dating of Quaternary bone material from Tasmanian caves—a comparison with ages determined by aspartic acid racemization and C^{14}. *Australian J. Earth Sci. 32*, 155–162.

Goede, A., D. C. Green, and R. S. Harmon (1986) Late Pleistocene palaeotemperature record from a Tasmanian speleothem. *Australian J. of Earth Sciences 33*, 333–342.

Gold, D. P., R. R. Parizek, and T. Giddings (1970) Water well explosions: An environmental hazard. *Earth and Mineral Sciences* (The Pennsylvania State University) *40*, 17–21.

Goldberg, P. S. and Y. Nathan (1975) The phosphate mineralogy of et-Tabun Cave, Mount Carmel, Israel. *Mineral. Mag. 40*, 253–258.

Goldsmith, J. R., D. L. Graf, and O. I. Joensuu (1955) The occurrence of magnesian calcites in nature. *Geochim. Cosmochim. Acta 7*, 212–230.

Goodchild, M. F. (1969) Stereographic cave mapping. *Natl. Speleol. Soc. Bull. 31*, 19–22.

Goodchild, M. F. and D. C. Ford (1971) Analysis of scallop patterns by simulation under controlled conditions. *J. Geol. 79*, 52–62.

Gospodarič, R. and P. Habič (1976) *Underground Water Tracing* (Institute for Karst Research, Postojna, Yugoslavia).

Gradzinski, R. and A. Radomski (1965) Origin and development of internal poljes "Hoyos" in the Sierra de los Organos. *Bull. Acad. Sci. Poland 13*, 181–186.

Graf, W. H. (1971) *Hydraulics of Sediment Transport* (McGraw-Hill, New York).

Grant, L. F. and L. A. Schmidt, Jr. (1958) Grouting deep solution channels under an earth fill dam. *J. Soil Mech. Foundations Div.*, *ASCE* Paper 1813.

Gray, D. M. (1970) *Handbook on the Principles of Hydrology* (Canadian Committee for the International Hydrological Decade, National Research Council of Canada, Ottawa, Ontario).

Greeley, R. (1971a) Geology of selected lava tubes in the Bend Area, Oregon. *Oregon Dept. Geol. Mineral Resources Bull. 71.*

Greeley, R. (1971b) Observations of actively forming lava tubes and associates structures, Hawaii. *Modern Geol. 2*, 207–223.

Greeley, R. (1971c) Lava tubes and channels in the lunar Marius Hills. *Moon 3*, 289–314.

Greeley, R. (1971d) Lunar Hadley Rille: Considerations of its origin. *Science 172*, 722–725.

Greeley, R. and J. H. Hyde (1972) Lava tubes of the cave basalt, Mount St. Helens, Washington. *Geol. Soc. Amer. Bull. 83*, 2397–2418.

Greene, F. C. (1908) Caves and cave formation of Mitchell limestone. *Proc. Indiana Acad. Sci. 18*, 175–184.

Greenfield, R. J. (1979) Review of geophysical approaches to the detection of karst. *Bull. Assoc. Eng. Geol. 16*, 393–408.

Grün, R. (1985) ESR dating speleothems: Limits of the method. In *ESR Dating and Dosimetry*, M. Ikeya and T. Miki, Eds. (Ionics Publishing Co., Tokyo), 61–72.

Grün, R. (1986) ESR-dating of a flowstone core from Cova de sa Bassa Blanca (Mammorca, Spain). *Endins*, 12, 19–23.

Grund, A. (1903) Die Karsthydrographie. Studien aus Westbosnien. *Geograph. Abhandl.* (Penck) *9*, 1–200.

Gulden, R. (1982) The 100 longest caves in the United States. *Natl. Speleol. Soc. News 40*, 261–264.

Gumbel, E. J. (1942) Statistical control curves for flood discharges. *Trans. Amer. Geophys. Union*, Pt. II, 489–509.

Gunn, J. (1981) Limestone solution rates and processes in the Waitomo District, New Zealand. *Earth Surface Processes and Landforms 6*, 427–455.

Gurnee, R. H. (1967) The Rio Camuy Cave Project, Puerto Rico. *Natl. Speleol. Soc. Bull. 29*, 27–34.

Gurnee, R. H. and J. Gurnee (1980) *Gurnee Guide to American Caves* (Zephyrus Press, Teaneck, NJ).

Gustavson, T. C., W. W. Simpkins, A. Alhades, and A. Hoadley (1982) Evaporite dissolution and development of karst features on the rolling plains of the Texas Panhandle. *Earth Surface Processes and Landforms 7*, 545–563.

Gvozdetskii, N. A. (1967) Occurrence of karst phenomena on the globe and problems of their typology. *Earth Res. 7*, 98–127.

Haas, J. L., Jr. (1959) Evaluation of ground water tracing methods used in speleology. *Natl. Speleol. Soc. Bull. 21*, 67–76.

Hack, J. T. (1957) Studies of longitudinal stream profiles in Virginia and Maryland. *U.S. Geol. Surv. Prof. Paper 294-B*, 45–97.

Hack, J. T. (1960a) Relation of solutional features to chemical character of

water in the Shenandoah Valley, Virginia. *U.S. Geol. Surv. Prof. Paper 400-B*, 387–390.
Hack, J. T. (1960b) Interpretation of erosional topography in humid temperate regions. *Amer. J. Science, 258-A*, 80–97.
Hack, J. T. (1973) Stream-profile analysis and stream-gradient index. *J. Res. U.S. Geol. Surv. 1*, 421–429.
Halliday, W. R. (1954) Ice caves of the United States. *Natl. Speleol. Soc. Bull. 16*, 3–28.
Halliday, W. R. (1955) A proposed classification of physical features found in caves. *Natl. Speleol. Soc. Bull. 17*, 32–33.
Halliday, W. R. (1960) Pseudokarst in the United States. *Natl. Speleol. Soc. Bull. 22*, 109–113.
Halliday, W. R. (1962) *Caves of California.* Western Speleol. Surv.
Halliday, W. R. (1963) *Caves of Washington.* Washington Dept. Conservation Inform. Circ. 40.
Halliday, W. R. (1966) Terrestrial pseudokarst and the lunar topography. *Natl. Speleol. Soc. Bull. 28*, 167–170.
Halliday, W. R. (1976) *Proc. Internatl. Symp. on Vulcanospeleology and Its Extraterrestrial Applications* (Western Speleol. Surv. Special Pub., Seattle).
Haq, B. U., J. Hardenbol, and P. R. Vail (1987) Chronology of fluctuating sea levels since the Triassic. *Science 235*, 1156–1167.
Hardie, L. A. (1967) The gypsum-anhydrite equilibrium at one atmosphere pressure. *Amer. Mineral. 52*, 171–200.
Harmon, R. S. (1979) An isotopic study of groundwater seepage in the central Kentucky karst. *Water Resources Res. 15*, 476–480.
Harmon, R. S., J. J. Drake, J. W. Hess, R. L. Jacobson, D. C. Ford, W. B. White, J. Fish, J. Coward, R. Ewers, and J. Quinlan (1973) Geochemistry of karst waters in North America. *Proc. 6th Internatl. Congress Speleol., Olomouc, 3*, 103–114.
Harmon, R. S., D. C. Ford, and H. P. Schwarcz (1977) Interglacial chronology of the Rocky and Mackenzie Mountains based upon ^{230}Th-^{234}U dating of calcite speleothems. *Canad. J. Earth Sci. 14*, 2543–2552.
Harmon, R. S., P. Thompson, H. P. Schwarcz, and D. C. Ford (1975a) Uranium series dating of speleothems. *Natl. Speleol. Soc. Bull. 37*, 21–33.
Harmon, R. S., P. Thompson, H. P. Schwarcz, and D. C. Ford (1978c) Late Pleistocene paleoclimates of North America as inferred from stable isotope studies of speleothems. *Quaternary Res. 9*, 54–70.
Harmon, R. S., W. B. White, J. J. Drake, and J. W. Hess (1975b) Regional hydrochemistry of North American carbonate terrains. *Water Resources Res. 11*, 963–967.
Harmon, R. S. and J. W. Hess (1982) Ground water geochemistry of the Burnsville Cove Area, Virginia. *Natl. Speleol. Soc. Bull. 44*, 84–89.
Harmon, R. S., H. P. Schwarcz, and D. C. Ford (1978a) Stable isotope geochemistry of speleothems and cave waters from the Flint Ridge–Mammoth Cave System, Kentucky: Implications for terrestrial climate change during the period 230,000 to 100,000 years B.P. *J. Geol. 86*, 373–384.

Harmon, R. S., H. P. Schwarcz, and D. C. Ford (1978b) Late Pleistocene sea level history of Bermuda. *Quaternary Res. 9*, 205-218.

Harmon, R. S., H. P. Schwarcz, D. C. Ford, and D. L. Koch (1979) An isotopic paleotemperature record for late Wisconsinan time in northeast Iowa. *Geology 7* 430-433.

Harned, H. S. and B. B. Owen (1958) *The Physical Chemistry of Electrolytic Solution* (Reinhold, New York).

Harpaz, Y. and J. Schwarz (1967) Operating a limestone aquifer as a reservoir for a water supply system. *Bull. Internatl. Assoc. Sci. Hydrol. 12*, 78-90.

Harvey, E. J., J. H. Williams, and T. R. Dinkel (1977) Application of thermal imagery and aerial photography to hydrologic studies of karst terrane in Missouri. *U.S. Geol. Surv. Water Resources Invest.* 77-16.

Hedges, J. (1969) Opferkessel. *Zeits. Geomorph. 13*, 22-55.

Hedges, J. (1979) The 1976 NSS Standard Map Symbols. *Natl. Speleol. Soc. Bull. 41*, 35-48.

Heine K. and M. A. Geyh (1983) Radiocarbon dating of speleothems from the Rössing Cave, Namib desert, and palaeoclimatic implications. *SASQUA Internatl. Symp.* 465-470.

Helwig, J. (1964) Stratigraphy of detrital fills of Carroll Cave, Camden County, Missouri. *Missouri Speleol. 6*, 1-15.

Hempel, J. C. (1975) Caves and karst of Monroe County, W.Va. *West Virginia Speleol. Surv. Bull. 4*.

Henderson, E. P. (1949) Some unusual formations in Skyline Caverns, Va. *Natl. Speleol. Soc. Bull. 11*, 31-34.

Hendy, C. H. (1971) The isotopic geochemistry of speleothems I. The calculation of the effects of different modes of formation on the isotope composition of speleothems and their applicability as palaeoclimatic indicators. *Geochim. Cosmochim. Acta 35* 801-824.

Hennig, G. J. and R. Grün (1983) ESR dating in Quaternary geology. *Quaternary Sci. Rev. 2*, 157-238.

Hensler, E. (1968) Some examples from the south Harz Region of cavern formation in gypsum. *Trans. Cave Res. Group Great Britain 10*, 33-44.

Herek, M. and V. T. Stringfield (1972) *Karst: Important Karst Regions of the Northern Hemisphere* (Elsevier, Amsterdam).

Herman, J. S. (1982) The dissolution kinetics of calcite, dolomite, and dolomitic rocks in the CO_2—water system. PhD Thesis, Geochemistry, The Pennsylvania State University.

Herman, J. S. and W. B. White (1985) Dissolution kinetics of dolomite: Effects of lithology and fluid flow velocity. *Geochim. Cosmochim. Acta 49*, 2017-2026.

Hess, J. W., Jr. (1974) Hydrochemical investigations of the Central Kentucky karst aquifer system. PhD Thesis, Geology, The Pennsylvania State University.

Hess, J. W., Jr. and W. B. White (1974) Hydrograph analysis of carbonate aquifers. Institute of Land and Water Resources, The Pennsylvania State Univ. Res. Pub. No. 83.

Hess, J. W., Jr. and W. B. White (1976) Analysis of karst aquifers from hydrographs of karst springs. *Proc. 6th Internatl. Congress Speleol. Olomouc, 4*, 115-120.

Hickman, H. L., Jr. (1969) Characteristics of municipal solid wastes. *Scrap Age 26*, 305–307.
High, C. and F. K. Hanna (1970) A method for the direct measurement of erosion on rock surfaces. *British Geomorph. Res. Group Tech. Bull. 5.*
Hill, C. A. (1976) *Cave Minerals* (National Speleological Society, Huntsville, AL).
Hill, C. A. (1981a) Origin of cave saltpeter. *J. Geol. 89*, 252–259.
Hill, C. A. (1981b) Origin of cave saltpeter. *Natl. Speleol. Soc. Bull. 43*, 110–126.
Hill, C. A. (1981c) Mineralogy of cave nitrates. *Natl. Speleol. Soc. Bull. 43*, 127–132.
Hill, C. A. (1981d) Speleogenesis of Carlsbad Caverns and other caves of the Guadalupe Mountains. *Proc. 8th Internatl. Congress Speleol.* (Bowling Green, KY), 143–144.
Hill, C. A., D. DePaepe, P. G. Eller, P. M. Hauer, J. Powers, and M. O. Smith (1981) Saltpeter caves of the United States. *Natl. Speleol. Soc. Bull. 43*, 84–87.
Hill, C. A. and R. C. Ewing (1977) Darapskite, $Na_3(NO_3)(SO_4) \cdot H_2O$, a new occurrence, in Texas. *Mineral. Mag. 41*, 548–550.
Hill, C. A. and P. Forti (1986) *Cave Minerals of the World* (National Speleological Society, Huntsville, AL).
Hill, C., W. Sutherland, and L. Tierney (1976) Caves of Wyoming. *Geol. Surv. Wyoming Bull. 59.*
Hogberg, R. K. and T. N. Bayer (1967) Guide to the Caves of Minnesota. *Minnesota Geol. Surv. Educat. Ser. No. 4.*
Holland, H. D., T. V. Kirsipu, J. S. Huebner, and U. M. Oxburgh (1964) On some aspects of the chemical evolution of cave waters. *J. Geol. 72*, 36–67.
Holsinger, J. R. (1966) A preliminary study of the effects of organic pollution of Banners Corner Cave, Virginia. *Internat. J. Speleol. 2*, 75–89.
Holsinger, J. R. (1975) Descriptions of Virginia caves. *Virginia Div. Mineral Resources Bull. 85.*
Horn, G. (1947) Karsthuler i Nordland. *Norges Geol. Undersokelse No. 165.*
Horton, R. E. (1945) Erosional development of streams and their drainage basins: Hydrophysical approach to quantitative morphology. *Geol. Soc. Amer. Bull. 56*, 275–370.
Hosley, R. J. (1971) *Cave Surveying and Mapping* (Crown Press. Indianapolis, IN).
Howard, A. D. (1963) The development of karst features. *Natl. Speleol. Soc. Bull. 25*, 45–65.
Howard, A. D. (1964) A model for cavern development under artesian ground water flow, with special reference to the Black Hills. *Natl. Speleol. Soc. Bull. 26*, 7–16.
Howard, A. D. and B. Y. Howard (1967) Solution of limestone under laminar flow between parallel boundaries. *Caves and Karst 9*, 25–38.
Howard, A. D. (1971) Quantitative measures of cave patterns. *Caves and Karst 13*, 1–7.
Howard, A. D., M. E. Keetch, and C. L. Vincent (1970) Topological and geometrical properties of braided streams. *Water Resources Res. 6*, 1674–1688.

Hubbard, D. A., Jr. (1984) Sinkhole distribution in the central and northern Valley and Ridge Province, Virginia. In *Sinkholes: Their Geology, Engineering, and Environmental Impact*, B. F. Beck, Ed. (A. A. Balkema, Rotterdam), 75–78.

Hubbert, M. K. (1940) The theory of ground-water motion. *J. Geol. 48*, 785–944.

IASH (1967) *Hydrology of Fractured Rock* (Unesco, Belgium).

Ikeya, M. (1975) Dating a stalactite by electron paramagnetic resonance. *Nature 255*, 48–50.

Inglis, C. C. (1949) The behavior and control of rivers and canals (with the aid of models). Poona, India, Central Waterpower Irrig. Navigation, *Res. Station Res. Pub. 13*, Pt. 1.

Irmay, S. (1958) On the theoretical derivation of Darcy and Forchheimer formulas. *Trans. Amer. Geophy. Union 39*, 702–707.

Jacobson, R. L. (1973) Controls on the quality of some carbonate ground waters: Dissolution constants of calcite and $CaHCO_3^+$ from 0 to 50°C. PhD Thesis, Geochemistry, The Pennsylvania State University.

Jacobson, R. L. and D. Langmuir (1970) The chemical history of some spring waters in carbonate rocks. *Ground Water 8*, 5–9.

Jacobson, R. L. and D. Langmuir (1972) An accurate method for calculating saturation levels of ground waters with respect to calcite and dolomite. *Trans. Cave. Res. Group Great Britain. 14*, 104–108.

Jacobson, R. L. and D. Langmuir (1974a) Dissociation constants of calcite and $CaHCO_3^+$ from 0 to 50°C. *Geochim. Cosmochim. Acta 38*, 301–318.

Jacobson, R. L. and D. Langmuir (1974b) Controls on the quality variations of some carbonate spring waters. *J. Hydrol. 23*, 247–265.

Jakucs, L. (1977) *Morphogenetics of Karst Regions* (John Wiley, New York).

James, A. N. and A. R. R. Lupton (1978) Gypsum and anhydrite in foundations of hydraulic structures. *Geotechnique 28*, 249–272.

James, J. M. (1977) Carbon dioxide in the cave atmosphere. *Trans. British Cave Res. Assoc. 4*, 417–429.

James, J. M., A. J. Pavey, and A. F. Rogers (1975) Foul air and the resulting hazards to cavers. *Trans. British Cave Res. Assoc. 2*, 79–88.

Jammal, S. E. (1984) Maturation of the Winter Park sinkhole. In *Sinkholes: Their Geology, Engineering, and Environmental Impact*, B. F. Beck, Ed. (A. A. Balkema, Rotterdam), 363–369.

Jenne, E. A. (1979) *Chemical Modeling in Aqueous Systems*. Amer. Chem. Soc. Symp. Series 93.

Jennings, J. E. (1966) Building on dolomites in the Transvaal. *The Civil Engineer in South Africa 8*, 41–62.

Jennings, J. E., A. B. A. Brink, A. Louw, and G. D. Gowan (1965) Sinkholes and subsidences in the Transvaal dolomite of South Africa. *Proc. 6th Internatl. Conf. Soil Mech.* 51–54.

Jennings, J. N. and M. J. Bik (1962) Karst morphology in Australian New Guinea. *Nature 194*, 1036–1038.

Jennings, J. N. (1966) The Big Hole near Braidwood, New South Wales, *J. Proc. Royal Soc. New South Wales. 98*, 215–219.

Jennings, J. N. (1972) The character of tropical humid karst. *Zeits. Geomorph. 16*, 336–341.

Jennings, J. N. (1983) Karst landforms. *Amer. Scientist 71,* 578–586.
Jennings, J. N. (1985) *Karst Geomorphology* (Basil Blackwell, Oxford).
Jennings, J. N. and M. M. Sweeting (1963) *The Limestone Ranges of the Fitzroy Basin, Western Australia.* Bonner Geograph. Abhand. No. 32, (Bonn).
Johnson, K. S. (1981) Dissolution of salt on the east flank of the Permian Basin in the southwestern U.S.A. *J. Hydrol. 54,* 75–93.
Johnston, W. D., Jr. (1933) Ground Water in the Paleozoic Rocks of Northern Alabama. *Geol. Surv. Alabama Spec. Report. 16.*
Joiner, T. J. and W. L. Scarbrough (1969) Hydrology of limestone terraines: Geophysical investigations. *Alabama Geol. Surv. Bull. 94-D.*
Jones, W. K. (1971) Characteristics of the underground floodplain. *Natl. Speleol. Soc. Bull. 33,* 105–114.
Jones, W. K. (1973) Hydrology of limestone karst in Greenbrier County, West Virginia. *West Virginia Geol. Surv. Bull. 36.*
Kastning, E. H. (1975) Cavern development in the Helderberg Plateau, East-Central New York. MS Thesis, Geology, University of Connecticut.
Kastrinos, J. R. and W. B. White (1986) Seasonal, hydrogeologic, and land-use controls on nitrate contamination of carbonate ground waters. *Proc. Environmental Problems in Karst Terranes and Their Solutions Conference* (Bowling Green, KY), 88–114.
Katzer, F. (1909) *Karst und Karsthydrographie. Zur Kunde der Balkanhalbinsel.* (Kajon, Sarajevo).
Kaye, C. A. (1957) The effect of solvent motion on limestone solution. *J. Geol. 65,* 35–46.
Kaye, C. A. (1959) Geology of Isla Mona, Puerto Rico, and notes on the age of Mona Passage. *U.S. Geol. Surv. Prof. Paper 317-C,* 141–178.
Keeling, C. D., R. B. Bacastow, and T. P. Whorf (1982) *Carbon Dioxide Review: 1982.* W. C. Clark, Ed. (Oxford University Press, New York), 377–385.
Kemmerly, P. R. (1982) Spatial analysis of a karst depression population: Clues to genesis. *Geol. Soc. Amer. Bull. 93,* 1078–1086.
Kemmerly, P. R. and S. K. Towe (1978) Karst depressions in a time context. *Earth Surface Processes 3,* 355–361.
Kempe, S. and C. Spaeth (1977) Eccentrics: Their capillaries and growth rates. *Proc. 7th Internatl. Congress Speleol., Sheffield,* 259–262.
Kennedy, J. M. (1968) A microwave radiometric study of buried karst topography. *Geol. Soc. Amer. Bull. 79,* 735–742.
Kern, D. M. (1960) The hydration of carbon dioxide. *J. Chem. Educat. 37,* 14–23.
Khan, D. H. (1960) Clay mineral distribution in some Rendzinas, red brown soils, and terra rossas on limestones of different geological ages. *Soil Science 90,* 312–319.
Kirk, K. G. (1974) Resistivity and gravity surveys applied to karst research. *Proc. 4th Conf. on Karst Geol. and Hydrol.* H. W. Rauch and E. Werner, Eds., West Virginia Geological Survey, 61–71.
Kirk, K. G. and E. R. Snyder (1977) A preliminary investigation of seismic techniques used to locate cavities in karst terrains. In *Hydrologic Problems in Karst Regions.* R. R. Dilamarter and S. C. Csallany, Eds. (Western Kentucky University, Bowling Green, KY).

Kirk, K. G. and E. Werner (1981) *Handbook of Geophysical Cavity-Locating Techniques.* U.S. Dept. Transportation Implementation Package FHWA-IP-81-3.
Kiver, E. P. and W. K. Steele (1975) Firn Caves in the Volcanic Craters of Mount Rainier, Washington. *Natl. Speleol. Soc. Bull. 37*, 45–55.
Klaer, W. (1957) Verwitterungsformen in Granit auf Korsika, *Erdkunde 11*, 150–156.
Knight, E. L., B. N. Irby, and S. Carey (1974) *Caves of Mississippi. Southern Mississippi Grotto* (National Speleological Society, Hattiesburg, MS).
Knight, F. J. (1971) Geologic problems of urban growth in limestone terrains of Pennsylvania. *Bull. Assoc. Eng. Geol. 8*, 91–101.
Knox, R. G. (1959) Land of the burnt out fires: Lava Beds National Monument, California. *Natl. Speleol. Soc. Bull. 21*, 55–66.
Kraft, R. and D. Yaakobi (1966) Some remarks on non-Darcy flow. *J. Hydrol. 4*, 171–181.
Krauskopf, K. B. (1956) Dissolution and precipitation of silica at low temperatures. *Geochim. Cosmochim. Acta 10*, 1–26.
Kritsky, S. N. and M. F. Menkel (1967) On principles of estimation methods of maximum discharge. In *Floods and Their Computation* (International Association of Scientific Hydrology) 29–41.
Kukla, J. and V. Ložek (1958) K problematice výzkumu jeskynních výplní (Problems of investigation of cave deposits). *Československy kras 11*, 19–83.
Kundert, C. J. (1952) The origin of palettes, Lehman Caves National Monument, Baker, Nevada. *Natl. Speleol. Soc. Bull. 14*, 30–33.
Kyrle, G. (1923) *Theoretische Späeologie* (Österreichischen Staatsdruckerei, Vienna).
LaMoreaux, P. E. and W. M. Warren (1973) Sinkhole. *Geotimes* (March), 18, 15.
Lange, A. L. (1954) Phreatic floor slot in Model Cave, Nevada. *Science 120*, 1099–1100.
Langmuir, D. (1971) The geochemistry of some carbonate ground waters in Central Pennsylvania. *Geochim. Cosmochim. Acta 35*, 1023–1045.
Larson, C., W. R. Halliday, and J. Nieland (1972) Selected caves of the Pacific Northwest. *Guidebook for 1972 Natl. Speleol. Soc. Convention.*
Latham, A. G., H. P. Schwarcz, D. C. Ford, and G. W. Pearce (1979) Paleomagnetism of stalagmite deposits. *Nature 280*, 383–385.
Latham, A. G., H. P. Schwarcz, and D. C. Ford (1986) The paleomagnetism and U-Th dating of Mexican stalagmite, DAS2. *Earth Planet. Sci. Lett. 79*, 195–207.
Latham, A. G., H. P. Schwarcz, D. C. Ford, and G. W. Pearce (1982) The paleomagnetism and U-Th dating of three Canadian speleothems: Evidence for the westward drift, 5.4–2.1 ka BP. *Canad. J. Earth Sci. 19*, 1985–1995.
Latham, E. E. (1969) *Soil Survey of Barren County, Kentucky* (U.S. Dept. of Agriculture).
Lattman, L. H. and R. R. Parizek (1964) Relationship between fracture traces and the occurrence of ground water in carbonate rocks. *J. Hydrol. 2*, 73–91.

Lauritzen, S.-E. and M. Gascoyne (1980) The first radiometric dating of Norwegian stalagmites—evidence of pre-Weichselian karst caves. *Norsk, Geogr. Tidsskr, 34* 77–82.
LaValle, P. (1967) Some aspects of linear karst depression development in south central Kentucky. *Ann. Assoc. Amer. Geograph. 57,* 49–71.
LaValle, P. (1968) Karst depression morphology in south central Kentucky. *Geografiska Ann. 50,* 94–108.
Lee, C.-H., and R. T.-S. Cheng (1974) On seawater encroachment in coastal aquifers. *Water Resources Res. 10,* 1039–1043.
LeGrand, H. E. (1952) Solution depressions in diorite in North Carolina. *Amer. J. Sci. 250,* 566–585.
LeGrand, H. E. and V. T. Stringfield (1971) Development and distribution of permeability in carbonate aquifers. *Water Resources Res. 7,* 1284–1294.
LeGrand, H. E. and V. T. Stringfield (1973) Karst hydrology—a review. *J. Hydrol. 20,* 97–120.
Lehmann, H. (1936) Morphologische Studien auf Java. *Geogr. Abhandl. A. Penck Series 3, 9.*
Lehmann, H. (1954a) Der tropische Kegelkarst auf den grossen Antillen. *Erdkunde 8,* 130–139.
Lehmann, H. (1954b) Das Karstphänomen in den verschiedenen Klimazonen. *Erdkunde 8,* 112–122.
Lehmann, H. (1960) International contributions to karst phenomena. *Zeits. Geomorph.* Suppl. 2.
Lehmann, O. (1932) Die Hydrographie Des Karstes. Enzyklopädie der Erdkunde (Franz Deuticke, Leipzig).
Leopold, L. B. and M. O'Brien-Marchand (1968) On the quantitative inventory of the riverscape. *Water Resources Res. 4,* 709–717.
Leopold, L. B. and M. G. Wolman (1960) River meanders. *Geol. Soc. Amer. Bull. 71,* 769–794.
Leopold, L. B., M. G. Wolman, and J. P. Miller (1964) *Fluvial Processes in Geomorphology* (W. B. Freeman, San Francisco).
Lesser-Jones, H. (1965) Confined fresh water aquifers in limestone, exploited in the north of Mexico with deep wells below sea level. *Internatl. Assoc. Sci. Hydrol. Proc. Symp. on Hydrology of Fractured Rocks, 2,* 526–539.
Liedtke, H. (1962) Eisrand und Karstpoljen am Westrand der Lukavicahochfläche (Westmontenegro). *Erdkunde 16,* 289–298.
Linsley, R. K., M. A. Kohler, and J. L. H. Paulhus (1949) *Applied Hydrology* (McGraw-Hill, New York).
Liu, S.-T. and G. H. Nancollas (1971) The kinetics of dissolution of calcium sulfate dihydrate. *J. Inorg. Nucl. Chem. 33,* 2311–2316.
Loewenthal, R. E. and G. v. R. Marais (1976) *Carbonate Chemistry of Aquatic Systems* (Ann Arbor Science Pub., Ann Arbor).
Lohman, S. W. (1972) Ground-water hydraulics. *U.S. Geol. Surv. Prof. Paper 708.*
Long, D. T. and Z. A. Saleem (1974) Hydrogeochemistry of carbonate groundwaters of an urban area. *Water Resources Res. 10,* 1229–1238.
Louis, H. (1955) Über die Entstehung der Karstpoljen auf Grund von

Beobachtungen im Taurus. *Bayerische Akad. Wissenschaften. 1955*, 309–317.
Louis, H. (1956) Die Entstehung der Poljen und ihre Stellung in der Karstabtragung auf Grund von Beobachtungen im Taurus. *Erdkunde 10*, 33–53.
Malott, C. A. (1929) Three cavern pictures. *Proc. Indiana Acad. Science 38*, 1–6.
Malott, C. A. (1932) Lost River at Wesley Chapel Gulf, Orange County, Indiana. *Proc. Indiana Acad. Sci. 41*, 285–316.
Malott, C. A. (1937) The invasion theory of cavern development. *Geol. Soc. Amer. Proc.*, 323.
Malott, C. A. (1949) A stormwater cavern in the Lost River Region of Orange County, Indiana. *Natl. Speleol. Soc. Bull. 11*, 64–68.
Malott, C. A. (1951) Wyandotte Cavern. *Natl. Speleol. Soc. Bull. 13*, 30–35.
Malott, C. A. (1952) The swallow-holes of Lost River, Orange County, Indiana. *Proc. Indiana Acad. Sci. 61*, 187–231.
Mandelbrot, B. B. (1983) *The Fractal Geometry of Nature* (W. H. Freeman, San Francisco).
Mangin, A. (1970) Contribution a l'étude des aquifères karstiques a partir de l'analyse des courbes de decrue et tarissement. *Ann. Speleol. 25*, 581–609.
Mangin, A. (1971) Étude des débits classés d'exutoires karstiques portant sur un cycle hydrologique. *Ann. Speleol. 26*, 283–329.
Mangin, A. (1973) Sur les transferts d'eau au niveau du karst noyé a partir de travaux sur la source de Fontestorbes. *Ann. Speleol. 28*, 21–40.
Manov, G. G., R. G. Bates, W. J. Hamer, and S. F. Acree (1943) Values of the constants in the Debye-Hückel equation for activity coefficients. *J. Amer. Chem. Soc. 65*, 1765–1767.
Marker, M. E. (1976) Note on some South African Pseudokarst. *Bol. Soc. Venezolana Espel. 7*, 5–12.
Markowicz, M., V. Popov, and M. Pulina (1972) Comments on karst denudation in Bulgaria. *Geog. Polonica 23*, 111–139.
Martel, E. A. (1921) *Nouveau traité des Eaux souterraines* (Doin, Paris).
Mathews, H. L., R. L. Cunningham, and G. W. Petersen (1973a) Spectral reflectance of selected Pennsylvania soils. *Soil Sci. Soc. Amer. Proc. 37*, 421–424.
Mathews, H. L., R. L. Cunningham, J. E. Cipra, and T. R. West (1973b) Application of multispectra remote sensing to soil survey research in southeastern Pennsylvania. *Soil Sci. Soc. Amer. Proc. 37*, 88–93.
Matson, G. C. (1909) *Water resources in the Blue Grass Region, Kentucky.* U.S. Geol. Surv. Water Supply Paper 233, 42–45.
Matthews, L. E. (1971) Descriptions of Tennessee Caves. *Tennessee Div. Geol. Bull. 69.*
Maurin, V. and J. Zötl (1967) *Specialists Conference on the Tracing of Subterranean Waters.* Steirische Beitr. Hydrogeol. Jahr. 1966/67.
McConnell, H. and J. M. Horn (1972) Probabilities of surface karst. In *Spatial Analysis in Geomorphology*, R. J. Chorley, Ed. (Harper & Row, New York), 111–133.
McDonald, R. C. (1976a) Hillslope base depressions in tower karst topography of Belize. *Zeits. Geomorph.* Suppl. 26, 98–103.

McDonald, R. C. (1976b) Limestone morphology in South Sulawesi, Indonesia. *Zeits. Geomorph.* Suppl. 26, 79–91.
McGregor, D. R., E. C. Pendery, and D. L. McGregor (1963) Solution caves in gypsum, north central Texas. *J. Geol.* 71, 108–115.
McGuinness, C. L. (1963) *The role of ground water in the natural water situation.* U.S. Geol. Surv. Water Supply Paper. 1800.
Mears, B. (1963) Karst-like features in badlands of the Arizona petrified forest. *Contr. Geol.* 2, 101–104.
Medville, D. M., J. C. Hempel, C. Plantz, and E. Werner (1979) Solutional landforms on carbonates of the southern Teton Range, Wyoming. *Natl. Speleol. Soc. Bull.* 41, 70–79.
Medville, D. M. and E. Werner (1974) Hydrogeology of the Death Canyon Limestone, Teton Range, Wyoming. *Proc. 4th Conf. on Karst Geol. and Hydrol.*, H. W. Rauch and E. Werner, Eds. (West Virginia Geol. Surv., Morgantown, WV), 95–102.
Merriam, D. F. and C. J. Mann (1957) Sinkholes and related geologic features in Kansas. *Trans. Kansas Acad. Sci.* 60, 207–243.
Merriam, P. (1950) Ice caves. *Natl. Speleol. Soc. Bull.* 12, 32–37.
Mijatović, B. F. (1968) A method of studying the hydrodynamic regime of karst aquifers by analysis of the discharge curve and level fluctuations during recession. *Vesnik Zavoda za Geološka i Geofizička Istraživanja*, Series B, No. 8, 41–81.
Mijatović, B. F. (1983) Karst poljes in Dinarides. In *Hydrogeology of Dinaric Karst*, B. F. Mijatović, Ed. (Geozavod, Belgrade).
Milanović, P. T. (1981) *Karst Hydrogeology* (Water Resources Publications, Littleton, CO).
Milanović, P. T. (1984) Some methods of hydrogeologic exploration and water regulation in the Dinaric Karst with special reference to their application in eastern Herzegovina. *Internatl. Contr. Hydrogeol.* 4, 160–200.
Miller, T. (1981) Hydrochemistry, hydrology, and morphology of the Caves Branch Karst, Belize. PhD. Thesis, Geography, McMaster University.
Mills, H. H. and D. D. Starnes (1983) Sinkhole morphometry in a fluviokarst region: eastern Highland Rim, Tennessee, U.S.A. *Zeits Geomorph.* 27, 39–54.
Milske, J. A., E. C. Alexander, and R. S. Lively (1983) Clastic sediments in Mystery Cave, southeastern Minnesota. *Natl. Speleol. Soc. Bull.* 45, 55–75.
Minton, M. (1984) Christmas in Huautla. *Assoc. Mexican Cave Studies Activities Newsletter* 14, 66–72.
Miotke, F.-D. (1972) Die Messung des CO_2-Gehaltes der Bodenluft mit dem Dräger-Gerat und die beschleunigte Kalklösung durch höhere Fliessgeschwindigkeiten. *Zeits. Geomorph.* 16, 93–102.
Miotke, F.-D (1974) Carbon dioxide and the soil atmosphere. *Abhand. Karst u. Höhlenkunde*, No. 9.
Miotke, F.-D. (1975a) Der Karst im zentralen Kentucky bei Mammoth Cave. *Jahrb. Geograph. Gesell. Hannover* 1973.
Miotke, F.-D. (1975b) Der CO_2-Gehalt in Bodenluft in seiner Bedeutung für die aktuelle Kalklösung in verschiedenen Klimaten. *Akad. Wissenscht. Göttingen Abhand.*, 51–67.

Miotke, F.-D. and A. N. Palmer (1972) *Genetic relationship between caves and landforms in the Mammoth Cave National Park Area* (Böhler, Wurzburg).

Moneymaker, B. C. (1941) Subriver solution cavities in the Tennessee Valley. *J. Geol.* 49, 74–86.

Moneymaker, B. C. (1948) Some broad aspects of limestone solution in the Tennessee Valley. *Trans. Amer. Geophys. Union* 29, 93–96.

Monroe, W. H. (1960) Sinkholes and towers in the karst area of North-Central Puerto Rico. *U.S. Geol. Surv. Prof. Paper 400-B*, 356–360.

Monroe, W. H. (1964) The zanjon, a solution feature of karst topography in Puerto Rico. *U.S. Geol. Surv. Prof. Paper 501-B*, 126–129.

Monroe, W. H. (1968) The karst features of Northern Puerto Rico. *Natl. Speleol. Soc. Bull.* 30, 75–86.

Monroe, W. H. (1976) The karst landforms of Puerto Rico. *U.S. Geol. Surv. Prof. Paper 899*.

Moore, D. G. (1954) Origin and development of sea caves. *Natl. Speleol. Soc. Bull.* 16, 71–76.

Moore, G. W. (1952) Speleothem—a new cave term. *Natl. Speleol. Soc. News* 10, [6] 2.

Moore, G. W. (1954) The origin of helictites. *Natl. Speleol. Soc. Occasional Paper No. 1*.

Moore, G. W. (1958) Role of earth tides in the formation of disc-shaped cave deposits. *Proc. 2nd Internatl. Congress Speleol.* 1 (Bari), 500–506.

Moore, G. W. (1962) The growth of stalactites. *Natl. Speleol. Soc. Bull.* 24, 95–106.

Moore, G. W. (1981) Manganese deposition in limestone caves. *Proc. 8th Internatl. Congress Speleol.* (Bowling Green, KY), 642–644.

Moore, J. D., F. Hinkle, and G. F. Moravec (1977) High-yield wells and springs along lineaments interpreted from LANDSAT imagery in Madison County, Alabama. *Internatl. Assoc. Hydrogeol. Mem.* 12, 477–486.

Moravec, G. F. (1974) Development of karren karst forms on the Newala limestone in the Cahaba Valley, Alabama. *Proc. 4th Conf. Karst Geol. and Hydrol.* H. W. Rauch and E. Werner, Eds. (West Virginia Geol: Surv. Morgantown, WV, 113–121.

Morehouse, D. F. (1968) Cave development via the sulfuric acid reaction. *Natl. Speleol. Soc. Bull.* 30, 1–10.

Morey, G. W., R. O. Fournier, and J. J. Rowe (1962) The solubility of quartz in water in the temperature interval from 20 to 300°C. *Geochim. Cosmochim. Acta* 26, 1029–1043.

Morey, G. W., R. O. Fournier, and J. J. Rowe (1964) The solubility of amorphous silica at 25°C. *J. Geophys. Res.* 69, 1995–2002.

Morisawa, M. (1971) Evaluating riverscapes. In *Environmental Geomorphology*. D. R. Coates, Ed. (State University of New York, at Binghamton), 91–106.

Moseley, C. M. (1975) The caves of Nova Scotia. *Canadian Caver* 7, 47–54.

Munsell Color (1976) *Munsell Book of Color* (Macbeth Div. Kollmorgen Corp., Baltimore, MD).

Murray, J. P., J. V. Rouse, and A. B. Carpenter (1981) Groundwater contamination by sanitary landfill leachate and domestic wastewater in carbonate terrain: Principal source diagnosis, chemical transport characteristics, and design implications. *Water Res. 15*, 745–757.

Murray, J. W. and R. V. Dietrich (1956) Brushite and taranakite from Pig Hole Cave, Giles County, Virginia. *Amer. Mineral. 41*, 616–626.

Myers, A. J. (1960a) An area of gypsum karst topography in Oklahoma. *Oklahoma Geol. Notes 20*, 10–14.

Myers, A. J. (1960b) Alabaster Caverns. *Oklahoma Geol. Notes 20*, 132–137.

Mylroie, J. E. (1977) Speleogenesis and karst geomorphology of the Helderberg Plateau, Schoharie County, New York. *New York Cave Surv. Bull. 2.*

Nair, V. S. K. and G. H. Nancollas (1958) Thermodynamics of ion association IV. Magnesium and zinc sulfate. *J. Chem. Soc., London*, 3706–3710.

Neuzil, C. E. and J. V. Tracy (1981) Flow through fractures. *Water Resources Res. 17*, 191–199.

Newton, J. G., C. W. Copeland, and L. W. Scarbrough (1973) Sinkhole problem along proposed route to interstate highway 459 near Greenwood, Alabama. *Alabama Geol. Surv. Circ. 83.*

Newton, J. G. and L. W. Hyde (1971) Sinkhole problem in and near Roberts Industrial Subdivision, Birmingham, Alabama. *Alabama Geol. Surv. Circ. 68.*

Noel, M. (1983) The magnetic remanence and anisotropy of susceptibility of cave sediments from Agen Allwedd, South Wales. *Geophys. J. Royal Astron. Soc. 72*, 557–570.

Noel, M. (1986a) The paleomagnetism and magnetic fabric of cave sediments from Pwll y Gwynt, South Wales. *Phys. Earth Planet. Interiors, 44*, 62–71.

Noel, M. (1986b) The paleomagnetism and magnetic fabric of sediments from Peak Cavern, Derbyshire. *Geophys. J. Royal Astron. Soc. 84*, 445–454.

Novak, I. D. (1973) Predicting coarse sediment transport: The Hjulstrom curve revisited. In *Fluvial Geomorphology*, M. Morisawa, Ed. (State University of New York, Binghamton, NY), 13–25.

Núñez-Jiménez, A. (1959) *Geografía de Cuba* (Editorial Lex, Havana).

Ogden, A. E. (1982) Karst denudation rates for selected spring basins in West Virginia. *Natl. Speleol. Soc. Bull. 44*, 6–10.

Ogden, A. and W. Ebaugh (1972) Development of and sedimentation in Milroy Cave. *Nittany Grotto Newsletter 20*, 5–22.

Olive, W. W.(1957) Solution-subsidence troughs, Castile formation of gypsum plain, Texas and New Mexico. *Geol. Soc. Amer. Bull. 68*, 351–358.

Omnes, G. (1977) High accuracy gravity applied to the detection of karstic cavities. *Internatl. Assoc. Hydrogeol. Mem. 12*, 273–284.

Ongley, E. D. (1968) An analysis of the meandering tendency of Serpentine Cave, N.S.W. *J. Hydrol. 6*, 15–32.

Otvos, E. G. (1976) Pseudokarst and pseudokarst terrains: Problems of terminology. *Geol. Soc. Amer. Bull. 87*, 1021–1027.

Palmer, A. N. (1975) The origin of maze caves. *Natl. Speleol. Soc. Bull. 37*, 56–76.

Palmer, A. N. (1981a) *A Geological Guide to Mammoth Cave National Park* (Zephyrus Press, Teaneck, NJ).

Palmer, A. N. (1981b) *The Geology of Wind Cave* (Wind Cave Natural History Assoc., Hot Springs, SD).

Palmer, A. N. (1981c) Hydrochemical factors in the origin of limestone caves. *Proc. 8th Internatl. Congress Speleol.* (Bowling Green, KY), 120–122.

Palmer, A. N. (1984) Geomorphic interpretation of karst features. In *Groundwater as a Geomorphic Agent*, R. G. LaFleur, Ed. (Allen and Unwin, London), 173–209.

Palmer, M. V. and A. N. Palmer (1975) Landform development in the Mitchell Plain of southern Indiana: Origin of a partially karsted plain. *Zeits. Geomorph.* 19, 1–39.

Palmer, R. (1985) *The Blue Holes of the Bahamas* (Jonathan Cape, London).

Palmquist, R. C., G. A. Madenford, and J. N. Van Driel (1976) Doline densities in northeastern Iowa. *Natl. Speleol. Soc. Bull.* 38, 59–67.

Panoš, V. and O. Štelcl (1968) Physiographic and geologic control in development of Cuban mogotes. *Zeits. Geomorph.* 12, 117–173.

Parizek, R. R. (1971) Impact of highways on the hydrogeologic environment. In *Environmental Geomorphology*, D. R. Coates, Ed. (State University of New York, Binghamton), 151–199.

Parizek, R. R. (1976) On the nature and significance of fracture traces and lineaments in carbonate and other terranes. In *Karst Hydrology and Water Resources.* V. Yevjevich, Ed. (Water Resources Publications, Fort Collins, CO), 47–108.

Parizek, R. R., L. T. Kardos, W. E. Sopper, E. A. Myers, D. E. Davis, M. A. Farrell, and J. B. Nesbitt (1967) *Waste Water Renovation and Conservation.* Penn State Studies, The Pennsylvania State University, No. 23.

Parizek, R. R. and S. H. Siddiqui (1970) Determining the sustained yields of wells in carbonate and fractured aquifers. *Ground Water* 8, 12–20.

Parizek, R. R. and W. B. White (1985) Application of Quaternary and Tertiary geological factors to environmental problems in Central Pennsylvania. *Guidebook to 50th Ann. Field Conf. of Pennsylvania Geologists*, (Harrisburg, PA), 63–119.

Parizek, R. R., W. B. White, and D. Langmuir (1971) *Hydrogeology and geochemistry of folded and faulted rocks of the central Appalachian type and related land use problems.* The Pennsylvania State University Earth and Mineral Sciences Experiment Station Circ. 82.

Parris, L. E. (1973) *Caves of Colorado* (Pruett Publishing Co., Boulder, CO).

Pearse, A. S., E. P. Creaser, and F. G. Hall (1936) *The Cenotes of Yucatan.* Carnegie Institute of Washington Pub. 457.

Pechorkin, I. A. (1966) *The geodynamics of the banks of the Kama River Reservoirs. Part I. The Engineering-Geological Conditions.* (Perm State University, Perm).

Pechorkin, I. A. (1969) *Geodynamics of Kama Water Storage Reservoir Shorelines. Part II Geological Processes* (Perm State University, Perm).

Perič, J. (1964) Underground reservoirs as hydrotechnical structures for the control of the water regime in karst and other springs. *Bull. Inst. Geol. Geophys. Res.* (Belgrade) Series B, No. 3, 67–97.

Peterson, D. N. and G. D. McKenzie (1968) Observations of a glacier cave in Glacier Bay National Monument, Alaska. *Natl. Speleol. Soc. Bull. 30*, 47–54.

Peterson, M. N. A. (1962) The mineralogy and petrology of upper Mississippian carbonate rocks of the Cumberland Plateau of Tennessee. *J. Geol. 70*, 1–31.

Pfeiffer, D. and J. Hahn (1972) Karst of Germany. In *Karst: Important Karst Regions of the Northern Hemisphere*, M. Herak and V. T. Stringfield, Eds. (Elsevier, Amsterdam), 189–223.

Pia, J. (1933) Die Theorien über die Löslichkeit des kohlensauren Kalkes. *Mitt. Geol. Gesells.*, Vienna *25*, 1–93.

Picknett, R. G. (1972) The pH of calcite solutions with and without magnesium carbonate present and the implications concerning rejuvenated aggressiveness. *Trans. Cave Res. Group Great Britain 14*, 141–150.

Pigott, C. D. (1962) Soil formation and development on the carboniferous limestone of Derbyshire. I. Parent materials. *J. Ecol. 50*, 145–156.

Piper, A. M. (1932) Ground water in north-central Tennessee. *U.S. Geol. Surv. Water Supply Paper 640*.

Pitman, J. I. (1978) Carbonate chemistry of groundwater from tropical tower karst in South Thailand. *Water Resources Res. 14*, 961–967.

Pitty, A. F.(1968) Calcium carbonate content of karst water in relation to flow-through time. *Nature 217*, 939–940.

Pitty, A. F. (1971) Rate of uptake of calcium carbonate in underground karst water. *Geol. Mag. 108*, 537–543.

Pluhar, A. and D. C. Ford (1970) Dolomite karren of the Niagara Escarpment, Ontario, Canada. *Zeits. Geomorph. 14*, 392–410.

Plummer, L. N. and E. Busenberg (1982) The solubilities of calcite, aragonite, and vaterite in CO_2-H_2O solutions between 0 and 90°C, and an evaluation of the aqueous model for the system $CaCO_3$-CO_2-H_2O. *Geochim. Cosmochim. Acta 46*, 1011–1040.

Plummer, L. N., B. F. Jones, and A. H. Truesdell (1978b) *WATEQF—A Fortran IV version of WATEQ, a computer program for Calculating Chemical equilibrium in natural waters*. U.S. Geol. Surv. Water Resources Invest. 76-13.

Plummer, L. N., D. L. Parkhurst, and T. M. L. Wigley (1979) Critical review of the kinetics of calcite dissolution and precipitation. In *Chemical Modeling in Aqueous Systems*, E. A. Jenne, Ed., Amer. Chem. Soc. Symp. Series 93, 537–573.

Plummer, L. N. and T. M. L. Wigley (1976) The dissolution of calcite in CO_2 saturated solutions at 25°C and 1 atmosphere total pressure. *Geochim. Cosmochim. Acta 40*, 191–202.

Plummer, L. N., T. M. L. Wigley, and D. L. Parkhurst (1978a) The kinetics of calcite dissolution in CO_2-water systems at 5 to 60°C and 0.0 to 1.0 atm CO_2. *Amer. J. Sci. 278*, 179–216.

Pohl, E. R. (1955) Vertical shafts in limestone caves. *Natl. Speleol. Soc. Occasional Paper No. 2*.

Pohl, E. R. and W. B. White (1965) Sulfate minerals: Their origin in the central Kentucky karst. *Amer. Mineral.* **50**, 1461–1465.

Popov, I. V., N. A. Gvozdetskii, A. G. Chikishev, and B. I. Kudelin (1972) Karst of the U.S.S.R. In *Karst: Important Karst Regions of the Northern Hemisphere*, M. Herak and V. T. Stringfield, Eds. (Elsevier, Amsterdam), 355–416.

Powell, R. L. (1961) Caves of Indiana. *Indiana Geol. Surv. Circ. 8*.

Powell, W. J. and P. E. LaMoreaux (1969) A problem of subsidence in a limestone terrane at Columbiana, Alabama. *Alabama Geol. Surv. Circ. 56*.

Prior, T. A., M. T. Mills, B. M. Ellis, C. Wood, E. J. Watkins, and P. G. Bowler (1971) Scientific investigations in Raufarhólshellir lava cave, southwest Iceland. *Trans. Cave Res. Group Great Britain*, **13**, 225–264.

Proctor, P. D., G. Kisvarsanyi, and E. Garrison (1977) Heavy metal additions to waters of the Joplin Area, Tri-State mining district, Missouri. In *Hydrologic Problems in Karst Regions*, R. R. Dilamarter and S. C. Csallany, Eds. (Western Kentucky University, Bowling Green, KY), 369–387.

Pulina, M. (1972) A comment on present-day chemical denudation in Poland. *Geogr. Polonica.* **23**, 45–62.

Quinlan, J. F. (1967) Classification of karst types: A review and synthesis emphasizing the North American literature, 1941–1966. *Natl. Speleol. Soc. Bull.* **29**, 107–109.

Quinlan, J. F. (1970) Central Kentucky karst. *Mediteraneé*, **7**, 235–253.

Quinlan, J. F. (1978) Types of karst, with emphasis on cover beds in their classification and development. PhD Thesis, Geology, University of Texas, Austin.

Quinlan, J. F. (1981) Hydrologic research techniques and instrumentation used in the Mammoth Cave Region, Kentucky. In *GSA Cincinnati '81 Field Trip Guidebooks*, T. G. Roberts, Ed. (Geological Society of America, Boulder, CO). Vol. III, 502–504.

Quinlan, J. F. (1984) Litigious problems associated with sinkholes emphasizing recent Kentucky cases alleging liability when sinkholes were flooded. In *Sinkholes: Their Geology, Engineering, and Environmental Impact*, B. F. Beck, Ed. (A. A. Balkema, Rotterdam), 293–296.

Quinlan, J. F. (1986) Legal aspects of sinkhole development and flooding in karst terranes: 1. Review and synthesis. *Environ. Geol. Water Sci.* **8**, 41–61.

Quinlan, J. F. and R. O. Ewers (1981) Hydrogeology of the Mammoth Cave Region, Kentucky. In *GSA Cincinnati '81 Field Trip Guidebooks*, T. G. Roberts, Ed. (Geological Society of America, Boulder, CO). In Vol. III, 457–506.

Quinlan, J. F. and R. O. Ewers (1985) Ground water flow in limestone: Rationale for a reliable strategy for efficient monitoring of ground water quality in karst areas. *Proc. 5th Natl. Symp. on Aquifer Restoration and Ground Water Monitoring*, 197–234.

Quinlan, J. F. and J. A. Ray (1981) *Groundwater Basins in the Mammoth Cave Region, Kentucky.* Map sheet, Friends of Karst Occas. Paper No. 1 (Mammoth Cave, KY).

Quinlan, J. F. and D. R. Rowe (1977) *Hydrology and Water Quality in the*

Central Kentucky Karst: Phase I. (Water Resources Center, University of Kentucky, Lexington, KY), Res. Report 101.

Quinlan, J. F. and D. R. Rowe (1978) *Hydrology and water quality in the central Kentucky karst: Phase II, Part A: Preliminary summary of the hydrogeology of the Mill Hole Sub-basin of the Turnhole Spring Groundwater Basin* (Water Resources Institute, University of Kentucky), Res. Report 109.

Quinlan, J. F., A. R. Smith, and K. S. Johnson (1986) Gypsum karst and salt karst of the United States of America. *Proc. Internatl. Symp. on Evaporite Karst*, Le Grotte d'Italia, Series 4, *13*, in press.

Raines, T. W. (1968) Sotano de las Golondrinas. *Assoc. Mexican Cave Studies Bull. 2*, 1–20.

Raines, T. W. (1972) Sotanito de Ahuacatlan. *Assoc. Mexican Cave Studies Cave Report Series No. 1.*

Rasmusson, G. (1959) Karstformen im Granit des Fichtelgebirges. *Die Höhle 10*, 1–4.

Rauch, H. W. and W. B. White (1970) Lithologic controls on the development of solution porosity in carbonate aquifers. *Water Resources Res. 6*, 1175–1192.

Rauch, H. W. and W. B. White (1977) Dissolution kinetics of carbonate rocks. 1. Effects of lithology on dissolution rate. *Water Resources Res. 13*, 381–394.

Reams, M. W. (1968) Cave sediments and the geomorphic history of the Ozarks. PhD Thesis, Geology (Washington University, St. Louis, MO).

Reardon, E. J. (1974) Thermodynamic properties of some sulfate, carbonate, and bicarbonate ion pairs. PhD Thesis, Geochemistry, The Pennsylvania State University.

Reardon, E. J. and D. Langmuir (1974) Thermodynamic properties of the ion pairs $MgCO_3°$ and $CaCO_3°$ from 10 to 50°C. *Amer. J. Sci. 274*, 599–612.

Reardon, E. J. and D. Langmuir (1976) Activity coefficients of $MgCO_3°$ and $CaSO_4°$ ion pairs as a function of ionic strength. *Geochim. Cosmochim. Acta 40*, 549–554.

Reckmeyer, V. (1963a) Radioactivity background in Hughs Cave, Alabama, *Huntsville Grotto Newsletter 4*, 32.

Reckmeyer, V. (1963b) Radioactivity in Alabama caves. *Huntsville Grotto Newsletter 4*, 82.

Reddell, J. R. (1964) *A Guide to the Caves of Texas.* (National Speleological Society, Huntsville, AL).

Reddell, J. R. (1977) Studies on the Caves and Cave Fauna of the Yucatan Peninsula. *Assoc. Mexican Cave Studies Bull. 6.*

Reddy, M. M., L. N. Plummer, and E. Busenberg (1981) Crystal growth of calcite from calcium bicarbonate solutions at constant P_{CO_2} and 25° C: a test of a calcite dissolution model. *Geochim. Cosmochim. Acta 45*, 1281–1289.

Reich, J. R., Jr. (1974) *Caves of Southeastern Pennsylvania.* Pennsylvania Geol. Surv. General Geol. Report 65.

Renault, P. (1967–1968) Contribution a l'étude des actions méchaniques et sédimentologiques dans la spéléogènese. *Ann. Speleol. 22*, 5–21, 209–267; *23*, 259–307, 529–596; *24*, 317–337.

Rhindress, R. C. (1971) Gasoline pollution of a karst aquifer. In *Hydrogeology and Geochemistry of Folded and Faulted Rocks of the Central Appalachian Type and Related Land Use Problems*, R. R. Parizek, W. B. White, and D. Langmuir, Eds. (The Pennsylvania State University), Earth and Mineral Sciences Experiment Station Circ. 82, 171–176.

Rhoades, R. and M. N. Sinacori (1941) Pattern of ground-water flow and solution. *J. Geol. 49*, 785–794.

Rickard, D. and E. L. Sjöberg (1983) Mixed kinetic control of calcite dissolution rates. *Amer. J. Sci. 283*, 815–830.

Rightmire, C. T. (1978) Seasonal variation in P_{CO_2} and ^{13}C content of soil atmosphere. *Water Resources Res. 14*, 691–692.

Robbins, C. and W. D. Keller (1952) Clay and other non-carbonate minerals in some limestones. *J. Sed. Petrol. 22*, 146–152.

Roberge, F. (1977) Another Teton? *New Engineer 6*, 17–22.

Robie, R. A., B. S. Hemingway, and J. R. Fisher (1978) Thermodynamic properties of minerals and related substances at 298.15 K and 1 bar (10^5 pascals) pressure and at higher temperatures. *U.S. Geol. Surv. Bull. 1452*.

Roglič, J. (1957) Quelques problèmes fondamentaux du karst. *L'Information Géographique 21*, 1–12.

Roglič, J. (1964) "Karst valleys" in the Dinaric Karst. *Erdkunde 18*, 113–116.

Roglič, J. (1972) Historical review of morphologic concepts. In *Karst: Important Karst Regions of the Northern Hemisphere*, M. Herak and V. T. Stringfield, Eds. (Elsevier, Amsterdam), 1–18.

Ross, S. H. (1969) *Introduction to Idaho Caves and Caving*. Idaho Bur. Mines Geol. Earth Sci. Series No. 2.

Rossi, M. G. (1974) Sur une series de mesures de teneurs en CO_2 de sols tropicaux. *Bull. Assoc. Geogr. Franc. 51*, 141–144.

Ruhe, R. V. (1975) *Geohydrology of karst terrain. Lost River Watershed, Southern Indiana*. Indiana Univ. Water Resources Ctr. Report Invest. No. 7.

Ruhe, R. V. (1977) Summary of geohydrologic relationships in the Lost River Watershed, Indiana applied to water use and environment. In *Hydrologic Problems in Karst Regions*, R. R. Dilamarter and S. C. Csallany, Eds. (Western Kentucky University, Bowling Green, KY), 64–78.

Rutherford, J. M. and R. M. Amundson (1974) Use of a computer program for cave survey data reduction. *Natl. Speleol. Soc. Bull. 36*, 7–17.

Saleem, Z. A. (1977) Road salts and quality of groundwater from a dolomite aquifer in the Chicago Area. In *Hydrologic Problems in Karst Regions*, R. R. Dilamarter and S. C. Csallany, Eds. (Western Kentucky University, Bowling Green, KY), 364–368.

Salomone, W. G. (1984) Part 1: The applicability of the Florida mandatory endorsement for sinkhole collapse coverage—legal aspects. In *Sinkholes: Their Geology, Engineering, and Environmental Impact*. B. F. Beck, Ed. (A. A. Balkema, Rotterdam), 319–328.

Schmid, E. (1958) *Höhlenforschung und Sedimentanalyse* (Schriften Inst. Ur u. Frühgeschichte Schweiz).

Schmidt, V. A. (1974) The paleohydrology of Laurel Caverns, Pennsylvania.

Proc. 4th Conf. Karst Geol. Hydrol., H. W. Rauch and E. Werner, Eds., (West Virginia Geological Survey), 123–129.

Schmidt, V. A. (1982) Magnetostratigraphy of sediments in Mammoth Cave, Kentucky. *Science 217*, 827–829.

Schmidt, V. A. and J. H. Schelleng (1970) The application of the method of least squares to the closing of multiply-connected loops in cave or geological surveys. *Natl. Speleol. Soc. Bull. 32*, 51–58.

Schumm, S. A. (1956) The evolution of drainage systems and slopes in badlands at Perth Amboy, New Jersey. *Geol. Soc. Amer. Bull. 67*, 597–646.

Schwarcz, H. and M. Gascoyne (1984) Uranium-series dating of quaternary deposits. In *Quaternary Dating Methods*, W. C. Mahaney, Ed. (Elsevier, Amsterdam), 33–51.

Schwarcz, H. P., R. S. Harmon, P. Thompson, and D. C. Ford (1976) Stable isotope studies of fluid inclusions in speleothems and their paleoclimatic significance. *Geochim. Cosmochim. Acta 40*, 657–665.

Shelley, M. B. (1954) Caves and karst of the U.S.S.R. *Natl. Speleol. Soc. Bull. 16*, 40–54.

Shepherd, R. G. (1985) Regression analysis of river profiles. *J. Geol. 93*, 377–384.

Shuster, E. T. (1970) Seasonal variation in carbonate spring water chemistry related to groundwater flow. MS Thesis, Geology, The Pennsylvania State University.

Shuster, E. T. and W. B. White (1971) Seasonal fluctuations in the chemistry of limestone springs: A possible means for characterizing carbonate aquifers. *J. Hydrol. 14*, 93–128.

Shuster, E. T. and W. B. White (1972) Source areas and climatic effects in carbonate ground waters determined by saturation indices and carbon dioxide pressures. *Water Resources Res. 8*, 1067–1073.

Siddiqui, S. H. and R. R. Parizek (1971) Hydrogeologic factors influencing well yields in folded and faulted carbonate rocks in Central Pennsylvania. *Water Resources Res. 7*, 1295–1312.

Siegel, F. R., J. P. Mills, and J. W. Pierce (1968) Aspectos petrograficos y geoquimicos de espeleothemas de opalo y calcita de la Cueva de la Bruja, Mendoza Republica Argentina. *Revista Assoc. Geol. Argentina. 23*, 5–19.

Siever, R. (1962) Silica solubility, 0–200°C, and the diagenesis of siliceous sediments. *J. Geol. 70*, 127–150.

Silar, J. (1965) Development of tower karst of China and North Vietnam. *Natl. Speleol. Soc. Bull. 27*, 35–46.

Sinclair, W. C. (1982) *Sinkhole development resulting from ground-water withdrawal in the Tampa Area, Florida.* U.S. Geol. Surv. Water Resources Invest. 81-50.

Sjöberg, E. L. (1976) A fundamental equation for calcite dissolution kinetics. *Geochim. Cosmochim. Acta 40*, 441–447.

Sjöberg, E. L. and D. T. Rickard (1984) Temperature dependence of calcite dissolution kinetics between 1 and 62°C at pH 2.7 to 8.4 in aqueous solutions. *Geochim. Cosmochim. Acta 48*, 485–493.

Skinner, A. F. (1983) Overestimate of stalagmitic calcite ESR dates due to laboratory heating. *Nature 304*, 152–154.

Smart, C. C. and M. C. Brown (1981) Some results and limitations in the application of hydraulic geometry to vadose stream passages. *Proc. 8th Internatl. Congress Speleol.* Bowling Green, KY, 724–726.

Smart, P. L. and I. M. S. Laidlaw (1977) An evaluation of some fluorescent dyes for water tracing. *Water Resources Res. 13*, 15–33.

Smith, B. W., P. L. Smart, M. C. R. Symons, and J. N. Andrews (1985) ESR dating of detritally contaminated calcites. *ESR Dating and Dosimetry*, M. Ikeya and T. Miki, Eds. (Ionics Publishing Co., Tokyo), 49–59.

Smith, D. I. (1962) *The solution of limestone in an arctic environment.* Institute of British Geographers. Special Pub. 4, 187–200.

Smith, D. I. and T. C. Atkinson (1976) Process, landforms, and climate in limestone regions. In *Geomorphology and Climate*, E. Derbyshire, Ed. (John Wiley, New York), 367–409.

Smith, D. I., T. C. Atkinson, and D. P. Drew (1976) The hydrology of limestone terrains. In *The Science of Speleology*. T. D. Ford and C. H. D. Cullingford, Eds. (Academic Press, London), 179–212.

Smith, D. I., D. P. Drew, and T. C. Atkinson (1972) Hypotheses of karst landform development in Jamaica. *Trans. Cave Res. Group Great Britain 14*, 159–173.

Smith, K. G. (1958) Erosional processes and landforms in Badlands National Monument, South Dakota. *Geol. Soc. Amer. Bull. 69*, 975–1008.

Smith, L. and F. W. Schwartz (1984) An analysis of the influence of fracture geometry on mass transport in fractured media. *Water Resources Res. 20*, 1241–1252.

Smith, R. M. and A. E. Martell (1976) *Critical Stability Constants* (Plenum Press, New York).

Smith, W. O. and A. N. Sayre (1964) *Turbulence in ground-water flow.* U.S. Geological Survey Prof. Paper 402-E, 1–9.

Soderberg, A. D. (1979) Expect the unexpected: Foundations for dams in karst. *Bull. Assoc. Eng. Geol. 16*, 409–425.

Sonderegger, J. L. (1970) Hydrology of limestone terranes: Photogeologic investigations. *Alabama Geol. Surv. Bull. 94C.*

Sowers, G. F. (1984) Correction and protection in limestone terrane. In *Sinkholes: Their Geology, Engineering, and Environmental Impact*, B. F. Beck, Ed. (A. A. Balkema, Rotterdam), 373–378.

Stearns, H. T. (1942) Hydrology of volcanic terranes. In *Hydrology*, O. E. Meinzer, Ed. (McGraw-Hill, New York), 678–703.

Steiner, R. S. (1975) Reinforced earth bridges highway sinkhole. *Civil Eng., Amer. Soc. Civil Eng. 45* 54–56.

Stone, B. (1983) A breakthrough at the Huautla resurgence. *Natl. Speleol. Soc. News 41*, 168–180.

Stone, B. (1984) The challenge of the Peña Colorada. *Assoc. Mexican Cave Studies Activities Letter, 14*, 46–55.

Strahler, A. N. (1952) Hypsometric (area-altitude) analysis of erosional topography. *Geol. Soc. Amer. Bull. 63*, 1117–1142.

Strahler, A. N. (1964) Quantitative geomorphology of drainage basins and channel networks. In *Handbook of Applied Hydrology*, Ven Te Chow, Ed., (McGraw-Hill, New York), 4-39–4-76.

Stringfield, V. T. and H. E. LeGrand (1969) Hydrology of carbonate rock terrains—a review. *J. Hydrol. 8*, 349–417.

Stumm, W. and J. J. Morgan (1981) *Aquatic Chemistry*, 2nd ed. (John Wiley, New York).
Suarez, D. L. (1983) Calcite supersaturation and precipitation kinetics in the lower Colorado River, All-American Canal and East Highline Canal. *Water Resources Res. 19*, 653–661.
Sunartadirdja, M. A. and H. Lehmann (1960) Der tropische Karst von Maros und Nord-Bone in SW-Celebes (Sulawesi). *Zeits. Geomorph.* Suppl. 2, 49–65.
Sutcliffe, A. J. (1973) Caves of the East African Rift Valley. *Trans. Cave Res. Group Great Britain 15*, 41–65.
Sweeting, M. M. (1950) Erosion cycles and limestone caverns in the Ingleborough District. *Geograph. J. 115*, 63–78.
Sweeting, M. M. (1958) The karstlands of Jamaica. *Geograph. J. 124*, 184–199.
Sweeting, M. M. (1972) *Karst Landforms* (Macmillan, London).
Sweeting, M. M. (1979) Weathering and solution of the Melinau limestones in the Gunong Mulu National Park, Sarawak, Malaysia. *Ann. Soc. Geol. Belgium 102*, 53–57.
Sweeting, M. M. and K. H. Pfeffer (1976) Karst processes. *Zeits. Geomorph.* Suppl. 26.
Sweeting, M. M. and G. S. Sweeting (1969) Some aspects of the Carboniferous limestone in relation to its landforms. *Méditerraneé, 7*, 201–209.
Swinnerton, A. C. (1932) Origin of limestone caverns. *Geol. Soc. Amer. Bull. 43*, 663–694.
Szczerban, E. and F. Urbani (1974) Carsos de Venezuela 4. Formas Carsicas en Areniscos Precambricas del Territorio Federal Amazonas y Estado Bolivar. *Bol. Soc. Venezolana Espeleol. 5*, 27–54.
Tennessee Valley Authority (1949) *Geology and Foundation Treatment* (Tennessee Valley Authority, Knoxville, TN).
Ternan, J. L. (1972) Comments on the use of a calcium hardness variability index in the study of carbonate aquifers: With reference to the Central Pennines, England. *J. Hydrol. 16*, 317–321.
Terzaghi, K. (1958) Landforms and subsurface drainage in the Gacka region in Yugoslavia. *Zeits. Geomorph. 2*, 76–100.
Thompson, G. M., D. N. Lumsden, R. L. Walker, and J. A. Carter (1975) Uranium series dating of stalagmites from Blanchard Springs Cavern, U.S.A. *Geochim. Cosmochim. Acta 39*, 1211–1218.
Thompson, P., D. C. Ford, and H. P. Schwarcz (1975) U^{234}/U^{238} ratios in limestone cave seepage waters and speleothem from West Virginia. *Geochim. Cosmochim. Acta 39*, 661–669.
Thompson, P., H. P. Schwarcz, and D. C. Ford (1976) Stable isotope geochemistry, geothermometry, and geochronology of speleothems from West Virginia. *Geol. Soc. Amer. Bull. 87*, 1730–1738.
Thrailkill, J. (1968a) Chemical and hydrologic factors in the excavation of limestone caves. *Geol. Soc. Amer. Bull. 79*, 19–46.
Thrailkill, J. (1968b) Dolomite cave deposits from Carlsbad Caverns. *J. Sed. Petrol. 38*, 141–145.
Thrailkill, J. (1970) *Solution geochemistry of the water of limestone terrains*. University of Kentucky Water Resources Inst. Res. Report No. 19.

Thrailkill, J. (1972) Carbonate chemistry of aquifer and stream water in Kentucky. *J. Hydrol. 16*, 93–104.
Tjia, H. D. (1969) Slope development in tropical karst. *Zeits. Geomorph. 13*, 260–266.
Tolson, J. S. and F. L. Doyle (1977) *Karst Hydrogeology*. Internatl. Assoc. Hydrogeol. Memoir 12.
Trainer, F. W. and R. C. Heath (1976) Bicarbonate content of groundwater in carbonate rock in eastern North America. *J. Hydrol. 31*, 37–55.
Tratman, E. K. (1969) *The Caves of Northwest Clare, Ireland* (David and Charles, Newton Abbot).
Tricart, J. and T. Cardoso da Silva (1960) Un exemple d'evolution karstique en milieu tropical sec: Le morne de Bom Jesus da Lapa (Bahia, Bresil). *Zeits. Geomorph. 4*, 29–42.
Troester, J. W. and W. B. White (1984) Seasonal fluctuations in the carbon dioxide partial pressure in a cave atmosphere. *Water Resources Res. 20*, 153–156.
Troester, J. W. and W. B. White (1986) Geochemical investigations of three tropical karst drainage basins in Puerto Rico. *Ground Water 24*, 475–482.
Troester, J. W., E. L. White, and W. B. White (1984) A comparison of sinkhole depth frequency distributions in temperate and tropical karst regions. In *Sinkholes: Their Geology, Engineering, and Environmental Impact*. B. F. Beck, Ed. (A. A. Balkema, Rotterdam), 65–73.
Trombe, F. (1952) *Traité de Spéléologie* (Payot, Paris).
Truesdale, G. A., A. L. Downing, and G. F. Lowden (1955) The solubility of oxygen in pure water and sea-water. *J. Appl. Chem. (London) 5*, 53–62.
Tschang, H.-L (1962) Some geomorphological observations in the region of Tampin, Southern Malaya. *Zeits. Geomorph. 6*, 253–259.
Tsui, P. C. and D. M. Curden (1984) Deformation associated with gypsum karst in the Salt River Escarpment, northeastern Alberta. *Canad. J. Earth Sci. 21*, 949–959.
Tullis, E. L. and J. P. Gries (1938) Black Hills caves. *Black Hills Engineer, 24*, 233–271.
Udden, J. A. (1925) *Etched potholes*. Univ. Texas Bull. No. 2509.
Ugolini, F. C. (1975) Ice-rafted sediments as a cause of some thermokarst lakes in the Noatak River delta, Alaska. *Science, 188*, 51–53.
Urbani, F. and E. Szczerban (1974) Venezuelan caves in non-carbonate rocks: A new field for karst research. *Natl. Speleol. Soc. News, 32*, 233–235.
Vandenberghe, A. (1964) Remarques sur les théories karstiques. *Bull. Bur. Rech. Géol. Minières* No. 2, 33–50.
Van Eysinga, F. W. B. (1983) *Geological Time Table* (Elsevier, Amsterdam).
Vanoni, V. A. (1963) Sediment transportation mechanics: Suspension of sediment. *J. Hydraulics Div. Proc. Amer. Soc. Civil Engineers 89*, 45–76.
Vanoni, V. A. (1966) Sediment transportation mechanics: Initiation of motion. *J. Hydraulics Div. Proc. Amer. Soc. Civil Engineers 92*, 291–314.
Varnedoe, W. W. Jr. (1973) *Alabama Caves and Caverns* (Privately published, Huntsville, AL).

Varnedoe, W. W. Jr. (1980) *Alabama Caves, 1980* Alabama Cave Survey.
Vennard, J. K. and R. L. Street (1982) *Elementary Fluid Mechanics*, 6th ed. (John Wiley, New York).
Verber J. L. and D. H. Stansbury (1958) Caves in Lake Erie Islands. *Ohio J. Sci. 53*, 358–362.
Verstappen, H. T. (1960) Some observations on karst development in the Malay Archipelago. *J. Tropical Geograph. 14*, 1–10.
Verstappen, H. T. (1964) Karst morphology of the Star Mountains (Central New Guinea) and its relation to lithology and climate. *Zeits. Geomorph. 8*, 40–49.
Vineyard, J. (1958) The reservoir theory of spring flow. *Natl. Speleol. Soc. Bull. 20*, 46–50.
Vineyard, J. and G. L. Feder (1974) Springs of Missouri. *Missouri Geol. Surv. and Water Resources*, WR 29.
Von Knebel, W. (1906) *Cave Science with Consideration of Karst Phenomena.* Summlung Naturwissenscraftlicher und Mathematischer Monographien, Heft 15 (Druck und Verlag von Friedrich Vieweg und Sohn, Braunschweig). (partial translation in *Cave Geology 1*, 139–162).
Von Wissmann, H. (1954) Der Karst der humiden Heissen und sommerheissen Gebiete Ostasiens. *Erdkunde 8*, 122–130.
Walker, A. C. (1961) *Contamination of underground water in the Bellevue area.* Ohio Div. of Water Report (Columbus, OH).
Walsh, J. (1972) The exploration of Rimstone River Cave, Perry County, Missouri. *Natl. Speleol. Soc. News 30*, 50–57.
Walton, W. C. (1962) *Selected analytical methods for well and aquifer evaluation.* Illinois State Water Surv. Bull. 49.
Ward, J. C. (1964) Turbulent flow in porous media. *J. Hydraulics Div., Proc. Amer. Soc. Civil Engineers 90*, 1–12.
Warren, W. M. and C. C. Wielchowsky (1973) Aerial remote sensing of carbonate terranes in Shelby County, Alabama. *Ground Water 11*, 14–26.
Watson, R. A. (1966a) Notes on the philosophy of caving. *Natl. Speleol. Soc. News 24*, 54–58.
Watson, R. A. (1966b) Central Kentucky karst hydrology. *Natl. Speleol. Soc. Bull. 28*, 159–166.
Watson, R. A. and P. M. Smith (1971) Underground wilderness. *Internatl. J. Environ. Studies 2*, 217–220.
Watson, R. A. and W. B. White (1985) The history of American theories of cave origin. *Geol. Soc. Amer. Centennial* Special Vol. 1, 109–123.
Webb, J. A. and B. L. Finlayson (1984) Allophane and opal speleothems from granite caves in south-east Queensland. *Australian J. Earth Sci. 31*, 341–349.
Wells, S. G. (1973) Geomorphology of the sinkhole plain in the Pennyroyal Plateau of the central Kentucky karst. MS Thesis, Geology, University of Cincinnati.
Werner, E. (1977) Chloride ion variations in some springs of the Greenbrier limestone karst of West Virginia. In *Hydrologic Problems in Karst Regions*, R. R. Dilamarter and S. C. Csallany, Eds. (Western Kentucky University, Bowling Green, KY), 357–363.
Weyl, P. K. (1958) The solution kinetics of calcite. *J. Geol. 66*, 163–176.

White, E. L. (1975) Role of carbonate rocks in modifying extreme flow behavior. PhD Thesis, Civil Engineering, The Pennsylvania State University.

White, E. L. (1976) Role of carbonate rocks in modifying flood flow behavior. *Water Resources Bull. 12*, 351–370.

White, E. L. (1977) Sustained flow in small Appalachian watersheds underlain by carbonate rocks. *J. Hydrol. 32*, 71–86.

White, E. L. and B. M. Reich (1970) Behavior of annual floods in limestone basins in Pennsylvania. *J. Hydrol. 10*, 193–198.

White, E. L. and W. B. White (1968) Dynamics of sediment transport in limestone caves. *Natl. Speleol. Soc. Bull. 30*, 115–129.

White, E. L. and W. B. White (1969) Processes of cavern breakdown. *Natl. Speleol. Soc. Bull. 31*, 83–96.

White, E. L. and W. B. White (1979) Quantitative morphology of landforms in carbonate rock basins in the Appalachian Highlands. *Geol. Soc. Amer. Bull. 90*, 385–396.

White, E. L. and W. B. White (1983) Karst landforms and drainage basin evolution in the Obey River basin, north-central Tennessee, U.S.A. *J. Hydrol. 61*, 69–82.

White, E. L. and W. B. White (1984) Flood hazards in karst terrain: Lessons from the Hurricane Agnes Storm. *Internatl. Contrib. Hydrogeology 1*, 261–264.

White, G. W. (1926) The limestone caves and caverns of Ohio. *Ohio J. Sci. 26*, 73–116.

White, P. J. (1948) The Devil's Sinkhole. *Natl. Speleol. Soc. Bull. 10*, 2–14.

White, W. B. (1960) Terminations of passages in Appalachian caves as evidence for a shallow phreatic origin. *Natl. Speleol. Soc. Bull. 22*, 43–53.

White, W. B. (1969) Conceptual models for limestone aquifers. *Ground Water, 7*, 15–21.

White, W. B. (1976a) *The Caves of Western Pennsylvania*. Pennsylvania Geol. Surv. General Geol. Report 67.

White, W. B. (1976b) Cave minerals and speleothems. In *The Science of Speleology*, T. D. Ford and C. H. D. Cullingford, Eds. (Academic Press, London), 267–327.

White, W. B. (1977a) Characterization of karst soils by near infrared spectroscopy. *Natl. Speleol. Soc. Bull. 39*, 27–31.

White, W. B. (1977b) Conceptual models for carbonate aquifers: Revisited. In *Hydrologic Problems in Karst Terrains*, R. R. Dilamarter and S. C. Csallany, Eds. (Western Kentucky University, Bowling Green, KY), 176–187.

White, W. B. (1977c) Role of solution kinetics in the development of karst aquifers. *Internatl. Assoc. Hydrogeol. Memoir 12*, 503–517.

White, W. B. (1979) Karst landforms in the Wasatch and Uinta Mountains, Utah. *Natl. Speleol. Soc. Bull. 41*, 80–88.

White, W. B. (1981) Reflectance spectra and color in speleothems. *Natl. Speleol. Soc. Bull. 43*, 20–26.

White, W. B. (1982) Mineralogy of the Butler Cave-Sinking Creek System. *Natl. Speleol. Soc. Bull. 44*, 90–97.

White, W. B. (1984) Rate processes: Chemical kinetics and karst landform

development. In *Groundwater as a Geomorphic Agent*, R. G. LaFleur, Ed. (Allen and Unwin, Boston), 227–248.
White, W. B. and G. H. Deike III (1962) Secondary mineralization in Wind Cave, South Dakota. *Natl. Speleol. Soc. Bull. 24*, 74–87.
White, W. B. and G. H. Deike III (1976) Hydraulic geometry of solution conduits. *Proc. 1976 Natl. Speleol. Soc. Convention*, E. Werner, Ed., 57–60.
White, W. B. and J. W. Hess (1982) Geomorphology of Burnsville Cove and the geology of the Butler Cave–Sinking Creek System. *Natl. Speleol. Soc. Bull. 44*, 67–77.
White, W. B., G. L. Jefferson, and J. F. Haman (1966) Quartzite karst in southeastern Venezuela. *Internatl. J. Speleol. 2*, 309–314.
White, W. B. and J. Longyear (1962) Some limitations on speleo-genetic speculation imposed by the hydraulics of groundwater flow in limestone. Nittany Grotto Newsletter 10, 155–167.
White, W. B., B. E. Scheetz, S. D. Atkinson, D. Ibberson, and C. A. Chess (1985) Mineralogy of Rohrer's Cave, Lancaster County, Pennsylvania. *Natl. Speleol. Soc. Bull. 47*, 17–27.
White, W. B. and V. A. Schmidt (1966) Hydrology of a karst area in east-central West Virginia. *Water Resources Res. 2*, 549–560.
White, W. B. and J. A. Stellmack (1968) Seasonal fluctuations in the chemistry of karst groundwater. *Proc. 4th Internatl. Congress Speleol. Ljubljana, 3*, 261–267.
White, W. B., R. A. Watson, E. R. Pohl, and R. Brucker (1970) The central Kentucky karst. *Geograph. Rev. 60*, 88–115.
White, W. B. and E. L. White (1970) Channel hydraulics of free-surface streams in caves. *Caves and Karst 12*, 41–48.
White, W. B. and E. L. White (1974) Base-level control of underground drainage in the Potomac River Basin. *Proc. 4th Conf. Karst Geol. and Hydrol.* H. W. Rauch and E. Werner, Eds. West Virginia Geological Survey, 41–53.
Whittemore, D. O. and R. R. Parizek (1976) Application of a VLF electromagnetometer to detection of karst underground channels. In *Karst Hydrology and Water Resources*, V. Yevjevich, Ed. (Water Resources Pub., Fort Collins, CO), 658–661.
Wigley, T. M. L. (1971) Ion pairing and water quality measurements. *Canad. J. Earth Sci. 8*, 468–476.
Wigley, T. M. L. (1977) *WATSPEC: A computer program for determining the equilibrium speciation of aqueous solutions*. British Geomorph. Res. Group Tech. Bull. 20.
Wigley, T. M. L. and M. C. Brown (1976) The physics of caves. In *The Science of Speleology*, T. D. Ford and C. H. D. Cullingford, Eds. (Academic Press, London), 329–358.
Wigley, T. M. L., J. J. Drake, J. F. Quinlan, and D. C. Ford (1973) Geomorphology and geochemistry of a gypsum karst near Canal Flats, British Columbia. *Canad. J. Earth Sci. 10*, 113–129.
Wigley, T. M. L. and L. N. Plummer (1976) Mixing of carbonate waters. *Geochim. Cosmochim. Acta 40*, 989–995.
Wilcock, J. D. (1968) Some developments in pulse-train analysis. *Trans. Cave Res. Group Great Britain 10*, 73–98.

Wilford, C. E. (1966) "Bell Holes" in Sarawak Caves. *Natl. Speleol. Soc. Bull.* *28*, 179–182.

Wilkening, M. H. and D. E. Watkins (1976) Air exchange and ^{222}Rn concentrations in the Carlsbad Caverns. *Health Physics 31*, 139–145.

Williams, J. H. and J. D. Vineyard (1976) Geologic indicators of subsidence and collapse in karst terrain in Missouri. Presentation at 55th Annual Meeting (Transportation Res. Board, Washington, DC).

Williams, P. W. (1963) An initial estimate of the speed of limestone solution in County Clare. *Irish Geograph. 4*, 432–441.

Williams, P. W. (1966a) Morphometric analysis of temperate karst landforms. *Irish Speleol. 1*, 23–31.

Williams, P. W. (1966b) *Limestone pavements with special reference to Western Ireland.* Inst. British Geographers. Trans. Paper No. 40, 155–172.

Williams, P. W. (1969) The geomorphic effects of ground water. In *Water, Earth, and Man*, R. J. Chorley, Ed. (Methuen, London), 269–284.

Williams, P. W. (1970) *Limestone morphology in Ireland* (Irish Geographical Studies, Dept. of Geography, Queens University, Belfast), 105–124.

Williams, P. W. (1971) Illustrating morphometric analysis of karst with examples from New Guinea. *Zeits. Geomorph. 15*, 40–61.

Williams, P. W. (1972a) The analysis of spatial characteristics of karst terrains. In *Spatial Analysis in Geomorphology*, R. J. Chorley, Ed. (Methuen, London), 135–163.

Williams, P. W. (1972b) Morphometric analysis of polygonal karst in New Guinea. *Geol. Soc. Amer. Bull. 83*, 761–796.

Williams, P. W. (1973) Variations in karstlandforms with altitude in New Guinea. *Geogr. Zeits. Beihefte 32*, 25–33.

Williams, P. W. (1983) The role of the subcutaneous zone in karst hydrology. *J. Hydrology. 61*, 45–67.

Williams, P. W. (1985) Subcutaneous hydrology and the development of doline and cockpit karst. *Zeits. Geomorph. 29*, 463–482.

Williams, P. W. and R. K. Dowling (1979) Solution of marble in the karst of the Pikikiruna Range, Northwest Nelson, New Zealand. *Earth Surface Processes 4*, 15–36.

Wilson, C. R. and P. A. Witherspoon (1974) Steady state flow in rigid networks of fractures. *Water Resources Res. 10*, 328–335.

Wilson, J. R. (1979) Glaciokarst in the Bear River Range, Utah. *Natl. Speleol. Soc. Bull. 41*, 89–94.

Witherspoon, P. A., C. H. Amick, J. E. Gale, and K. Iwai (1979) Observations of a potential size effect in experimental determination of the hydraulic properties of fractures. *Water Resources Res. 15*, 1142–1146.

Witherspoon, P. A., J. S. Y. Wang, K. Iwai, and J. E. Gale (1980) Validity of cubic law for fluid flow in a deformable rock fracture. *Water Resources Res. 16*, 1016–1024.

Wojcik, Z. (1961) Karst phenomena and caves in the karkonosze granites. *Die Höhle, 12*, 1–44.

Wolfe, J. A. (1978) A paleobotanical interpretation of Tertiary climates in the northern hemisphere. *Amer. Scientist 66*, 694–703.

Wolfe, J. A. (1979) *Temperature parameters of humid to mesic forests of eastern Asia and relation to forests of other regions of the northern hemisphere and Australia.* U.S. Geol. Surv. Prof. Paper 1106.

Wolfe, T. E. (1973) Sedimentation in karst drainage basins along the Allegheny Escarpment in southeastern West Virginia. PhD Thesis, Geography, McMaster University.

Wood, C. (1974) The genesis and classification of lava tube caves. *Trans. British Cave Res. Assoc. 1,* 15–28.

Wood, C. (1976) Caves in rocks of volcanic origin. In *The Science of Speleology,* T. D. Ford and C. H. D. Cullingford, Eds. (Academic Press, London), 127–150.

Wood, W. W. and M. J. Petraitis (1984) Origin and distribution of carbon dioxide in the unsaturated zone of the southern High Plains. *Water Resources Res. 20,* 1193–1208.

Woodward, H. P. (1961) A stream piracy theory of cave formation. *Natl. Speleol. Soc. Bull. 23,* 39–58.

Wright, H. E. Jr. and D. G. Frey (1965) *The Quaternary of the United States* (Princeton University Press, Princeton, NJ).

Wulkowicz, G. M. and Z. A. Saleem (1974) Chloride balance of an urban basin in the Chicago area. *Water Resources Res. 10,* 974–982.

Yarborough, K. A. (1976) Investigation of radiation produced by radon and thoron in natural caves administered by the National Park Service. *Proc. Natl. Cave Management Symp.,* T. Aley and D. Rhodes, Eds. (Speleobooks, Albuquerque, NM), 59–69.

Yarborough, K. A. (1977) Airborne alpha radiation in natural caves administered by the National Park Service. *Proc. Natl. Cave Management Symp.* R. Zuber, J. Chester, S. Gilbert, and D. Rhodes, Eds. (Adobe Press, Albuquerque, NM), 125–132.

Yarborough, K. A. (1978) The National Park Service cave radiation research and monitoring program. *Proc. Natl. Cave Management Symp.* R. C. Wilson and J. J. Lewis, Eds. (Pygmy Dwarf Press, Oregon City, OR), 27–40.

Yevjevich, V. (1976) *Karst Hydrology and Water Resources* (Water Resources Publications, Fort Collins, CO).

Yonge, C. J., D. C. Ford, J. Gray, and H. P. Schwarcz (1985) Stable isotope studies of cave seepage water. *Chemical Geol. 58,* 97–105.

Zötl, J. G. (1961) Die Hydrographie des nordostalpinen Karstes. *Steirische Beit. Hydrogeol.,* 1960/61, 53–183.

Zötl, J. G. (1974) *Karsthydrogeologie* (Springer-Verlag, Vienna).

Zuffardi, P. (1976) Karsts and economic mineral deposits. In *Handbook of Strata-Bound and Stratiform Ore Deposits,* Vol. 3 *Supergene and Surficial Ore Deposits: Textures and Fabrics,* K. H. Wolf, Ed. (Elsevier, Amsterdam), 175–212.

Illustration credits

Donald W. Ash: Fig. 9.18
Association for Mexican Cave Studies: Fig. 3.17
William Back: Fig. 9.19
Barry F. Beck: Fig. 3.20
George Brook: Fig. 7.5
Derek C. Ford: Figs. 9.1, 9.3, 9.4, 9.5, and 9.6
Ronald Greeley: Fig. 12.6
John Gunn: Fig. 8.2
Kenneth W. Johnson: Fig. 11.2
William K. Jones: Fig. 4.2
Ernst Kastning: Fig. 3.12
Keith Kirk: Fig. 13.5
Charlie and Jo Larson: Fig. 12.5
Brainerd Mears: Fig. 12.7
Missouri Speleological Survey: Fig. 3.11
T. P. O'Holleran: Fig. 8.15
Arthur N. Palmer: Fig. 9.7
James F. Quinlan: Fig. 11.1
Victor A. Schmidt: Fig. 10.7
Bernard L. Smeltzer: Fig. 3.9
John A. Stellmack: Fig. 4.7
Jerry D. Vineyard: Fig. 13.6
Paul Williams: Fig. 2.14
West Virginia Speleological Survey: Figs. 3.10 and 3.14

Subject Index

Activity coefficient, 124, 125, 127
Adriatic karst, 39
Aeolian deposits, 221, 238
Aerial photography, 4
Alkalinity measurements, 134, 135
Allochthonous sediments, 220, 221
Allogenic recharge, 183, 190f, 201, 213, 279, 280f, 281
Allogenic runoff, 172
Allogenic sediments, 159f
Allogenic streams, 157f, 172
Anastomoses, 92
Anatase, 240
Anhydrite, 260
Anion measurements, 136
Anthodites, 222, 247, 255, 256f
Appalachian Highlands, 15
Appalachian Plateau, 21
Aquifers, 149–151, 173f
   coastal karst, 402–405, 403f
   conduit, 171, 284
   diffuse flow, 171, 174, 284
   perched, 172
   response, 208-210
Aragonite, 247, 256, 257f, 301
Ardealite, 229
Area ratio, doline, 33, 34
Arrhenius equation, 142
Artesian aquifer, 151f
Artesian caves, 278
Aspect ratio, 33
Aults, 44
Authigenic recharge, 279, 280f

Autochthonous sediments, 220, 221, 228, 238

Barite, 240
Base flow, 174, 189, 190
Basins, surface and subsurface, 153
Bauxite, 225
Bedforms, 241
Bedrock geology, 213
Beekmantown dolomite, 202, 284
Beekmantown wells, 203f
Bellefonte dolomite, 147f
Bernoulli equation, 160, 161, 180
Bifurcation ratio. *See* Stream, ordering
Big Sunk Cane, 17
Biotite, 240
Biphosphammite, 229
Birnessite, 263
Blind valley, 13–14
Blister caves, 336–337
Blödite, 259
Boehmite, 225
Bogaz, 44
Boiling Hole, 211f, 212
Bom Jesus de Lapa, 46
Bond number, 250
Borneo, 42
Botryoidal forms, 222, 255, 257f
Bouguer anomaly, 374
Breakdown, 221, 229–233
   features, 236–237
   mechanisms, 233–235
Brookite, 240

Brush Valley, 25f
Brushite, 229

Ca/Mg ratio, 140, 213
Calcite, 247, 283f
  dissolution kinetics, 142, 213
  reaction rate, 144
Canyon, 72–76, 72f
Carbon dioxide
  atmosphere, 193, 194f
  availability, 202
  cave atmosphere, 196, 197f
  climate control, 213
  denundation, 219
  enrichment, 211f
  global model, 198, 199f
  in soils, 193–196, 195f, 210, 215
  karst waters, 201
  measurements, 132
  partial pressure, 138
  production, 196–198
  sources, 198–200
  vertical shafts, 296f
Carbonate aquifers, 171–183
  closed, 203, 204, 205f, 211, 213, 214
  flow types, 286
  open, 203, 204, 205f, 206, 213, 214
Carbonate mineral solubility, 128
Carbonate minerals, 247
Carbonate rocks, 3, 221
  composition, 221–224
  dissolution, 119–148, 212
Cation measurements, 135
Cave
  Agen Allwedd, 316
  Alabaster, 333
  Anvil, 83, 282
  Big Hole, 25
  Big Ridge, 21
  Blue Grotto, 352
  Brady's Bend, 78, 83f
  Bungonia, 200
  Butler-Sinking Creek, 91, 206, 262
  Cameron, 78
  Carlsbad Caverns, 71, 260, 298, 299f, 308; radon levels, 403
  Carroll, 242f
  Cascade Pit, 296f
  Casement, 345
  Castleguard, 90, 278, 403
  Clearwater, 65
  Cottonwood, 260
  Coumo d'Hyouernedo, 65
  Crevice, 65
  Crystal, radon levels, 403
  Cumberland Caverns, 65, 261f
  Dachstein Ice, 263
  Dante's Descent, 23
  Dent de Crolles, 65
  Devil's Sinkhole, 23
  Dingo Donga, 229
  Doolin System, 182
  Drum, 200
  Easegill-Lancaster System, 65
  Eisriesenwelt, 263
  El Sotano, 23
  Et-Tabun, 229
  Fighiera-Corchia-Farolfi, 65
  Fingal's, 352
  Fisher, 78, 81f
  Fisher Ridge System, 65
  Fleming, 78, 79f
  Fletcher, 78, 80f, 374, 375f
  Forrat Mico, 337f, 338f
  Fossil Mountain Ice, 263
  Friers Hole, 65, 179
  Grand Caverns, 254
  Greenville Saltpetre, 83, 84f
  Grill, 200
  Gua Air Jernih, 65
  Gunong Mulu, 264
  Hellhole, 23, 237
  Hirlatzhöhle, 65
  Hölloch, 65, 273
  Horn Hollow, 97f
  Hostermans Pit, 251f
  Howards Waterfall, 399
  Howe's, 14
  Huautla, 89
  Indian, radon levels, 403
  Jewel, 65, 263
  Kananda Atea, 65
  Kane, 260, 298
  L'Alpe, 65
  La Guangola, 229
  Lehman, 254, 403
  Lilburn, radon levels, 403
  Little Trimmer, 321f
  Luray Caverns, 374
  McFail's, 14
  Mammoth, aquifers, 156f, 175; breakdown, 233–237; CO_2 levels, 196; cave levels, 85, 86f, 272, 318; Cleaveland Avenue, 74, 75f; glacial damming, 65; length, 65; meander bends, 176; minerals, 259–260; New Discovery, 74, 75f, 94f; paleomagnetism, 316, 317f; radon

levels, 403; runoff character, 175; scallops, 178, 179; sediments, 239, 239f; Turnhole, 209f; vertical shafts, 296f; water balance, 174
Mark Twain, 78
Marvel, 237
Milroy, 242f
Murra-el-elevyn, 229
Mystery (Minnesota), 242f
Mystery (Missouri), 200
Niah Great, 229
Ojo Guareña, 65
Optimistceskaya, 316, 334f
Organ System, 65
Owl, 210
Ozernaja, 65
Paradise Ice Cave, 344, 345f
Parker, 14
Peak Cavern, 65
Peña Colorado, 278
Penns, 187
Petrogale, 229
Pierre St. Martin, 65
Pig Hole, 229
Poulnagollum, 179
Purificacion, 65
Putrid Pit, 200
Pwll y Gwynt, 316
Ragge Jarre Raige, 88
Red Del Rio Silencio, 65
Rickwood Caverns, 95f
Rio Camuy System, 23
Rössing, 322
Roudsea Wood, 179
Runaway Bay, 101f
Sedom, 338
Siebenhengste-Hohgant, 65
Sierra de El Abra, 273
Skipton, 229
Skull, 14, 78, 82f
Skylight, 335f
Skyline Caverns, 256f
Smullton Sink, 25f
Soldiers, radon levels, 403
Sotano Golondrinas, 23
Swago Pit, 90, 296f
Thorn Mountain, 254
Tytoona, 197f
Unknown, 229
White Lady, 179
Wilgie Mia, 229
Wind, 65, 71
Wyandotte, 237
Zbrasov, 200
Zolushka, 65
Cave passage, paleodischarge, 179
Cave volume, chemical and lithologic control, 285f
Caverns. *See* Cave
Caves, 60–64
  angulate, 79f
  artesian, 273, 278
  bad-air, 200
  Black Hills, 240, 257
  branchwork, 78, 81f, 82f
  classification, 353
  conduit, 78, 80f
  coral, 256, 257f
  definition, 60
  Demanova Valley, 271
  entrances, 61, 62f
  Guadalupe Mountains, 260
  high gradient, 87–91
  length distributions, 61–64
  levels, 318–319, 319f
  long cave list, 65
  maps, 64–65
  maze and network, 78, 83f, 84f, 282f
  Mona Island, 229
  nonintegrated, 278
  patterns, 69f, 77–91, 280f
  pearls, 252f, 257
  radon levels, 403
  regional surveys, 66, 67
  sediment classification, 221
  shapes, 68–77
  surveys, 66, 67f
  terminations, 61–64
  tiered, 85
Ceiling channels, 92
Ceiling pockets, 99, 101f
Celestite, 240, 259
Cenote, 26
Chalcedony, 223, 225
Champlainian limestone, 285f
Channel features, 92
Chemical character, karst water, 201
Chemical equilibrium, 121–130, 206, 220
Chemical evolution, 203
Chemical hydrographs, 209f
Chemical kinetics, 140–147
Chemical reactions, 120
Chemical sediments, 221
Chlorite, 223, 224
Chromite, 240
Cibao formation, 49f

Classification
  cave sediments, 221
  caves, 353
  karst aquifers, 286
  karst landforms, 20
  speleothems, 222
Classification by cover, 117–118
Clay
  in carbonate rocks, 223
  unctuous, 237–238
Clay minerals, 224, 283f
Cleft karren, 50, 55, 57f
Closed depressions, 22f, 23, 32, 33, 225, 227f. See also Sinkholes
Coburn formation, 284
Cockpits, 31, 32, 33
Coconino Plateau, 23
Colluvium, 21
Colossal Dome, 296f
Commercial caves, 377–378
Compositional tetrahedron, 283
Compound sinks, 31, 33, 40
Concentration, 290
Concretions, 223
Conductivity measurements, 132, 133, 208. See also Specific conductance
Conduit aquifers, 171, 287. See also Carbonate aquifers
Conduit flow, 160, 171, 172
Conduit rate curve, 287f
Conduit springs, 202, 206, 207f, 211f, 293
Conduit systems, 174, 182, 206, 207f, 213
Cone karst, 47, 48f, 49f
Crandallite, 229, 262
Critical thresholds, 290–293
Crusts, 222, 247, 259
Crystal linings, 223
Cuba, 42
Cube law, 171
Cumberland Plateau, 17
Cutter, 5, 42–44, 58. See also Grike
Cutter-and-pinnacle karst, 357, 358f

D'Arcy flow, 154
D'Arcy's law, 169, 189
D'Arcy-Weisbach equation, 163, 164, 169, 271
Dahllite, 229
Dams, 5, 369–372
Darai Hills, 32f, 33f
Darapskite, 262

Dating
  amino acid recemization, 316–317
  carbon-14, 308–310, 316
  clastic sediment record, 326–327
  electron-spin resonance, 312–313, 313f, 314f
  magnetic records, 315–316
  techniques, 309f
  thermoluminescence, 313–315
  uranium-series, 310–312
Debye-Hückel equation, 124, 125
Deckenkarren, 50, 94
Density
  gypsum, 333
  limestone, 216, 217, 231, 232f
  water, 162
Denudation rate
  gypsum, 333
  limestone, 215–219, 216f, 218f
Depositional sequences, 240, 241
Depression density, doline, 33, 34, 36, 37, 110f, 172
Detritus, weathering, 228
Diaspore, 225
Diffuse flow springs, 174, 202, 206, 207f, 210, 211f
Diffuse infiltration, 172, 183, 201, 213, 218
Discharge parameters, 187
Dissociation constants, 128
Dissolution kinetics, 142, 146, 213, 335
Dissolution rate, 143f
Dissolution reactions, 121
Distributions
  caves, 62f, 63f
  sinkholes, 35f
Diving gear, 5
Dolines
  definition, 20, 24f, 25f, 26f, 27f, 29f, 30f. See also Sinkholes
  measures, 33–39
  sediment deposition, 225, 227
Dolomite, 247, 283f
  dissolution kinetics, 146
Drager apparatus, 132
Drainage, Hidden River, 156f
Drainage basins, 8, 279–282
Drainage divide, 8
Drainage network, 8
Drapery, 222, 253
Dripstone, 222, 247, 250, 259
Dry valley, 12f, 13, 17
Dunn method, 180
Dye tracing, 180

Elliptical tube, 72f, 73–76, 81f, 83, 84f
Endellite clays, 228
Engineering foundation, 5
Enthalpy, 126, 127
Entrance talus, 221, 237
Entrances, 61
Epidote, 240
Epsomite, 259, 260
Equilibrium constants, 121, 122, 124–127, 128
Erratic forms, 222
Etched features, 99–101
Etchpits, 99
Evaporites
  beds in U.S., 329f
  chemistry, 259, 260
  deposits, 221, 258–262, 329
  interstratal karst, 329
  karst types, 328
  mineralogy, 258, 259
  rock solubility, 329–331
  speleothems, 260–262
Evapotranspiration, 174, 175, 198, 217, 218
Exhaustion coefficient, 186, 188, 189
Exploration, 5

Fall velocity, 246
Fence diagram, 167f
Filiform, 255
Fitzroy Basin, 44
Flexural stress, 231
Flood pulse, 192
Floods, 190–192, 385–387
Floor slots, 92
Flow pattern, initialization, 289f
Flowstone, 222, 247, 253, 254f
Flutes, 95, 178
Fluvial deposits, 221, 238
Fluvial drainage, 9
Fluviokarst, 10, 112, 113, 157f
Flysch, 41
Foot-cave, 47
Forchheimer equation, 170
Ford-Ewers model, 272, 278, 294
Foundation engineering problems, 356–362
Foundation engineering solution, 361–362
Fountain of Vaucluse, 19
Fracture aquifers, 287. *See also* Conduit aquifers
Fracture flow, 171

Fracture systems, 380–383, 381f, 382f
Friction factor, 163, 171
Frost wedging, 234
Froude number, 165, 166, 167f, 244, 297

Gage point, 8
Galena dolomite, 298
Garnet, 240
Gasconade dolomite, 81f, 283
Gatesburg dolomite, 202, 211f
Gatesburg wells, 203f
Geomorphology, 4
Geophysical methods, 373f. *See also* Remote sensing
Geothermal ice caves, 346
Ghyben-Herzberg principle, 402
Gibbs free energy, 125, 126
Gibbsite, 225
Glacial alluviations, 18
Glacial debris, 18
Glacial deposits, 221, 238
Glacial drift, 21
Glacial fill, 18
Glacier caves, 343–346
Glasgow Upland, 14
Globulites, 257f
Goethite, 223, 240
Green River, 14, 85, 174, 175
Greenbrier limestone, 13f, 80f, 84f
Grike, 42, 44, 116. *See also* Cutter
Groove karren, 50, 55, 57f
Grottos, 6
Groundwater, 4, 171, 265–266
Groundwater flow, 160
Growth rate curve, 321f
Guanine, 229
Guano, 228, 229, 262
Gypsum
  caves, 333–335, 334f
  dissolution rate, 335
  formula, chemistry, 259
  in caves, 261f
  karst landforms, 331–333
  needles, 262
  shallow doline, 332f
  speleothems, 260

Hagen-Poiseuille equation, 162, 271
Halite, 259
Hannayite, 229
Hardness
  conduit flow, 207f–208
  definition, 136–138, 137f

452 / Subject index

Hardness (continued)
   denundation rate, 215–218
   diffuse flow, 207f, 208, 211, 211f, 212
   karst waters, 201–203
   specific conductance, 137f, 138
   springs and wells, 137f
   time variation, 206, 207f
Hazards
   explosions, 398–399, 401
   floods, 385–387
   industrial wastes, 399–401
   radiation, 401–402
Helderberg limestone, 21, 79f, 82
Helderberg Plateau, 14
Helictites, 222, 255, 255f
Hematite, 223, 225, 228
Henry's law, 388
Hexahydrite, 259
High storage aquifers, 190f
Highland Rim, 17, 37, 38f, 103
Hjulstrom curve, 244, 245f
Hohlkarren, 50, 54
Holokarst, 10, 11f
Honey Creek, 19
Horizontal grooves, 93
Hornblende, 240
Hot springs, 200
Huautla Plateau, 264
Hums, 46
Huntite, 247
Hydraulic conductivity, 152, 153f, 183, 186
Hydraulic features, 93–99
Hydraulic geometry, 174
Hydraulic jump, 166
Hydraulic radius, 161, 164–166, 168, 245
Hydrogen sulfide, 260
Hydrographs, 12f, 183–186, 191, 192, 206, 217
Hydrologic cycle, 155
Hydrologic models, 271, 272
Hydrologic system, 171
Hydrology, 4
Hydromagnesite, 247
Hydroxyapatite, 229

Ice, 221, 263
Icy Cove, 30
Igneous rock karst, 340
Illite, 223, 224, 228
Incised meanders, 92
Index of pitting, doline, 34
Indiana, 43
Infiltrates, 221, 237, 238

Infrared imagery, 4
Ingleborough, 15f
Inorganic species, 120
Internal runoff, 172, 201, 213
Ion activity product, 138
Ion pairs, 127–128
Ireland, 42
Isotope compositions, 322–324
Isotope ratios, 324

Jamaica, 32, 42

Kaminitza, 50, 53, 54f, 332
Kaolinite, 223, 224, 228
Karren, 46, 49–59
   cleft karren, 50, 55, 57f
   deckenkarren, 50, 94
   groove, 50, 55, 57f
   hohlkarren, 50, 54
   kaminitza, 50, 53, 54f, 332
   kluftkarren, 43, 50, 55, 56f, 115f, 332, 343
   meänderkarren, 50, 54
   pedistal, 50
   pinnacle, 50, 58f, 59
   pit-and-tunnel, 50, 56, 57f, 58
   rillenkarren, 50, 51, 52f, 116, 332, 338f
   rinnenkarren, 50, 53, 332, 343
   rundkarren, 50, 54
   spitzkarren, 46, 50, 59, 338f
   trench, 55, 57f
   trittkarren, 50, 53f
   wandkarren, 50
Karst
   Adriatic, 106
   alpine, 58, 103, 182
   aquifer flow system, 173
   buried, 117
   cave, 116
   characteristics, 109
   chemical reactions, 120
   cockpit, 108, 111f
   cone and tower, 46–49, 109, 111f, 112f, 212
   constant, 64
   covered, 117
   cycle, 10
   definition, 3
   denudation, 215–219, 216f
   Dinaric, 42
   doline, 103, 107, 108f
   drainage, 9–11
   drainage basins, 4, 9

Subject index / 453

exhumed, 118
exposed, 117
fenster, 25f
fluvio, 112
hydrology, 4, 149-192
interstratal, 117
kegalkarst, 46-47, 49f
labyrinth, 116
landforms/landscapes, 20, 103-118
mantled, 117
naked, 117
pavement, 113, 115f, 212
phytokarst, 59
plains, 319-320
plateau margin, 114f
polje, 29f, 116
polygonal, 32, 32f, 34
process, 104f, 105
relict, 118
science, 7
springs, 9, 18-19, 384-385, 397f
subaqueous, 117
subsoil, 117
temperate climate, 59
terrain, 3, 5
time scale, 303, 304, 305f
tropical, 103
turmkarst, 46
valleys, 11-18
water chemistry, 201
water climatic controls, 213
water regional controls, 212
water table, 181-183
Karst aquifer, central Kentucky, 175
Karst process, 104f, 105
  evaporites, 328
Karstification, 10, 304
Kentucky, southcentral, 18, 174, 175, 212
Kinetic models, 271, 272
Kinetics, dissolution, 142-147
Kluftkarren, 43, 50, 55, 56f, 115f, 332
Knobs, 46
Kyanite, 240

LANDSAT, 4, 384
Laminar flow, 162, 164f, 189, 290
Landfills, 392-393, 393f
Landform characteristics, 3, 8-59
Lapiez. *See* Karren
Laplace equation, 169
Lares limestone, 49f
Lava tube
  Ape Cave, 348, 350f

Raufarholshellir, 348
Saddle Butte System, 349f
Surtshellir, 348
Lava tube distribution map, 350f
Lava tubes, 347-351
Leachate, 224
Legal questions, 367-368
Length distributions, 61-64
Leucoxene, 240
Limestone pedestal, 215, 216f
Limestone/dolomite ratio, 213
Limonite, 223, 240, 263
Lithology, 213
Low flow, 174
Low storage aquifers, 190f
Loyalhanna limestone, 106, 283

MDCP, 37
Madagascar soils, 194
Magnesite, 247
Magnetite, 240
Maligne River, 192
Mammoth Spring, 19f
Manning equation, 168, 180
Manning's n, 168, 245
Mauna Loa Observatory, 194f
Mazes, 178
Mean annual discharge, 175
Meander bend spacing, 177
Meänderkarren, 50, 54
Mechanical transport, 3, 10
Mendip Hills, 37
Merokarsts, 10
Meteoric water line, 324-325f
Meyer-Peter formula, 245
Milroy formation, 284
Mineral deposition and wedging, 235
Mirabilite, 259, 260
Mischungskorrosion, 300
Mitchell Plain, 39
Mixing solutions, 298, 299
Mogotes, 47, 48f
Monetite, 229
Monohydrocalcite, 247
Montmorillonite, 223, 224
Moonmilk, 222, 247, 256, 257f
Moraine, 21
Mundrabillaite, 229

Nesquehonite, 247
Neversink Shaft, 296
Newberyite, 229
Niahite, 229

Nitrate levels, karst springs, 397f
Nitrate minerals, 221, 262
Nolin River, 174, 175

Obey River, 17, 18
Oolite, 43
Open channel flow, 164, 212
Opferkessel. *See* Solution basins
Organic debris, 221, 228
Oulopholite, 222, 260, 261f

pH measurements, 129, 133, 143, 204, 224, 225
Paleokarst, 220
Paragenesis, 276f
Passages
  anastomotic, 68, 69f, 71, 83
  angulate, 68, 82f
  cross section, 71–77
  linear, 68, 82f
  maze, 68, 77
  network, 68, 71
  orientation, 70f
  single conduit, 68, 69f, 77, 78
  sinuous, 69, 176, 177f
  spongework, 68, 69f, 71, 99
Patterns, cave, 69f, 77–91
Pedestal karren, 50
Pendants, 94, 95f
Pennyroyal Plain, 15f, 37, 38f, 103
Pennyroyal Plain, 15f, 37, 38f, 103
Permeability, 151, 153f, 172
Phosphammite, 229
Phosphate minerals, 221, 262
Phreatic storage, 173f
Phreatic zone, 149, 150f, 268
Piezometric surface, 149, 273
Pinnacle karren, 46, 49–51, 58f
Pinnacles, 5, 42, 45, 46, 48f
Pit. *See* Cave
Pit-and-tunnel karren, 50, 56, 57f, 58
Pleistocene, 19, 302–327
Polje, 5, 39–42, 116
Pollution, 5
Pollution sources, 389
Pollution threats, 387–402
Ponor, 18
Popovo Polje, 40f
Porosity, 151, 154f, 154
Porous media flow, 168
Prandl-von Karman equation, 163
Pseudokarst, 347–354
Pseudokarst landform elements, 352f
Pyrite, 259, 260, 263

Quartzite karst, 341, 343

Radon levels, 403
Rate constant, 142, 144, 145
Rate equation, 141, 142, 335
Recreational caving, 378–379
Reflectance ratio, 226f
Remote sensing, 372–377
Reservoir engineering, 368–372
Residual features, 101–102
Resistate minerals, 221, 263
Resurgence, Rio Camuy, 211f, 212
Reynolds number, 97, 161–165, 167, 170, 244, 291, 297
Rillenkarren, 50–52f, 116, 332, 338f
Rillenstein, 55
Rimstone dams, 222, 257
Rinnenkarren, 50, 53, 332, 343
Rise, Orangeville, 207f
Rooms, 76, 77
Rossette, 70f
Rundkarren, 50, 54
Runoff, 8
Runoff characteristics, 183–192
Runoff intensity, 175
Rutile, 240

SMOW, 324, 325f
Salona formation, 284
Salt karst landforms and caves, 337–339
Saltpetre, 262
Saltwater intrusion, 404
Saturation, time variation, 206
Saturation Index
  definition, 138–139
  karst waters, 201
  shafts, 296f
  travertine, 257–258
Saturation ratio, 144
Sauter mean, 98, 179
Scallops, 96–100f, 178, 179, 179f, 180
Sea caves, 352, 353
Sediments
  balance, 158–160
  clastic, 239f, 241f
  fluvial, 238–247
  gravitational, 237, 238
  mineralogy, 247. *See also* Carbonate minerals
  transport, 241–247
Shenandoah Valley, 15
Shields, 222, 254
Shields curve, 244, 245
Sierra de los Organos, 47

Sigmoid Hills, 47, 48f
Silica, 224
Sinkhole collapse, 364f, 365, 366f, 367f
Sinkhole fillings, 225–227
Sinkhole ponds, 23
Sinkhole Plain, 15f, 109, 212, 278
Sinkholes, 3, 17, 20, 28, 35f, 220, 338.
    See also Dolines
Sinking streams, 3, 103, 172, 201. *See also* Swallow holes
Sinks, 28, 30, 30f, 31f
Soil leaching, 224
Soil piping, 42, 357–359, 359f, 362–367
Soil water, 201
Soils
    Beekmantown, 203
    Gatesburg, 202
    karst, 224–225
    terra rosa, 224
Solubility
    aragonite, 248–249, 249f
    calcite, 131f, 249f
    carbon dioxide, 123f
    carbonates, 128–131
    curves, 131f, 249f, 331f
    product constant, 121, 129, 138, 331
    silica, 342f
Solution basins, 341f
Solution canyons, 44
Solution chemistry, 119–148, 224
Solution chimneys, 21
Solution corridors, 45f
Solution sinks, 22
Solutional cavities, 5
Solutional cavity collapse, 359–361, 360f
Solutional sculpturing
    caves, 91–102
    surface, 49–51, 52f
Solutional transport, 3, 343
Solutional weathering, 3
Sotanito, Ahuacatlan, 21
Spattermites, 256, 257f
Specific conductance, 132, 137f, 138, 218
Specific yield, 187
Speleo Digest, 7
Speleology, 4
Speleothem, record, 320–326
Speleothems, 220, 222, 250–257
Spitzkarren, 46, 50, 59, 338f
Spongework, 68, 69f, 71, 99
Spore-tracing, 181
Spring
    Aghia Eleousa, 187

Big, 211, 211f, 384
Davis, 187
Graham, 211f, 212
Obod, 384
Ras-el-Ain, 187
Rock, 185f, 187, 207f
San Marcos, 187
San Pedro, 211f, 212
Silver, 187
Spruce Creek, 207
Thompson, 185f, 187
Turnhole, 209, 209f, 211f, 212
Tuscumbia, 187
Spring Creek basin, 182, 191f
Spring discharge, 174
Springs
    alluviated, 18
    Beekmantown, 203f
    conduit, 203f, 206, 211f
    diffuse, 174, 202, 203f, 206, 211f
    discharge, 187
    Fountain's Fell, 208f
    Gatesburg, 202, 203f
    hardness, 137f
    near shore, 19
    offshore, 19
Stalactite, 222, 251f, 252
Stalagmite, 222, 252
Staurolite, 240
Stokes law, 246
Storativity, 186–189
Stream
    numbering, 9, 14
    ordering, 8, 9, 15, 15f
    profile, 16, 17f
Struvite, 229
Subaqueous forms, 222
Subcritical flow, 166
Subcutaneous zone, 172
Suffosion caves, 353
Sulfuric acid solution, 298
Sunk Cane Branch, 17f
Supercritical flow, 166
Survey, 4, 67
Swallet ordering. *See* Swallow holes
Swallow holes, 9, 10, 14, 15f, 18, 191

Talus caves, 353, 354
Taranakite, 229, 262
Tectonic caves, 352–353
Temperature
    measurements, 132, 133
    time variation, 207, 307
Tendrils, 288

Tension dome, 231f
Thermal equilibrium, 206
Thermodynamic quantities, 126
Thermokarst, 346–347
Thornthwaite potential, 174
Threshold
  critical, 290–293, 292f
  hydraulic, 291
  kinetic, 292–293
  transport, 291–292
Tinajita, 53, 54f, 332
Topography
  clint, 42
  pinnacle, 5, 42
Tourmaline, 240
Tower karst, 46–49
Transmission efficiency, 215
Transmissivity, 186–189
Travertine, 213, 221, 257, 258f
Trench karren, 55, 57f
Trittkarren, 50, 53f
Tufa, 257
Turbulent flow, 162, 163, 164f, 165
Turnhole Spring, 209f, 211f

Underground wilderness, 379
Urea, 229
Uricite, 229
Uvalas. *See* Compound sinks

Vadose zone, 149, 150f, 155, 172, 200
Valhalla Shaft, 296f
Valley sinks, 31
Van't Hoff equation, 127
Vanport limestone, 83
Vaughns Dome, 296f
Velocity profile, 164f

Vermiculite, 223, 224
Vermiform, 255
Vertical drops, 5
Vertical grooves, 95
Vertical rills, 93, 94f
Vertical shafts, 21, 77, 91, 172, 233, 295, 296f
Viscosity, water, 162

Wall pockets, 99
Wandkarren, 50
Wasatch Mountains, 55
Waste composition, 392
Waste injection, 404–405
Water
  balance, 155, 157f, 158, 174
  chemical characteristics, 200, 201, 212–219
  hardness, 202
  properties, 162
  quality standards, 390
  supply, 5, 187, 380–385
  table, 150, 181–183, 294, 295f
  tracing, 180
  wells, 380–383. *See also* Aquifers
  yield, 383–384
Weathering detritus, 221, 228
Well data, 202, 203f
Wells, hardness, 137f
Wet suits, 5
Wetted perimeter, 164, 166f
Whitlockite, 229

Yucatan, 26

Zanjones, 44
Zircon, 240

Author Index

Acree, S. F., 124
Ahlstrand, G. M., 402, 403
Alexander, E. C., 242f
Aley, T., 180, 369, 391
Alhades, A., 328
American Public Health Association, 136
Amick, C. H., 171
Amundson, R. M., 65
Anderson, C. H., 344, 345
Andrews, J. N., 313, 322
Andrieux, C., 250, 255
Apgar, M. A., 393
Aronis, G., 188
Ash, D. W., 299
Ashton, K., 192
Atkinson, S. D., 262
Atkinson, T. C., 151, 172, 198, 217, 218, 322, 403
Aubert, D., 225

Bacastow, R. B., 194
Back, W., 301, 405
Bada, J. L., 313, 317
Baes, C. F., Jr., 194
Bagnold, R. A., 246
Bakalowicz, M., 189, 206
Baker, G., 257
Baker, V. R., 14, 177
Balch, E. S., 263
Balenzano, F., 229
Barnes, H. H., Jr., 168
Baron, G., 247
Barr, T. C., Jr., 67

Bartrum, J. A., 340
Bassett, J. L., 206, 207
Bates, R. G., 124, 133
Bauer, F., 51
Baumgardner, R. W., 363
Bayer, T. N., 67
Bear, J., 160
Beck, B. F., 26, 363
Benjamin, G. J., 322
Berner, R. A., 143, 249
Bik, M. J., 46
Black, D. F., 399
Blount, C. W., 330, 331
Blumberg, P. N., 98
Bögli, A., 7, 49, 51, 54, 299
Bosnak, A. D., 401
Bowler, P. G., 348
Box, E. O., 198, 199, 215
Bradbury, J. C., 220
Braker, W. L., 224
Brandt, A., 328
Bretz, J H., 67, 238, 269, 270, 333
Bridge, P. J., 229
Brink, A. B. A., 367
Brod, L., 66
Broecker, W. S., 309, 310
Brook, G. A., 42, 116, 198, 199, 214, 215, 384
Brown, M. C., 178, 192, 263, 384
Brucker, R. W., 236, 296, 297
Buhmann, D., 293
Bull, P., 327
Burdon, D. J., 171, 186, 188

Burwell, E. B., Jr., 370
Busenberg, E., 145, 146
Butcher, A. L., 66
Butler, D. K., 362
Butler, J. H., 356
Buzio, A., 337

Caillere, S., 247
Calandri, G., 337
Campbell, N. P., 67, 113
Cardosa da Silva, T., 46
Carey, S., 67
Carmi, I., 172
Carpenter, A. B., 393
Carroll, D. E., 224
Carroll, R. W., Jr., 354
Carter, J. A., 310
Carwile, R. H., 240
Castillo, E., 171
Cavaillé, A., 266
Chabert, C., 63f, 65
Chalcraft, D., 343
Chao, E. C. T., 239, 240, 241
Charlton, R. E., Jr., 65
Cheng, R. T.-S., 404
Cherry, J. A., 151, 187, 402
Chess, C. A., 262
Chico, R. J., 374
Chikishev, A. G., 328
Chinese Academy, 112
Chow, V. T., 168
Christ, C. L., 116, 125
Cipra, J. E., 225
Clark, W. C., 193
Clarke, R. M., 229
Clausen, E. N., 351
Colby, B. R., 245
Cole, L. J., 26
Coleman, J. C., 96
Colveé, P., 343
Compton, R. G., 145
Conn, H., 67
Conn, J., 67
Copeland, C. W., 364
Corbel, J., 119, 215, 216, 219
Courbon, P., 65, 88
Coward, J., 201
Cowell, D. W., 214
Crawford, N. C., 114, 181, 386, 401
Creaser, E. P., 26
Crowther, J., 196
Csallany, S. C., 6
Cullingford, C. H. D., 7
Cunningham, R. L., 225
Curden, D. M., 336

Curl, R. L., 60, 61, 64, 98, 142, 178, 249, 250, 253
Cvijič, J., 40, 268

Dalton, R. F., 67
Daly, P. J., 145
Daugherty, R. L., 163
Davies, W. E., 23, 54, 59, 67, 116, 228–230, 236, 239, 240, 241, 270
Davis, D. E., 395
Davis, D. G., 260, 298
Davis, N. W., 65
Davis, S. N., 151
Davis, W. M., 265, 268, 270
Day, M., 32, 33
Deal, D. E., 257, 263
DeBellard Pietri, E., 351
Debenham, N. C., 315
Deike, G. H., III, 86, 177, 178, 179, 228, 257, 318
Deike, R. G., 69
Delecour, F., 132, 196
Dell'Anna, L., 229
DePaepe, D., 262
DesMarais, D. J., 403
DeWiest, R., 151
Dickson, F. W., 330, 331
Dietrich, R. V., 229
Dilamarter, R. C., 6
Dinkel, T. R., 384
DiPierro, M., 229
Doehring, D. O., 356
Donini, G., 337
Douglas, H. H., 67
Dowling, R. K., 218
Downing, A. L., 388
Doyle, F. L., 6
Dragovich, D., 341
Drake, J. J., 37, 148, 200, 201, 214, 218, 328
Dratnal, E., 396
Dreiss, S., 284
Drever, J. I., 224
Drew, D. P., 151
Dreybrodt, W., 293, 299
Droppa, A., 271
Dublyanskii, V. N., 334
Dunkley, J. R., 259
Durov, S. A., 298

Eagon, H. B., Jr., 383
Ebaugh, W. F., 242, 376
Ege, J. R., 363
Egemeier, S. J., 260, 298
Ehhalt, D. H., 324

Author index / 459

Einstein, H. A., 244–246
Ek, C., 132, 196
Eller, P. G., 262
Elliott, L. P., 401
Ellis, B. M., 348
Ellwood, B. B., 316
Endo, H. K., 171
Environmental Protection Agency
    (EPA), 388, 390
Ericson, D. B., 306
Even, H., 172
Ewers, R. O., 201, 272–279, 288, 289, 294, 400
Ewing, M., 306
Ewing, R. C., 262

Fantidis, J., 324
Farrell, M. A., 395
Feder, G. L., 18
Fellows, L. D., 43
Fenneman, N. M., 303
Ferguson, G. E., 18, 187
Finlayson, B. L., 340
Fiore, S., 229
Fischbeck, R., 247
Fish, J. E., 201, 273
Fisher, J. R., 126
Fletcher, M. W., 180
Flint, R. F., 308
Folk, R. L., 59
Folkoff, M. E., 198, 199, 215
Foose, R. M., 362, 364, 367
Ford, D. C., 7, 37, 42, 49, 51, 55, 56, 60, 97, 113, 116, 201, 214, 218, 220, 272–279, 310, 312, 316, 322–326, 328
Forti, P., 247, 262, 337
Fournier, R. O., 224, 342
Frank, R. M., 239, 240
Franke, H. W., 253
Franklin, A. G., 362
Franz, R., 67
Freeman, J. P., 259
Freeze, R. A., 151, 187, 402
French, H. M., 347
Fretwell, J. D., 404
Frey, D. G., 306
Frink, J. W., 371
Frondel, C., 223
Frostick, A. C., 257
Fry, P. L., 402

Gale, J. E., 171
Gale, S. J., 179
Gams, I., 42, 116

Gardner, J. H., 265, 268
Garrels, R. M., 119, 125
Garrison, E., 399
Garza, S., 187
Gascoyne, M., 253, 310–312, 322, 323
George, A. I., 259
Gerson, R., 172
Gerstenhauer, A., 110
Geyh, M. A., 322
Géze, B., 255, 266
Giddings, M. T., Jr., 182, 398
Giggenbach, W. F., 346
Gilewska, S., 196
Gillieson, D., 327
Giusti, E. V., 377
Glazek, J., 218
Glew, J. R., 51, 52
Glover, R. R., 181
Goede, A., 313, 317, 321
Goeller, H. E., 194
Gold, D. P., 398
Goldberg, P. S., 229
Goldsmith, J. R., 221
Goodchild, M. F., 97
Gospodaric, R., 180
Gowan, G. D., 367
Gradzinski, R., 42
Graf, D. L., 221
Graf, W. H., 243, 245, 246
Grant, L. F., 370
Gray, D. M., 174
Gray, J., 324, 325
Greeley, R., 67, 348, 350
Green, D. C., 321
Greene, F. C., 267
Greenfield, R. J., 372, 376
Gries, J. P., 257
Grün, R., 313, 314
Grund, A., 41, 265, 268
Gulden, R., 65
Gumbel, E. J., 385
Gunn, J., 218
Gurnee, J., 378
Gurnee, R. H., 23, 378
Gustavson, T. C., 328, 363
Gvozdetskii, N. A., 3, 328

Haas, J. L., Jr., 180
Habič, P., 180
Hack, J. T., 15, 16, 17, 319
Hahn, J., 328
Hall, F. G., 26
Halliday, W. R., 67, 263, 344, 345, 347, 348, 350, 352
Haman, J. F., 343

Hamer, W. J., 124
Hanna, F. K., 217
Hanshaw, B. B., 301, 405
Haq, B. U., 308
Hardenbol, J., 308
Hardie, L. A., 330
Harmon, R. S., 200, 201, 206, 214, 310, 312, 321, 322, 324, 326
Harned, H. S., 121
Harpaz, Y., 384
Harris, S. E., Jr., 67
Harvey, E. J., 384
Hauer, P. M., 262
Hawkinson, E. F., 240
Health, R. C., 215
Hedges, J., 66, 340, 341
Heine, K., 322
Helwig, J., 240, 242
Hemingway, B. S., 126
Hempel, J. C., 80, 84, 113
Henderson, E. P., 256
Hendy, C. H., 324
Hennig, G. J., 313
Hensler, E., 334
Herek, M., 6, 46
Herman, J. S., 145, 146, 147
Hess, J. W., 201, 206, 209, 211, 214, 296, 297, 322
Hickman, H. L., Jr., 392
High, C., 217
Hill, C., 63f, 67, 262
Hill, C. A., 247, 259, 262, 298
Hinkle, F., 381
Hoadley, A. D., 328, 363
Hogberg, R. K., 67
Holland, H. D., 248
Holsinger, J. R., 60, 67, 396
Horn, G., 40
Horn, J. M., 39
Horton, R. E., 8
Hosley, R. J., 66
Howard, A. D., 42, 178, 278, 290
Howard, B. Y., 290
Hubbard, D. A., Jr., 108
Hubbert, M. K., 169, 404
Huebner, J. S., 248
Humphreville, J. A., 362
Hyde, J. H., 348, 350f
Hyde, L. W., 364
Hynes-Griffin, M. E., 362

IASH, 6
Ibberson, D., 262
Ikeya, M., 313

Ingersoll, A. C., 163
Inglis, C. C., 177
Irby, B. N., 67
Iwai, K., 171

Jacobson, R. L., 133, 139, 148, 185f, 201, 203
Jakucs, L., 6
James, A. N., 335
James, J. M., 200
Jammal, S. E., 363
Jefferson, G. L., 343
Jenne, E. A., 148
Jennings, J. E., 367, 367f
Jennings, J. N., 6, 25, 44, 46, 47, 49, 51, 116, 215, 217
Joensuu, O. I., 221
Johe, D. E., 383
Johnson, K. S., 328, 329f, 330f
Johnston, W. D., Jr., 267
Joiner, T. J., 374
Jones, B. F., 143
Jones, W. K., 108, 241

Karadi, G. M., 171
Kardos, L. T., 395
Kasprzak, K., 396
Kastning, E. H., 82
Kastrinos, J. R., 396, 397
Kaszowski, L., 196
Katzer, F., 265
Kaye, C. A., 229, 271
Keeling, C. D., 194f
Keetch, M. E., 178
Keller, W. D., 223
Kemmerly, P. R., 37, 38
Kempe, S., 255, 328
Kennedy, J. M., 376
Kern, D. M., 142
Khan, D. H., 224
Kirk, K. G., 372, 374, 375f, 376
Kirsipu, T. V., 248
Kisvarsanyi, G., 399
Kiver, E. P., 346
Klaer, W., 341
Knight, E. L., 67
Knight, F. J., 356
Knox, R. G., 348
Kobylecki, A., 196
Koch, D. L., 326
Kohler, M. A., 186
Kraft, R., 170
Krauskopf, K. B., 342
Kritsky, S. N., 385

Krizek, R. J., 171
Kudelin, B. I., 328
Kukla, J., 237
Kundert, C. J., 254
Kyrle, G., 263

Lagrange, R., 247
Laidlaw, I. M. S., 181
LaMoreaux, P. E., 364, 366
Lange, A. L., 92
Langmuir, D., 128, 133, 137f, 139, 204, 205, 393
Larson, C., 345f
Latham, A. G., 310, 316
Latham, E. E., 224
Lattman, L. H., 381
Lauritzen, S.-E., 322
LaValle, P., 39
Lawson, T. J., 322
Lee, C.-H., 404
LeGrande, H. E., 171, 341
Lehmann, H., 6, 112
Lehmann, O., 266
Leopold, L. B., 176, 177, 243, 377
Lesser-Jones, H., 405
Lindsley, R. K., 186
Lingham, C. W., 18
Liu, S.-T., 335
Lively, R. S., 242f
Loewenthal, R. E., 119
Lohman, S. W., 152
Long, D. T., 394
Long, J. C. S., 171
Longyear, J., 271
Louis, H., 41
Louw, A., 367, 367f
Love, S. K., 18
Lowden, G. F., 388
Ložek, V., 247
Lumsden, D. N., 310
Lupton, A. R. R., 335

Madenford, G. A., 38
Magaritz, M., 172
Malott, C. A., 268
Mandelbrot, B. B., 64
Mangin, A., 189
Mann, C. J., 108
Manov, G. G., 124
Marais, G. v. R., 119
Marker, M. E., 343
Markowicz, M., 218f
Markowicz-Lohinowicz, M., 218f
Martel, E. A., 266

Mason, A. P., 340
Massello, J. W., 369
Mathews, H. L., 225
Matson, G. C., 267
Matthews, L. E., 67
Maurin, V., 180
McConnell, H., 39
McDonald, R. C., 110, 112
McGregor, D. L., 333
McGregor, D. R., 333
McKenzie, G. D., 345
Mears, B., 352f
Medville, D. M., 44, 113
Menkel, M. F., 385
Merriam, D. F., 108
Merriam, P., 263
Mijatović, B. F., 41, 42, 189
Milanović, P. T., 6, 151, 155, 188, 384
Miller, J. P., 243
Miller, T., 211f, 218f
Mills, H. H., 37
Mills, J. P., 343
Mills, M. T., 348
Milske, J. A., 242f
Minton, M., 88
Miotke, F.-D., 85, 132, 195f, 196, 318
Moneymaker, B. C., 370
Monroe, W. H., 44, 111
Moore, C. H., 59
Moore, D. G., 352
Moore, G. W., 220, 228, 250, 254, 255, 263, 352
Moore, J. D., 381
Moravec, G. F., 51, 381
Morehouse, D. F., 298
Morey, G. W., 224, 342f
Morgan, E. L., 401
Morgan, J. J., 119
Morisawa, M., 377
Morse, J. W., 139, 142, 143f
Moseley, C. M., 334
Müller, G., 247
Murray, J. P., 393
Murray, J. W., 229
Myers, A. J., 328
Myers, E. A., 333, 395
Mylroie, J. E., 238

Nair, V. S. K., 128
Nancollas, G. H., 128, 335
Nathan, Y., 229
Nesbitt, J. B., 395
Neuzil, C. E., 171
Newton, J. G., 364

Nieland, J., 345f
Noel, M., 316
Novak, I. D., 244, 245f
Núñez-Jiménez, A., 111

O'Brien-Marchand, M., 377
Ogden, A. E., 218, 218f, 252f
Oleksynowna, B., 196
Oleksynowa, K., 196
Olive, W. W., 332
Olson, E. A., 309, 310
Olson, J. S., 194f
Omnes, G., 374
Ongley, E. D., 71
Ormay, S., 170
Orr, P. C., 310
Otvos, E. G., 347
Owen, B. B., 121
Oxburgh, U. M., 248

Palmer, A. N., 67, 68, 71, 85, 86, 107, 178, 257, 279, 280–282, 290, 291f, 293, 308, 318, 334
Palmer, M. V., 107
Palmer, R., 308
Palmquist, R. C., 38
Panoš, V., 111
Papakis, N., 171, 186, 188
Parizek, R. R., 303, 376, 381, 382f, 383, 394, 395, 398
Parkhurst, D. L., 144
Parris, L. E., 67
Patrick, D. M., 362
Paulhaus, J. L. H., 186
Pavey, A. J., 200
Pearce, G. W., 316
Pearse, A. S., 26
Pechorkin, I. A., 372
Pendery, E. C., 333
Perič, J., 384
Peterson, D. N., 345
Peterson, G. W., 225
Peterson, M. N. A., 224
Petraitis, M. J., 200
Pfeffer, K. H., 6
Pfeiffer, D., 6, 328
Pia, J., 265
Picknett, R. G., 300
Pierce, J. W., 343
Piper, A. M., 265, 267
Pitman, J. I., 206
Pitty, A. F., 206
Plantz, C., 113
Pluhar, A., 49, 51, 55, 56

Plummer, L. N., 144, 145, 146, 148, 300, 301
Pobeguin, T., 247
Pohl, E. R., 21, 235, 259
Popov, I. V., 328
Popov, V., 218f
Poulson, T. L., 259
Powell, R. L., 67
Powell, W. J., 364
Powers, J., 262
Prior, T. A., 348
Proctor, P. D., 399
Pulina, M., 216, 218f
Pye, K., 343
Pyle, T. E., 301

Quinlan, J. F., 14, 117, 156, 181, 182, 201, 220, 278, 289f, 294, 328, 329f, 334, 368, 387, 400

Radomski, A., 42
Railton, C. L., 66
Raines, T. W., 21, 23
Rasmusson, G., 341
Rauch, H. W., 284, 285f
Ray, J. A., 156, 182, 278
Reams, M. W., 238–240
Reardon, R. J., 128
Reckmeyer, V., 402
Reddell, J. R., 26
Reddy, M. M., 145
Reich, B. M., 191f, 192
Reich, J. R., Jr., 67
Renault, P., 275
Rhindress, R. C., 398
Rhoades, R., 271
Rickard, D. T., 144, 145
Rightmire, C. T., 194
Robbins, C., 223
Roberge, F., 369
Roberts, H. H., 59
Robie, R. A., 126
Robinson, B. W., 229
Rogers, A. F., 200
Roglič, J., 41, 44, 113
Ross, S. H., 67, 348
Rossi, G., 337
Rossi, M. G., 194, 195f
Rotty, R. M., 194f
Rouse, J. V., 393
Rowe, D. R., 14, 156, 181, 400
Rowe, J. J., 224, 342f
Ruhe, R. V., 206, 207f, 225
Rutherford, J. M., 65

Safadi, C., 188
Salamone, W. G., 368
Saleem, Z. A., 394
Sayre, A. N., 170
Scarbrough, W. L., 364, 374
Scheetz, B. E., 263
Schelleng, J. H., 65
Schmid, E., 238, 240, 241
Schmidt, L. A., Jr., 370
Schmidt, V. A., 65, 283, 316, 317f, 319
Schuum, S. A., 9
Schwarcz, H. P., 310, 312, 316, 322, 323f, 324, 325f, 326
Schwartz, F. W., 171
Schwarz, J., 384
Seeger, M., 328
Shelley, M. B., 328, 347
Shepherd, R. G., 16
Shuster, E. T., 171, 203, 206, 207f, 211f
Siddiqui, S. H., 382f, 383
Siegel, F. R., 343
Siever, R., 342f
Silar, J., 47, 112
Simpkins, W. W., 328
Sinacori, M. N., 271
Sinclair, W. C., 364
Sjöberg, E. L., 144, 145
Skinner, A. F., 313f, 313
Slifer, D., 67
Smart, C. C., 178
Smart, P. L., 181, 278, 313, 322, 403
Smeltzer, B., 79
Smith, A. R., 328, 329f
Smith, B. W., 313
Smith, D. I., 151, 217, 218f
Smith, G. L., 259
Smith, K. G., 351
Smith, L., 171
Smith, M. O., 262
Smith, P. M., 379
Smith, R. M., 127
Smith, W. O., 170
Snyder, E. R., 376
Soderberg, A. D., 369
Sonderegger, J. L., 381
Sopper, W. E., 395
Sowers, G. F., 362
Spaeth, C., 255
Stansbury, D. H., 337
Starkey, H. C., 224
Starnes, D. D., 37
Stearns, H. T., 349
Steele, W. K., 346
Steiner, R. S., 362

Štelcl, O., 111
Stellmack, J. A., 115f
Stewart, M. T., 404
Stone, B., 278
Strahler, A. N., 8, 14
Street, R. L., 160
Stringfield, V. T., 6, 46, 171
Strohm, W. E., Jr., 362
Stumm, W., 119
Suarez, D. L., 258
Sunartadirdja, M. A., 112
Sutcliffe, A. J., 348
Sutherland, W., 63f, 67, 262
Sweeting, G. S., 284
Sweeting, M. M., 6, 31, 42, 44, 51, 116, 182, 218f, 270, 284
Swinnerton, A. C., 265, 269, 270
Symons, M. C. R., 313
Szczerban, E., 343

Tennessee Valley Authority, 370
Ternan, J. L., 207
Terzaghi, K., 30, 41
Thompson, G. M., 310
Thompson, P., 310, 312, 324
Thrailkill, J., 139, 148, 201, 247, 271, 372f
Tierney, L., 63f, 67, 262
Tjia, H. D., 47, 112
Tolson, J. S., 6
Towe, S. K., 37
Tracy, J. V., 171
Trainer, F. W., 215
Tratman, E. K., 182
Tricart, J., 46
Troester, J. W., 34, 35f, 197f, 211f
Trombe, F., 139, 266
Truesdale, G. A., 388
Truesdell, A. H., 148
Tschang, H.-L., 341
Tsui, P. C., 336
Tullis, E. L., 257

Udden, J. A., 53
Ugolini, F. C., 347
Urbani, F., 343

Vail, P. R., 308
Vandenberghe, A., 266
Van Driel, J. N., 38
Van Eysinga, F. W. B., 305
Vanoni, V. A., 244, 245f, 246
Varnedoe, W. W., Jr., 62f, 63f, 67
Vennard, J. K., 160

Verber, J. L., 337
Vernon, R. O., 18
Verstappen, H. T., 112
Vincent, C. L., 178
Vineyard, J. D., 18, 365, 366f
Vladi, F., 328
Von Knebel, W., 266
Von Wissmann, H., 47, 112

Walker, A. C., 395
Walker, R. L., 310
Walsh, J., 200
Waltham, A. C., 322
Walton, W. C., 187, 383
Wang, J. S. Y., 171
Ward, J. C., 170
Warren, W. M., 366, 376
Watkins, D. E., 402
Watkins, E. J., 348
Watson, P. J., 259
Watson, R. A., 264, 265, 378, 379
Webb, J. A., 340
Weidie, A. E., 301
Weissen, F., 132, 196
Werner, E., 44, 113, 372, 374, 376, 394
West, T. R., 225
Weyl, P. K., 142, 288
White, E. L., 34, 35f, 63f, 176, 179, 180, 189, 190f, 191f, 192, 243, 270, 318, 386
White, G. W., 67
White, P. J., 23
White, W. B., 34, 35f, 63f, 68, 83, 113, 146, 172, 176, 177, 178, 179, 180, 195f, 197f, 201, 203, 206, 209f, 211f, 214, 218, 218f, 225, 235, 243, 247, 253, 257, 259, 262–265, 270, 271, 282f, 284, 285f, 293, 296f, 297, 303, 318, 343, 386, 396, 397
Whittemore, D. O., 376
Whorf, T. P., 194f
Wielchowsky, C. C., 376
Wigley, T. M. L., 144, 148, 214, 259, 263, 290, 300, 328, 331, 403
Wilcock, J. D., 192
Wilford, C. E., 99
Wilkening, M. H., 402
Williams, J. H., 365, 366f, 369, 384
Williams, P. W., 14, 15, 32, 33, 34, 42, 47, 112, 113, 216, 218f
Wilson, C. R., 171
Wilson, J. R., 113
Witherspoon, P. A., 171
Wojcik, Z., 341
Wolfe, J. A., 306, 307f
Wolfe, T. E., 240
Wollin, G., 306
Wolman, M. G., 176, 177, 243
Wood, C., 348
Wood, W. W., 200
Woodward, H. P., 268
Wright, H. E., Jr., 306
Wulkowicz, G. M., 394

Yaakobi, D., 170
Yarborough, K. A., 402, 403
Yevjevich, V., 6
Yonge, C. J., 324, 325f

Zeris, K., 188
Zotl, J. G., 6, 180–182
Zuffardi, P., 220